ndations of Information Science
Buckland, Jonathan Furner, and Markus Krajewski

Retrieval

ation: The Culture of Wikipedia
Reagle Jr.

ut Cards & Catalogs, 1548–1929
i, translated by Peter Krapp

gue: From Index Cards to Dewey Decimals to Alger Hiss

Information and Intrigue

History and Fou
Edited by Michael

Human Informatio
by Julian Warner

Good Faith Collabo
by Joseph Michael

Paper Machines: Ab
by Markus Krajewsk

Information and Intri
by Colin B. Burke

Information and Intrigue

From Index Cards to Dewey Decimals to Alger Hiss

Colin B. Burke

The MIT Press
Cambridge, Massachusetts
London, England

MIT Press books may be purchased at special quantity discounts for business or sales promotional use. For information, please email special_sales@mitpress.mit.edu.

This book was set in Stone Sans and Stone Serif by the MIT Press. Printed and bound in the United States of America.

Library of Congress Cataloging-in-Publication Data

Burke, Colin B., 1936-
Information and intrigue : from index cards to Dewey decimals to Alger Hiss / Colin B. Burke.
pages cm
Includes bibliographical references and index.
ISBN 978-0-262-02702-1
1. Field, Herbert Haviland, 1868-1921. 2. Bibliographers—Biography. 3. Concilium Bibliographicum—History. 4. Classification—Books—Science. 5. Information storage and retrieval systems—Science. 6. Field, Noel Haviland, 1904-1970. 7. Diplomats—Biography. 8. Information science—History. 9. Science—Political aspects—History—20th century. 10. Science and state—History—20th century. I. Title.
Z1004.F54B87 2014
020.9—dc23

2013035293

10 9 8 7 6 5 4 3 2 1

To my wife, Rose, who shared with me the joy and privilege of creating and nurturing beautiful and unique human life.

Contents

Preface

A Compelling Historical Puzzle

About a decade ago, when I was focusing on the emerging field of the history of information, I saw a brief mention of a science information service run by an American in Switzerland. That report said his late nineteenth century Concilium Bibliographicum was sending millions of bibliographic index cards to scientists in Europe and America and that he was using radically new information methods. A hint that the Concilium's founder might have been the father of an alleged American Communist spy and turncoat (and friend of men such as Alger Hiss) led me to temporarily put aside most of my other projects as I began a search for answers to the puzzles of the extraordinary lives of Herbert and Noel Field. The search did not remain temporary. It took me to archives on both coasts, to numerous requests for classified information, and to sending letters and emails to dozens of institutions and people around the Western world. At each step, I became more enthralled with the history of the Fields and more impressed with Herbert Field's information system and its ties to the development of science and to America and Europe's politics.

The Concilium's history is one of the most fascinating stories I have ever encountered. This volume traces the Fields and science information systems to the beginnings of the Cold War. The next work will tell the amazing stories of the Fields and science information to our own time and the era of Open Access, when new types of information reformers demanded that all scientific articles be provided to readers without charge.

I have done my best to avoid using the specialized language of information science, and I have organized the book so that readers can skip over the few technical sections and, if they wish, concentrate on the compelling story of three generations of a Quaker family dealing with a modernizing America, a world in conflict, Communism, and even espionage.

Unfortunately, the many people and institutions I should acknowledge cannot be saluted here. Space constraints mean the "thanks" will be found on my web page (http://userpages.umbc.edu/~burke) where there is also an illustrative bibliography.

Introduction

Modern Science Information Emerges from a Calling, a Complex Family, and America's Changing Liberalism

Herbert Haviland Field, an idealistic American intellectual phenom, experienced a Quaker's version of a Calling in the 1890s. He devised a plan to revolutionize the world's science information systems—a plan that would further internationalism, the grand cause of his era. His struggle to create and maintain his daring system led him back and forth across two continents and into numerous entanglements: in nationalistic struggles over the control of science information; in the new system of American millionaires' philanthropy; in the politics of emerging American professional science (and "social science"); in a tug-of-war between types of science practice; and in the efforts of Paul Otlet, another great information visionary, to create a worldwide database for all subjects. (Like Herbert Field's system, Otlet's contained aspects that were ahead of their time; Otlet's scheme even had logical features common to the World Wide Web.)

Herbert Field's life and his Swiss-based information adventure, the Concilium Bibliographicum, continued to unfold with much drama. He became a paid intelligence agent for the United States in World War I, even sending coded warnings to a young American spymaster, Allen Dulles, about the Red Menace that had seized power in Russia. After the war, Herbert struggled to revive his great information experiment. But just when his ultimate (and at times demeaning) pleas for financial help from friends and American philanthropies were heard, and thus might have allowed him to reestablish the Concilium, he fell victim to the last bursts of the great influenza pandemic.

That did not end the story of Herbert's information crusade or of his family's adventures. American and European advocates continued the battle to save his system. Then, less than a decade after Herbert's death, his son Noel experienced his own Calling. It differed from Herbert's, however, leading him into the American diplomatic corps. Then, along with his friends Alger Hiss and Laurence Duggan, Noel Field became a source for Soviet intelligence during the 1930s (and certainly in Noel's case, the 1940s). Noel also served as a secret informant for Allen Dulles during World War II. After that, his life turned into a nightmare. Beginning in 1949, the Soviets imprisoned Noel, his wife, his foster daughter, and his brother Hermann. All of them spent a half-decade in the torture dungeons of Communist Europe while Noel became the center-point of Stalin's schemes, which led to horrendous purges in Eastern Europe. Hundreds were executed, thousands imprisoned. Meanwhile, on the other side of the Atlantic, many of Noel's friends and associates became targets during the hunt for spies in America.

Noel and his father were not the only Fields to live unusual and influential lives. Herbert Field's brother Hamilton propelled the introduction of modern art and a bohemian lifestyle in America. Herbert's youngest son, Hermann, took a leading role in the new holistic architecture movement in the United States while Herbert's eldest daughter became an important Sangerite birth-control advocate and practitioner. Even Herbert's wife became a Communist fellow traveler (at least) when she was a close friend of two of the more significant Soviet agents in America during the 1930s.

All of these connected players and events tie into the history of scientific information. A strange continuity marked the more than century-long histories of Herbert Field, his family, and science information due to the rise and morphing of cultural, political, and ideological liberalism and its institutions. Whether in the nineteenth century, when science was viewed as a progressive, liberating, and modernizing force, or in the Cold War era when science information was vital to the struggle against totalitarianism, or in our own era when the more vibrant part of the Open Access movement was linked to science-distrusting postmodernism (and to post–New Leftism), science information's fate has been shaped by ideas and values as well as by technical and economic forces. The ever-changing liberal modernism, and reactions to it, have been and continue to be important to the history of information.

The story of the Fields and information science is not "monographic." Simple explanations cannot account for the Fields' contributions and mistakes. Although science information systems are its focus, this is not a technical work or a "report." It is, and has to be, an intertwined historical biography of people, ideas, institutions, and technologies.

The complex histories of science information, Herbert Field's adventures, and his family's traumas require readers to deal with many people, institutions, and ideas—and with changing vocabularies related to information systems. I have spared readers most of the specialized terminology of librarians and information scientists, but practitioners will recognize technical challenges whenever they are discussed. I have also consciously avoided the vocabularies of such fields as Science-Culture Studies, Modernity, and Deconstructionism, although those approaches are relevant to the Fields' story.

Other rewards compensate for the absence of a technical vocabulary and postmodern terminology. The histories of Herbert Field's life and system, and the development of science information, are exciting stories.[1]

I To Revolutionize Information, but Not the World

1 Raising a Perfectly Modern Herbert

The Fields' American presence dates from the colonial era when the family chose to settle in New England rather than Pennsylvania, the expected destination of displaced English Quakers. Despite locating in a region the Puritans tried to keep as a haven for Congregationalists, at times by hanging Quakers, the Fields prospered. By the eighteenth century, they held large amounts of valuable land in Connecticut and were establishing themselves in New York. They were not alone, in either Greenwich or in New York State's eastern countryside. A number of their denomination's fellows joined them. They formed a Quaker community that helped maintain faith and culture as the Fields made a decision critical to young Herbert Field's life—to move to the rising urban commercial center, New York City.[1]

In the mid-nineteenth century, Herbert's family was flourishing in the city's competitive business community. They prospered as dry goods merchants and importers. Herbert's grandfather was a millionaire with a beautiful home on Willow Street in Brooklyn, just a few blocks from where he would help his sons Charles and Aaron build their own impressive townhomes. His boys constructed their houses next to each other on the heights over the East River. From that handsome setting, they could oversee their company's wharf and warehouse. Columbia Heights was a place-to-be for the ambitious. Rising shipping magnates made it their home. The clipper-ship builders, the Chapmans, selected it. Even William R. Grace, whose companies came to dominate Peru and became America's conduit for trade with Central America, lived there while leading a political reform movement in New York City.[2]

By the 1860s, the Fields were leaders of the influential Quaker group that made the Heights the most desirable upscale neighborhood in Brooklyn. Its promenade along the river, its rows of sumptuous townhouses, and its freedom from pollution and city crime meant it was a true upper middle-class haven. Even the wealthy Roebling family that constructed the Brooklyn Bridge built a residence on the Heights. From the top floor of the townhome at 110 Columbia Heights, the disabled Washington Roebling supervised the completion of the engineering marvel of the era.[3]

A New Type of Quaker

Austere living and avoidance of secular pleasures were not hallmarks of Brooklyn Heights' Quakers. Even before the American Civil War, Herbert's father Aaron and his religious fellows, although Hicksites, were no longer unadorned and simple country Quiet People. The Quaker families in Brooklyn still spoke using "thee" and "thou," but few continued to wear the plain coats and dresses of the old farm Quakers. Even the older Friends' strict rules about marrying only Quakers of the same social background were somewhat relaxed. In addition, the Fields had many contacts with those outside the faith and they enjoyed many worldly pleasures.[4]

The family had joined the social world of the Quaker international capitalists. The Fields were financially able to enjoy European travel, and they took advantage of the cultural attractions in New York City and the surprising number of musical and intellectual offerings in Brooklyn. However, the Fields did not forsake the fundamental ideals of the early Friends' communities. They did not drift into the arms of traditional Christian churches, or even that of their neighbor, Brooklyn's famed Henry Ward Beecher. To ensure their religiosity, the Fields helped to establish the celebrated Brooklyn Friends School in the community's newly constructed Quaker meeting house. Even the liberalized Quakers, who had abandoned much of Christian doctrine and relied upon their inner light for guidance, felt a need to separate their young from the outside influences of an increasingly diverse city. The public schools held too many dangers for Quaker children.[5]

Wealth, Tradition, and Reform

By the time Herbert was born in 1868, his father Aaron Field was a well-to-do capitalist merchant. Of significance for Herbert's life, all the Fields went unscathed through the traumatic economic and financial upheavals of the nineteenth century. As a result, Herbert grew up in what twenty-first-century Americans would interpret as a setting of secure privilege, financially and psychologically. His father's household, for example, always had three or four live-in servants. But his family was oriented toward much more than self and materialism. Herbert was raised in a home and community pledged to social reform, pacifism, and at least the fundamentals of the old Quakerism. To help maintain such loyalties the Fields, like most other Quakers, continued to try to marry within the faith.

Aaron first married when he was in his twenties to Charlotte Cromwell, a woman of the same age and religion but from a family of unassuming farmers with different social connections. The couple quickly became important members of the most modern Quaker organizations. Aaron was a leader in Brooklyn's historical and charitable causes while being a devoted family man. Aaron and Charlotte followed tradition and soon had three children, a girl and two boys. Then, as the Civil War began, Aaron's family was rocked by his wife's death, just as the couple's last child, Edward, was born. Aaron shouldered his responsibility and raised his family by himself for three years. Then, he broke with tradition. He took a bride who was of the faith but who was a near decade his junior. She was almost the Quaker "girl next door," living only a block or so away, but she was of a more sophisticated social breed than his first wife.

A Daughter, Wife, and Mother—Of Intellect, Influence, and Means

Aaron's new spouse, Lydia, came from an elite Quaker family, the Havilands. Some of her relatives had built a prosperous chinaware business in New York City; others created a famous ceramic company in France. None were strangers to influential groups in America or Europe. Contacts included the presidents of the United States and luminaries in the arts on both continents. The French Havilands, who were also married into the European Fields, mingled with the continent's royalty.[6]

Lydia's father was an influential man. He had made a fortune as a drug merchant, allowing him to retire at a relatively early age. He then concentrated on raising his children and completing his impressive townhome in Brooklyn Heights, and on managing his funds. However, the American Civil War changed his plans. His investments in the South became worthless. Although still well-off, he was no longer among the very rich. Despite that, his children had the best and became accustomed to a pampered life.

Whereas Charlotte came from Cornwall, a quiet small town fifty miles up the Hudson River, Lydia was an urban girl who "had it all," at least for a while. Her milieu was cosmopolitan and wealthy. But she had not been a society belle and had few acceptable young men courting her. The youngest of seven children, she remained at home and unmarried until she was close to thirty, two years after her mother died, and just after her father suddenly remarried. Lydia had not been a recluse spinster lady, however. She was active in the many Quaker and social reform activities in Brooklyn. She also was one of the founding members of Swarthmore College, a trustee of New York and Brooklyn homes for the destitute, an active builder of the YMCA and YWCA, and a strong but not militant figure in the women's suffrage movement.[7]

Living a block or so away from the Field's townhomes and being a member of his Quaker Meeting, Lydia had caught Aaron's attention. When an acceptable amount of time passed after Charlotte's death, Aaron began a courtship. It took much convincing, but Lydia consented to become a Field—on her own terms.

An important part of the successful courtship was Aaron's career after leaving his father's dry goods business. He became an independent auctioneer and went from being well-off to being rich. Lydia was not, as Charlotte had been, a young and innocent bride, and she recognized a relationship between wealth and happiness. In addition, she had her own ambitions and expectations. One indication of her demand for a decisive role in the family surfaced when Aaron and his children moved into Lydia's home at 72 Columbia Heights, rather than having her move down the street into

Aaron's nearby Hicks Street home. Within a decade, she convinced Aaron to build a magnificent new multilevel townhome with an indoor pool at 106 Columbia Heights. All that increased Lydia's influence over Herbert and Charlotte's children and allowed her to put her peculiar stamp on the lives of her own. Her father's wealth, then his misfortunes, would have consequences for her children. Lydia did not want to change her life too much—or too suddenly. Although her father's economic standing had declined, she continued to have social ambitions. She expected to be able to afford all that wealth could bring after she married. Aaron would always be under pressure to provide the money. Another indication of her ambitions was that Lydia did not rush into family life after the marriage. It took three years for her and Aaron to produce their first child, Herbert, and another five years before their second son, Hamilton, was born. Three years year later came their third and last child, Anna.

Raising a "Better Sort" through Devotion and International Exposure

Perhaps it was genetics, or perhaps it was because Lydia was an older first-time mother, that led her children to have some physical difficulties. Unlike Charlotte's robust children, Lydia's showed signs of ill health. They required attention, if not indulgence. But even before she discovered that she had needy children, Lydia was determined that her offspring would have everything necessary to make them part of "the better sort."

There was more that was important for the future of Lydia's natural children. She viewed child rearing as a more distant and formal task than had Charlotte. In addition, giving birth to three sickly children put a lenient cast on her mothering. Lydia's devotion to her sons deepened after a family tragedy. Anna had suffered through a lingering and painful illness for years and went blind before her death in 1883, at age seven.

Other influences molded Herbert and Hamilton's lives. Of great significance for her boys' futures were Aaron's international business connections and Lydia's European Haviland relatives. The Havilands had established thriving businesses and a place in the continent's art and social worlds. Their exciting lives matched Lydia's dreams. The Haviland name had become a byword in Europe and America for artistic, even collectable, ceramics and dinnerware. Their products turned them into millionaires and they became socially well connected on the continent. Significantly, they were patrons and friends of the most advanced artists of the nineteenth and early twentieth century. Renoir was a neighbor and painted the Haviland children's portraits. The Havilands favored Picasso with their patronage. Later, with the direct help of the Paris branch of the Havilands, Herbert and Hamilton Field became part of a new social type, well-to-do American expatriates.

An Ideal Enlightened Life, Even in Brooklyn

Despite their international links and wealth, Lydia and Aaron remained Brooklyn Meeting people. Both Herbert and Hamilton (and the other children in the house) were surrounded by liberal Quakers and immersed in their culture. Almost all of Herbert's boyhood friends were of the Inner Light and he, like his siblings, was sent to the local Friend's School where, except for a run home for lunch, the day was spent learning an integrated mix of religion, ethics, and, notably, the latest secular knowledge. Unlike the German Mennonites in America, or some of the Quakers who remained in the countryside, many urban Friends were able to embrace the new learning, even Darwinism, without damage to their religious convictions. Such liberal and pacifist Friends were finding, for example, ways to accommodate the Quaker belief that Love guides the world to the Darwinian "struggle for survival." The Fields were examples of the American Whig so well described by Daniel Walker Howe.[8]

Lydia made sure that Herbert did not grow up as an American "urban provincial." She and Aaron took every step to make him more than just a Brooklyn Quaker boy. His parents loosened many cultural restraints. Many of the prohibitions against worldliness and secular enjoyments, such as piano music and literature, dissolved. All the Field children learned to play several musical instruments; Herbert was the star. He amazed family and friends by being able to play, on various instruments, scores he had heard only once at New York City's concert halls. The Fields and their neighbors also collected the best paintings, attended to world events, and, especially important for Herbert and Hamilton, they were aware of cultural and intellectual trends in Europe. Both boys became multilingual and were able to read the literature of most European nations. That, and Herbert's near photographic memory, led to his ability to recite French and German works during family gatherings. Aaron's extensive dealing in imported Japanese goods and art had a profound influence on his children, especially Hamilton.

Aaron's import and auction businesses and letters from the French Havilands kept the family attuned to European tastes and politics. They knew of the rise and consequences of Germany's new nationalism. They were aware of the dangerous ethnic and nationalistic urges as the Austrian Empire was disintegrating and losing its ethnic diversity battles. Aaron and Lydia told the young Herbert and Hamilton details of the Franco-Prussian War, of the horrible siege of Paris, and of the agonies of the Paris Commune.

The Underground Railroad (and Other Movements) toward a Liberal America

Significant was the family's longtime involvement in a broadening range of social reforms. Herbert's ancestors were among the first in New England to free their slaves after the American Revolution. But the majority of the family did not become rabid abolitionists, politicos, or socialists during the nineteenth century. Rather, they were moderates and focused on expanding civil rights and on the creation of new voluntary charitable and advocacy associations. Herbert's mother and her mother (and their friends) were early leaders in the women's suffrage movement and had been close to the feminist and antislavery advocate Lucretia Mott. The Havilands were so important to reform movements they had two midwestern towns, one in Ohio and one in Kansas, named after one of their more enthusiastic and unusually assertive relatives. That woman, Laura Haviland, became an activist and engaged in illegal activities before the Civil War as she supported stations on the underground railroad that aided slaves fleeing from the South. She became so enthusiastic that she abandoned the Quakers and became a Methodist because they had declared for active, not passive, opposition to slavery.

The Columbia Heights Quaker ladies of the later nineteenth century shared the early reform impulse, but expressed it in a tempered way. They remained pacifist Quakers while founding institutions for indigents, homeless women and children, and ladies needing employable skills. Well into the twentieth century, Herbert's female relatives were also leaders in establishing YWCAs and uplifting literary groups for those of their own class.

The Quaker Meetings' menfolk sided with the middle-class political reforms of the era and looked forward to a nonpartisan, scientific, civil service government. They believed that reason and science could change society, avoiding what many saw as impending class warfare. Aaron Field was among them, devoting much time and money to historical research on Brooklyn and acting as an advocate for the local hospitals.[9] All the Fields became supporters of the new American Library Association. There were limits to their involvements in reform, however.

Progressive but Not Red Radicals

Wealth and culture shaped the Fields' perspectives and values. They saw capitalism as a liberating force. The evidence of that surrounded them. An example: They and their children received Tiffany-made personal invitations to the 1883 opening of the astounding engineering accomplishment, the Brooklyn Bridge. Their next-door neighbors, the bridge-building Roeblings, personally signed the invitations. Although liberal, the nineteenth-century Fields did not approve of the radical agrarian populism of the American West, nor did they join the ranks of those who supported America's version of socialism or militant unions. For Aaron's generation, left-leaning politicians such as Eugene Debs and his Socialist Party represented a danger, not an opportunity.[10]

A Modern Liberal, Indulged and Educated, Dealing with Impediments

Lydia and Aaron provided all their children with marvelous childhoods. Endless progress and growth surrounded the boys as America and Brooklyn reaped the benefits of industrialization and economic expansion. Aaron and Lydia hoped that Herbert would be content to remain in Brooklyn and share in managing his father's auction business. But his mother's devotion to her firstborn, the boy's temperament, and a decision by the overseers of the Brooklyn Friends School led Herbert onto a different path. Neither Herbert nor Hamilton were handed the responsibilities for the family's economic welfare. Rather, Charlotte's sons were asked to forego their possible artistic ambitions and move into the business world. Edward, Herbert's older half-brother, as had his brother Henry, went into the family business. Edward soon found himself with great responsibilities for the family interests after Henry unexpectedly died in middle age.

Herbert might have followed a business career, but his still somewhat worrisome health, Lydia's indulgence, and the Brooklyn Meeting's decision not to establish its own secondary school until 1907 put him on a different trajectory. The lack of a Friends high school left a serious gap to fill for Quaker families that had money to spare for advanced education. There had been no worries about Henry and Edward's educations, but Herbert increasingly seemed temperamentally ill suited to follow the traditional Field career pattern.

His health declined during his teens and he suffered from migraine headaches. He stammered, and on the eve of his manhood declared a dislike of the business world. Throughout life he was a chronic complainer about his physical condition and was a "last minute Louie," putting off any mundane or unpleasant task until the final minutes. Neither trait fit with a rigorous and pressure-filled business life. There were other problems. Herbert was not gregarious and he had that speech difficulty. Although he had worked at controlling that impediment, even in his manhood it could be detected no matter what language he was speaking.[11]

Thus, Aaron and Lydia sadly recognized that an apprenticeship within the family business would not suit him. As well, it was expected of caring middle-class parents to provide a high school education, as they had tried to do for Henry and Edward; and, college attendance was no longer just a destination for those entering the ministry. Despite the Quakers' distrust of colleges, Lydia became determined that her favored son Herbert would join the some 2 percent of young people pursing education into their early twenties. The question was where to send him?

The first step toward his advanced education was to find an appropriate high school. There were many private and nurturing boarding schools on the East Coast, but Lydia feared the available high-status ones might force conflicting religious beliefs on her son. At minimum, attendance at any of them meant sending him out of the neighborhood's Quaker surroundings during his critical juvenile years. New York and Brooklyn's new public schools certainly posed some threats. Might it be better to educate the fragile boy at home with private tutors, she asked Aaron? Then, recommendations by Herbert's teachers at the Friends' school, and by businessmen such as the Brooklyn publishing magnate Albert S. Barnes, tipped the scales. Aaron and Lydia decided their son should, as had Edward and Henry, move out of the Quaker's educational fold but remain in his Brooklyn neighborhood.

The Museum and That New Kind of School, Just down the Street

Aaron and Lydia were careful about the choice of schools. They selected one near their home that offered courses which might keep their precocious boy's interest and that might provide the basis for an occupation. In respect to interest, Lydia and Aaron had sensed how much Herbert loved the newly emerging sciences. They had seen his eyes sparkle during the frequent trips to New York's American Museum of Natural History, which was then as much a research center for zoology and biology as a pleasant stop on family excursions. There were lectures as well as exhibits, and the "professors" at the museum were glad to speak with respectable visitors. They passed on an enthusiasm that impressed Herbert. It was an exciting place and an exhilarating time for science. By the mid-nineteenth century, the intellectual framework of modern science was emerging. There was one amazing discovery after another, especially in the natural sciences. The museum was on the cutting edge.[12]

So, it seemed, was a Brooklyn private high school on Livingston Street. Aaron's business friends and neighbors' assurances about the suitability of the nearby Brooklyn Collegiate and Polytechnic Institute quieted some of Aaron and Lydia's concerns about the school. They had sent Henry and Edward there in the mid-1870s; but neither proved to be a scholar. Henry stayed for only a year; Edward had to repeat a grade, and then left the school before graduating. But, Lydia felt that her own boys could do well at "Poly." One thing in its favor was that it was in walking distance of home. Another was that its science offerings might keep Herbert's attention. Yet another was its acceptance of technology, as implied in its name. Perhaps Herbert could become another Roebling?[13]

Brooklyn Collegiate and Polytechnic Institute was born in the 1850s when New York and Brooklyn's most respectable men committed funds to establish what became one of the nation's best college preparatory schools for boys between the ages of nine and seventeen. The founders had sought more than a replication of traditional elite schools. They wanted to create an urban institution to train leaders for the new age of industry and commerce (and for "metropolitan life") while maintaining high academic standards. Well before a Progressive reform ideology developed, they had faith that such well-educated businessmen would make capitalism a socially responsible force.[14]

When the fourteen-year-old Herbert entered the school, it already had a national reputation as one of the best centers for pre-college science, commercial, and technical preparation. As well, it was nonsectarian and had a splendid liberal arts track that prepared students for the best colleges. Added attractions for the Fields: it was less than a mile from the Heights' sumptuous townhomes; many of its students were from the neighborhood; and, it admitted only the most well-qualified young men.

Like many of its other bright students, Herbert blossomed at Poly, but not by focusing on what later became the school's forte, engineering. He concentrated on general science and language studies, telling his parents he was not ready to select an occupation. Rather, he was preparing for entrance into an emerging type of American college—one imitating the new universities of Germany that were leading the world in modern scientific research. Herbert wanted to continue to explore his general interests in science.

That New Creature, the American Research University

Herbert's outstanding performance at Poly, combined with Lydia's ambitions for her sons and Aaron's increasing wealth, led to another critical decision. Aaron was to support Herbert through college. Unlike Edward who went to nearby Columbia College for a quick law degree, Herbert was to have his pick of four-year institutions, even if such an expensive education did not have a clear link to making a living. Although Lydia again worried about sending him away from his Quaker community, she and Aaron bowed to Herbert's wishes and the recommendations of the "professors" at Poly. Lydia and Aaron decided that he could pick among the very first American schools that had found the resources for undergraduate and graduate education in the sciences.

There were few such colleges in America because science education was expensive and because there were few academically justified science courses. Traditional liberal arts colleges needed only a small library, a dining hall, lodging for the students, and bare classrooms. Good science and technology schools needed all that, and much, much more. Apparatus had to be purchased and constantly replaced; supplies and samples were required; special buildings were often essential; explorations had to be funded; and, science professors who were able to command higher salaries than teachers in the humanities had to be supported. In addition, college leaders were struggling to define viable curricula for science classes. All that meant that only a few institutions could afford to try to imitate the well-funded German research universities.

Three such rare institutions were high on Herbert's list: Columbia, Yale, and Harvard. New York City's Columbia had some scientific schools and America's first library school (founded by the young Melvil Dewey) but it seemed to be lagging behind the others in achieving the university ideal. It was not until 1896 that Columbia changed its name from *college* to *university* and even then its new departments leaned more toward technical training than the pure sciences. The college also seemed to the Fields to have too much of a connection to Episcopalian traditions.

Connecticut's Yale was relatively close to home. On the other hand, the distant Harvard in Massachusetts did not seem to pose a threat to the boy's religion. Yale had remained committed to many of the old Congregationalist ways. It had a prescribed course of study and was yet to be recognized as a modern university. In contrast, the once conservative Harvard had shifted to a more liberal religious orientation, Unitarianism, in the previous century. Unitarianism was so theologically liberal that many thought Harvard's leaders had "pushed God off of the campus."[15]

Unitarians prided themselves on having adapted to the Enlightenment and the use of reason. Their symbol became a flame, rather than a cross. For Lydia, that seemed a plus. She and Aaron's liberal Quaker beliefs shared much with Unitarianism. At least Harvard would not evangelize Herbert during compulsory chapel attendance, she concluded. Important for Herbert, Harvard's dynamic president had instituted the elective system so that undergraduates selected many of their classes. And that reforming educator, Charles William Eliot, had invested heavily in graduate schools and scientific "museums" that were theory oriented.[16]

Herbert Chooses a University and a Career

Eliot had more than amplified the university's earlier commitments to liberalism and science. He made sure the university carried on the work of the world-famous Swiss zoologist and geologist, Louis Agassiz. Eliot also spent much money to attract and retain energetic and distinguished researchers in several of the other specializing scientific fields. Geology, biology, and zoology had been favored in Harvard's post–Civil War budgets.

The Agassiz family had joined in with generous financial support, especially for the university's zoology museum, making it perhaps the greatest zoological research center in America. That was particularly attractive to Herbert because his days at Brooklyn's Friends School and Poly had sparked great interest in what was then the queen of the natural sciences, zoology. Harvard had not sacrificed the liberal arts, however. The best scholars in philosophy, history, and the arts brought academic blessings to political liberalism and causes such as educational reform and internationalism. Few Harvard students left the school without having classes in all the "new learning."[17]

By 1885 the seventeen-year-old Herbert had set his mind on Harvard, and more. Herbert decided on a career. He informed his parents that he was intent on becoming something new to the world, a professional scientist. Aaron was troubled by that. He knew there were few well-paying jobs for such men. Although he had been able to put aside money for all his children's futures, he worried about Herbert because he knew that most scientists worked in not-too-well-paying colleges. Herbert, however, did not fret; perhaps he was too young to be concerned about money. Certainly, he was confident in his academic abilities.

2 An Unexpected Library Revolution, at an Unexpected Place, by an Unusual Young Fellow Named Melvil Dewey

In 1885, as he was preparing to leave Brooklyn for Cambridge, Herbert Field had no idea the head of the nation's first library school at New York's Columbia University would soon have a profound influence on his life's work. In the mid-1870s, a decade before Herbert entered Harvard, a very unusual student at a small rural Massachusetts school began what might be seen as an information revolution. The school, Amherst College, was an institution that had remained a close-knit and warmly paternal liberal arts school with a relatively small budget, while Harvard was beginning to be a research-oriented institution with the funds to meet a growing need for a vast expansion of its facilities, including its libraries. Besides attending an "old fashioned" college, the young man had a far different social, educational, and religious background than Herbert. Melvil Dewey was an unlikely candidate for leading any intellectual or even a library revolution—or having any links to the likes of a Herbert Field.

Amherst College was founded in the 1820s to provide education for devout, usually poor, young men bound for the ministry, as well as to fill an educational gap in what was then a near frontier area. At that time, being ninety miles west of Boston meant the college was far from the centers of culture and learning. After an early growth spurt, the school remained small and resource poor and, unlike the liberalizing Unitarian Harvard, kept its commitment to Congregationalism. While the students in Cambridge were moving toward secularism, during the pre–Civil War years Amherst's became involved in the waves of emotionally charged revival movements that sought to restore faith, piety, and deep personal religious commitment. As in the other fervent revivals of the era, the students' turn toward a more heart-based religion was accompanied by obligations to a spectrum of reform causes, ranging from temperance to the abolition of slavery. Those commitments came with greater degrees of passion and absolutism than those of the liberal reformers in Boston or New York City.[1]

By the end of the Civil War, Amherst and most of its students had calmed and the school, like other of New England's small private colleges, was beginning to serve a more prosperous and less religiously centered population.[2] However, it would be decades before the college deserved anything like the elitist "preppie" reputation of later New England private colleges. The school remained religiously oriented and concentrated on undergraduate general education. But as its income and endowments increased it tried to accommodate to the new learning and to the financial demands that accompanied it.[3]

Those economic burdens included financing a growing library collection. The college's small assortment, less than fifteen hundred books in the 1820s, grew to six thousand by the eve of the Civil War. Then, it began to grow at a much faster rate. The school was soon adding over a thousand volumes a year. As well, those gifts and purchases included more subjects than when the collection focused on theology.

Organizing the collection, finding the means to protect the books from theft, and providing guarded access to the library tapped always-scarce resources. Although tiny compared to Harvard's collections, Amherst's was becoming a burden. Adding to the cost was the need for additional rooms, as well as expensive shelving and furniture. A faculty member had to be appointed as a part-time librarian. Just as the Civil War was ending, an investment was made in a search for ways to satisfy faculty and student needs while, at the same time, reducing the demands on the librarian as he responded to questions about the library's holdings. An experiment was begun. It took much time and effort to start building a card/slip catalog of the collection, one that could be consulted by the librarian and his helper and, perhaps, trusted students. The experiment was not that revolutionary, however. Following the format of earlier printed catalogs, the books were listed on cards arranged by the author's name along with an indicator of the physical location of the book, if there was a fixed one for the work. While of some help, the new catalog did not solve the college's library problem.

Within a decade after the war, keeping track of a collection with close to twenty thousand books, reshelving them, processing the paperwork on four or more new books a day, and finding places for them in the stacks, was overwhelming the librarian and his student helper. Then, more financial burdens had to be taken on. In response to student demand, in 1871 the library was opened for six hours a day, not just three hours a week. That called for additional staff time.

There were other pressures on the school's personnel and budget. As the range of subjects in the collection increased, the librarian faced more and more demands on his memory. Patrons who did not know the names of authors when they sought books on various topics asked the librarian for help. Even if a relevant author or two was suggested, a user could not be sure that he would get other related works because the books were, at best, cataloged and stored on the shelves by authors' names. Works on the same subject were scattered over the library's rooms. As a result, "search time" was becoming unbearable. As well, when a significant number of new works was added to the collection there had to be a costly labor-intensive repositioning of books. That task was especially onerous in libraries that used room-shelf-row types of identifiers for book locations. In such situations the catalog entry for each book had to be changed to point to the new absolute physical location of an item.

A Determined Young Man of Economy from the Revival District

As Amherst's library problem was deepening, an eighteen-year-old from the area of western New York called the Burnt-Over region decided to switch colleges.[4] In 1870, Melvil Dewey left the college sponsored by his rather fundamentalist Seventh Day Baptist denomination and entered Amherst to attempt to gain the education and credentials he needed to become a crusader for education for the common man. Soon, his experiences at Amherst shifted his target from democratic schooling to libraries as the medium for egalitarian education. The shift, however, did not mean a loss of his near evangelical drive for reform. In that and much else, Dewey was more like the Amherst students of the 1830s than those of the 1870s.

Melvil was raised in a small town on the far western edge of the state of New York. Adams Center was in an area not too far removed from frontier times and had been swept by revivals and social experiments for decades. The Deweys' later residence, Oneida, New York, was the site of one of the most radical religious-based communities in American history. The family was not part of John Humphrey Noyes' perfectionist communal experiment, but the Deweys raised their children to be religiously committed. That, his small town background, and the lower middle-class economic status of his family, made Melvil's life and perspectives very different from those of Herbert Field and his rather elite liberal urban friends. However, Melvil's education and encounters at Amherst led him to share some interests with those men despite his becoming an unusual mixture of the intellectual and cultural old-and-new.

To help pay for his college education, Melvil took a job working as a library assistant at Amherst, just as the college was entering its "library crisis" years. The library work shifted his youthful focus on school improvement to library problems. Dewey observed the frustrations at the college and began to think of ways to solve what he saw as unnecessary disorganization and waste. He called on his past experiences and interests, as well as his evangelical-like energy, to find solutions. Since childhood he had been fascinated by mathematics and orderly ways of problem solving. Even in his early teenage years he was applying them to develop means to bring economy and efficiency to the organization of his and others' lives.

Included in Dewey's list of ideas for reordering society were three that he wrapped in revivalist enthusiasm. One was shared by the likes of Herbert Field: Dewey spent a lifetime advocating the adoption of the metric system that scientists and academics were urging as a universal standard. In contrast, the second idea seemed odd: Throughout much of his life Dewey labored for something few academics approved. He spent time and a great amount of money urging the adoption of his system of simplified spelling.

In the name of Economy he sought to reduce the number of letters in words through his version of phonetic spelling. For example, he changed his first name to have single *l*s, altered *relative* to *relativ*, changed *rhyme* to *rime*, and asked that his last name be spelled *Dui*. Few university types joined in that crusade. The third idea received a mixed reception: Dewey targeted another "uneconomic" habit. That campaign would gain the support of some American intellectuals, although many humanists resisted his pleas. Dewey argued the Roman numerical system wasted time and space and it suffered from not having a zero. Why write, for example, MMXXXIV when 2034 would do, he asked?

In the 1870s, Amherst showed no interest in Dewey's spelling or number reforms, but his library work impressed the school's administration. Upon gaining his undergraduate degree when he was twenty-three (a relatively late age for all but financially poor college graduates of the era) he was hired as the school's full-time librarian. The administration placed great faith in him and in his promise to bring order and efficiency to the library. They believed that he could fulfill his pledge of a money- and time-saving radical reorganization of the school's collection.

Melvil had readied himself for the challenge. Soon after he entered college he had begun surveying the conditions in other libraries. He found that his small school was certainly not alone in its state of wasteful disorganization. At best, he later described, some libraries shelved their works within a few gross "subjects," ones so broad that books were hard to find. That was accompanied, in many libraries, by cataloging systems that gave the location of a book by what was termed a "press mark." A typical mark had "A" for the shelf area containing Chemistry, "B" for the particular shelf, and a number, say "5," to indicate the book was the fifth book on that shelf. Such systems did more than tie a book to a specific location. They made it difficult for users to find works and they called for expensive redoing of catalogs if a collection had to be moved. As bad to the economy-minded Dewey, each of the libraries he surveyed had a different classification-identification-location system. That prevented savings in effort and money. If all libraries could agree on a standard, then a book could be given the same classification-location number and users would know where to look for a volume no matter what library they visited.[5]

An Evangelical Enthusiasm and the Goal of Economy

Dewey had also been investigating various library remedies proposed by others, hoping to put them together in a useful combination. His evangelical-like approach is reflected in his statement of how a major part of his eventual solution occurred to him: "For months I dreamd night and day that there must be somewhere a satisfactory solution. After months of study, one Sunday during a long sermon by Pres. Stearns, while I lookt stedfastly at him without hearing a word, my mind absorbed in the vital problem, the solution flasht over me so that I jumpt in my seat and came very near shouting 'Eureka'! It was to get absolute simplicity by using the simplest known symbols, the

Arabic numerals as decimals, with the ordinary significance of nought, to number a classification of all human knowledge in print."[6]

Central to his evolving plans for creating a new library system were: a subject approach; the use of Arabic numbers to classify/identify books (and, he later stated, clippings and business documents); and, an improved card catalog. The young Dewey paid little attention to the problems of a new form of communication that was developing at the time, scientific journals.

Cataloging and shelving works by subjects, he believed, might solve many interconnected problems. "Subjects" would facilitate the placement of intellectually related works close to each other and would allow finding many works of interest by browsing. A "subject" organization could also be more efficient and less costly for library management. Books might not have to be identified by a physical location but, perhaps, by the "relativ" location within the subject area. Those "subject" identifiers might also serve another Economy goal. When a collection had to be moved the identifier could remain the same.

There was a suggested special version of Arabic numbers that Dewey explored in detail. He thought it might facilitate a flexibility vitally needed for the identifiers of items in expanding collections. Combined with card files they also might allow patrons to search for information in the catalog rather than in the shelves. Moreover, Dewey thought that his search would lead to a practical classification scheme (an outline of knowledge) that would satisfy all the needs of a general library and its users.

Dewey honed his ideas while he worked on the Amherst collection. After two years, he made decisions critical to the history of libraries and the organization of information. He left Amherst; he decided not to pursue an advanced academic degree; and, he founded a for-profit company to serve libraries and businesses. In addition, he did something that was strange to academics like Herbert Field and his scientific colleagues. Dewey copyrighted his work. He treated his publications, if not ideas, as property. There was something else that was important to Herbert Field's own information Calling: Dewey would not rush to create foreign-language versions of his systems, as he foresaw few customers outside of North America.

A Published System and a Library Bureau

Believing that he had put together all that was needed for his library organization system, he published his first edition of his *Dewey Decimal Classification and Relativ Index* in 1876, just after he left Amherst. At the same time, he helped found the American Library Association and he established a company to sell products designed for offices as well as libraries.[7] He and his for-profit firm, soon called the Library Bureau, created and sold methods and equipment for Economy (later known as "Efficiency").[8] Dewey continued his library work, however. In 1883, Columbia University in New York City hired him to attempt to bring its library up to the new academic standards. It also allowed him to found America's first library school. That school was not oriented to research. Dewey viewed it as a tool for thorough vocational education in library "Economy." After five years and some conflicts with Columbia's administrators, he moved on to modernize New York's state library in Albany and then spearheaded some reforms in the state's educational system. He had moved his library school with him and continued viewing it as a place for training a workforce for the expanding general library market.

By that time, despite hints of the emergence of more sophisticated classification and cataloging systems, he was finding that his classification was on the way to becoming a standard, at least in America's growing number of public libraries. Dewey made an additional move in 1895, one that was predictable given his social and religious background. He founded his own version of a utopian colony (at least for the rising Christian middle class) at the then remote Lake Placid, New York. It was a far different social environment than what Herbert Field and his friends were accustomed to.

A System for the People, Not the Researcher—Papermaking Advances, the Post Office, and A New Information Technology

Dewey's library system had emerged out of a merger of a practical challenge at Amherst's library, his youthful interest in mathematics and "order," and his sense of potential markets. Dewey was not a speculative or abstract thinker. His goal had been to devise a useful system that classified all knowledge, but one that allowed a busy librarian in a typical library to catalog and manage the books in growing and ever-changing collections. As Dewey had been modifying and merging old and new ideas about library technology and classification systems he had simplified his task by focusing on the needs of a common user and on a method for bound books—not articles, newspapers, or archival documents.

Especially important was the creation of a classification system comprehensible to typical patrons, as well as to usually harried librarians who did not know the details of every topic in a collection—but who would have to classify books as they came into their library. To Dewey, the existing approaches, such as the great intellectual efforts by the British Museum and England's Royal Society, which were aimed at scholars and scientists, suffered from classifications that were unfriendly to "the people" and library personnel.[9]

Dewey had confronted more than an intellectual problem. He knew that libraries needed a hardware revolution. For Dewey, and even other library modernizers such as Ezra Abbot at Harvard University's large library, the elegant and expensive commemorative printed catalogs universities sent to their supporters and other libraries, had outlived their usefulness. They certainly did not fit the needs or resources of a town or school library.[10] Melvil made a commitment to card catalogs as the solution, although existing card and "slip" systems had many flaws.

There had been several attempts, such as in Germany during the 1820s, and at least at Amherst and Harvard in the United States beginning in the 1860s, to use those technologies for a library's permanent catalog. As well, librarians had used cards as a means to record acquisitions until they posted them in a new edition of a book catalog. However, there were sincere objections by those who saw the card catalog as uninviting, hard to use, easily destroyed, and, importantly, as a technology that prevented sharing of information among libraries. Quality, durability, and other technical problems had also limited the adoption of cards. As a result, no card system had proven to be a practical tool for the public use in a library. Melvil had bravely decided to find ways overcome the technical problems and make cards practical and affordable.

He soon decided that standardization was essential. He refashioned the special postcard recently introduced by the United States Post Office. The postcard was the result of new chemical and papermaking advances that soon produced another marvel, cardboard. Dewey was able to convince the Post Office to change the size of its card to one better suited to library use. With the government and private contractors making a standard size card, he predicted, manufacturers would improve quality and lower prices. Satisfied with his cards, Dewey then designed an innovative piece of furniture, a wooden library card-cabinet. His "Dewey cabinets" were ingenious and were a practical alternative to the high-cost automated card-display systems that were soon proposed by other library innovators.[11]

Dewey's cabinets had separate drawers to hold the cards in the correct "subject" order and a clever security device to ensure (he hoped) that cards would not spill when a drawer was being used. He devised his own version of a long rod that ran through the specially punched holes in the cards to prevent accidents—those who had experience with earlier paper-slip and card systems deeply feared catastrophes when drawers were opened. With the cabinet, rod, and cards, he thought his catalog could easily be updated, and most importantly, it could provide a convenient and efficient way to search through a library's current holdings.

Melvil went further. With Thomas Edison's help, he devised a standard for the handwriting on the cards.

More than technology was required for success. Because there was no organization in the nineteenth century to provide a central classification service, Dewey knew he had to create a schema that was intellectually flexible and that a harried and perhaps a not-well-trained local librarian could use to classify his own collection. It had to be one that was simple enough for patrons and staff to understand. As well, the system had to have a notation that satisfied the needs of shelving and retrieval of books. Melvil also sought one that facilitated standardization across libraries so cost saving could be shared. The instruction manual for any future classification system was also an important consideration: he was sure that unless it was friendly and affordable, the system would fail.

One of the keys to the success of Dewey's system was his construction of the *Relativ Index*. It did make his system friendly. He alphabetically listed, and in terms common to Americans, all the names, synonyms, and related terms for the "subjects" in his system. Beside them were the identifiers. A librarian, even a user, would look for those common names and be pointed to the indicators. If interested in tariffs, one could go to the "T" section, look down the list, and easily find the identifier "337."

Identifiers and Decimals, but What Kind Please?

One of the most economic aspects of Dewey's library system was his reliance on numbers, not words or letters, as integrated identification, classification, and shelf-arrangement tools. His "numbers only" identifiers served as indicators of the subject covered by a work and as the pointers to its physical location.

Dewey, as had others before him, recognized that numbers had many practical advantages over words or letters. Words were informative but impractical to use when marking books. Although a combination of letters and numbers (which more elaborate systems would later use) allowed more combinations, pure numbers took much less writing effort; they were unambiguous; they were near universal symbols; and, they took relatively little space so they could be written on the spines of books. With the visible marks, librarians could efficiently shelve items in the correct places and patrons could easily locate them.[12]

Dewey did believe the type of numbers he selected would soon become an efficient, universal "language." His numbers were to be decimal Arabic, not Roman. And, he put a special and important interpretation to his numbers. They were not to be arithmetic. They were not to be used for addition or any other mathematical purpose. His system used each number as a category indicator. That was new and strange, even though others had previously tried using such "Decimal" numbers for library work. Even in the 1920s, numbered decimal classifications in libraries seemed odd in many European countries whose managers also typically liked words and had a dislike for all catalogs.[13] In addition, to make his system practical for use in regular libraries, Dewey wanted as few numbers as possible for classifying item. Initially, he hoped that three numbers would prove satisfactory.

Orderly + Parsimonious + Familiar = Information Breakthrough

Sparse numbering depended on an efficient and well-ordered classification system. The one Dewey used was orderly, uncluttered, and parsimonious—and it was friendly to the "common librarian." It reflected the way knowledge was hierarchically organized in the Western world in the mid-nineteenth century. Moreover, to a great degree, his system matched the relative attention paid to various topics at the time. The understandable nature of his taxonomy (based on Francis Bacon's schema and the division of subjects in colleges) proved to be a major attraction—but it also was a limitation of the system. On one hand, its familiarity, its relative simplicity, and its hierarchy of knowledge allowed his users to search from the general to the particular within a subject. On the other hand, those features made it difficult to adapt the system to later changes in knowledge structures. Also, the system was, at its most detailed levels, oriented to the work of traditional academics who saw categorization, not the newer empirical research, as their fundamental activity.[14]

As with some other aspects of his system, Dewey's knowledge structure was not the most sophisticated of the era. Library professionals, and those responsible for classification in specialized fields, soon began to devise alternatives. An elegant one was a system that emerged from the work of experts in Boston's large libraries. However, Charles Ammi Cutter's system was complex and called for skilled professional classifiers—and much time-consuming effort.[15]

Dewey's classification scheme was intentionally limited. In his Decimal system, all knowledge was to fit into ten highest-level categories. Each category was subdivided, hierarchically, into ten subdivisions, and once more, if necessary, into another ten. However, Dewey's hope that a cost-effective three-digit subject code would be sufficient for organizing a library's holdings withered as libraries encountered exploding subject specialization, and as librarians demanded a system that fulfilled logistical and subject specialists needs.

Classifications for the Public

Dewey, however, had begun with great hopes for a successful sparse classification for small libraries. An early version of his schema yielded these ten highest-level categories:

000. Generalities (Encyclopedias, Periodicals, Journals, Bibliographies)
100. Philosophy, Esthetics, Psychology
200. Religion
300. Social Sciences
400. Linguistics
500. Pure Science
600. Applied Sciences
700. Arts and Recreation
800. Literature
900. Geography, History

The most inclusive, highest level, Decimals were sensible, as were the first attempts at the hierarchical subdivisions of the major categories. The subdivisions fit well with the nature of the collections of general libraries and with the concepts and needs of the public. For example, the division of the "Pure Sciences" into the following ten categories appeared adequate to respond to the needs of an educated gentleman of the nineteenth century, even when, as Dewey envisioned, only one subject code was assigned to a work. Moreover, the categories were understandable enough to allow a librarian to assign the classifications.

500. Pure Sciences
510. Mathematics
520. Astronomy
530. Physics
540. Chemistry
550. Earth Sciences
560. Paleontology
570. Biological Sciences
580. Botany
590. Zoology

Dewey realized he could not meet all the library challenges with his basic classification system, but he favored usability over complexity. There were also some weaknesses that became evident as knowledge expanded. Some of his classifications seemed a bit strained, even in the nineteenth century. He placed, for example, much of medical knowledge in the general category "Technology." Then, when the policies of libraries changed from holding only literature that was classic and uplifting to providing the public with popular fiction, the Dewey categories were overwhelmed. Much later, it was difficult to find a sensible intellectual home within the system for new fields such as Computer Science. Dewey's separation of science and technology would prove especially troublesome.

However, Melvil's classification scheme had great advantages. For example, its original description/instruction book fit into one small volume that sold at an affordable price—unlike some later systems that required many huge, expensive, and intimidating volumes of rules and codes. That, and his enthusiastic sales campaigns, soon made his system the popular choice among American librarians.

The Consequences of Specialization—Where Should We Cast Our Bibliographic Fate?

Three digits yields only one thousand classifications, but that seemed enough for storing and recovering information in a public library of the nineteenth century. In fact, some librarians complained that was far too many for their needs. However, one thousand "subjects" soon proved inadequate, even for the small college libraries of Dewey's time. Dewey responded by deepening his classifications. To allow for more subcategories, Dewey added a decimal-point divider to his scheme. Subject classification for butterflies, for example, called for 595 for the class "Other Invertebrates," then 7 for "Insects," 8 for "Lepidoptera," and 9 for "Butterflies," yielding a classification of 595.789.

There were more practical demands that tore away at Dewey's minimalist goals during the twentieth century. Inventory management called for many appended codes to allow the unique identification of a book. Dewey would later be convinced to add codes for secondary subjects, types of works, places, and dates. By the mid-twentieth century a "Dewey number" might be a dozen places long.

There was another force working against his economic goals. A demand for catalogs organized by authors, not just subjects, quickly arose. That placed more demands on Dewey's originally bare system. To adjust the use of words to his numeric system and the need to place identifiers on the spines of books, he added "Cutter" identification codes. They turned names into just a single letter and a digit or two. Dewey also approved his catalog entries being sorted by name, not his numeric subject identifiers. In reaction, accommodating to the expansion of knowledge and the logistical demands, within a generation his instruction booklet grew from forty-four to some eight hundred pages. He also had to expand his innovative *Relativ Index*. In addition, he soon allowed words to be printed on catalog cards such that users could find items by alphabetic "subject terms," not just numbers, titles, and authors' names.[16]

By the early 1890s, more than a decade after Dewey first published his *Decimal*, a few visionaries in America and Europe began thinking of using his system for classifying articles and other types of specialized documents. They wanted a world bibliographic system that could organize all knowledge, not just arrange books in a small library. They sought a system for specialists, not for the common man or an undergraduate student in a liberal arts college.

But much about Dewey and his structure was worrisome. Could a classification built by a liberal arts graduate from a small semirural college ever be adequate? Could well-placed intellectuals and scientists work with a for-profit entrepreneur who spelled *Relative* without an *e*? Would Dewey charge them fees?

3 The Great Men at Harvard and Herbert's Information "Calling"

When Herbert Field boarded the train for Cambridge, Massachusetts, in 1885 he had no direct knowledge of Melvil Dewey, but he was psychologically, intellectually, and morally well armed for university life. He felt that he had at last found effective ways to minimize his stammer, and he had developed tricks to manage his painful migraine headaches. He already had command of several languages, including Russian, Dutch, and Latin. He had a remarkable ability to concentrate on a problem, he had a photographic memory, and Poly had attended to his liberal studies. As well, Herbert's Quaker upbringing, and perhaps shyness, led him to avoid the temptations of undergraduate life. There is no sign that he drank, got into "girl trouble," or became involved in any town-gown conflicts.

His discipline and talents allowed him to finish his special undergraduate natural science program within less than three years. But Herbert was not a scientific nerd. He studied much European history and literature while in Cambridge, writing some impressive student essays and exhibiting an interest in one of the great causes of the period, internationalism.[1]

His attainments pleased Lydia and Aaron. Their Herbert was the first-ever true college man in the Fields' history, and they were especially proud to have a son who was a distinguished college graduate at age twenty. It did not take much to persuade Lydia and Aaron to support him through another three years at Harvard, which he needed to obtain a rare thing in America, a PhD degree, and to become, he hoped, a path-breaking researcher.

Herbert, Meet Edward Laurens Mark—Mentor, Innovator, and Star of a New Academic Breed

Herbert had a smooth transition to graduate studies because of his talent and because of his mentor's guidance. Herbert's advanced work was with a caring and attentive man who was one of the first professional academic stars of the era. Edward Laurens Mark was of a new breed in American higher education. He earned his living through his scientific specialty. Most American academics had been ministers turned professors, or gentlemen of means who went into university teaching out of intellectual interest, not to gain a livelihood.[2] Mark had a different background and career. He graduated from a publicly funded institution that was emerging from its small-town, small-college roots as a result of the nation's drive for economic modernization in the 1850s, the University of Michigan. Although its administration was moving the college toward the new university model, Michigan's primary goals were not, like Harvard, the nurturing of well-educated gentlemen or the pursuit of theoretical knowledge, but the production of scientific and technical specialists who could apply their learning to practical problems.

Mark was so impressive at Michigan that he received funding for advanced study in zoology in Germany. The experience changed him and he moved from a pragmatic view of research to a more theoretical orientation. He did so well in Leipzig that he soon received a great honor. He gained a regular faculty appointment at America's rich and powerful Harvard University. He was "at the top" at the beginning of his career. He spent the remainder of his life in Cambridge—where else would an American naturalist want to go?

Harvard had carefully selected Mark to carry on the work of the revered Louis Agassiz, the founder of the Harvard's Museum of Comparative Zoology. With the Harvard appointment, Mark had a rare opportunity. He had the setting and means to pursue a self-determined research agenda in pure science. He did not have to tailor his work to fit any politically driven and perhaps transient interest. He was not even pressured to accept the elder Agassiz's so-called racist, anti-Darwinian beliefs.

Mark did not let Harvard down. He kept his attention on "pure science," becoming an innovator in the study of zoology, which was quite a broadly defined and central field of study in the nineteenth century. Zoology then included what the twentieth century would call medical anatomy, paleontology, biology, and a host of other subjects that later became freestanding specialties.

Mark continued to study zoology from a theoretical perspective, but zoology's findings were becoming relevant to agriculture and public health with the result that politicians considered it valuable enough to begin to support research in it at the new state colleges. It was a fundamental and vital subject. Moreover, since the appearance of Darwin's theory of evolution, scholars and theologians linked significant and politically charged theological and social questions to zoology.[3]

By the 1890s, colleges throughout America had established programs in zoology—but Harvard and Mark remained at the summit of the field, partly because they kept their focus on advancing the science and leaving applications to others. Consequently, Mark's graduates had the pick of the best academic positions in America.

Edward Mark remained at the forefront as he moved zoology into experimentation, not just classification. He put the newly improved microscope to use on a wide range of problems, making zoology "scientific." He became world famous in what became known as "microscopy." It was a subfield that was making significant discoveries about the structure and development of insects, animals, and fish. Mark was also an institution builder and a trendsetter. He founded the path-breaking Bermuda Biological Research station where his students could study fish and other creatures in, importantly, their natural context.[4]

"Paperwork" and Networks in the Growing Science Professions

Mark was also a style setter in one of the most critical parts of the quickly professionalizing sciences, the formal scientific paper. He developed formats that became standards for the articles submitted to academic journals. Mark was interested in bibliography and passed (in fact, forced) that interest onto his students (significantly, to Herbert Field). Mark insisted that at all times his students carry small cards to record citations to books and articles they encountered. They were to share the cards with their colleagues. Mark also had a hand in modernizing Harvard's library, including its Ezra Abbot–led shift to a card-based catalog, and in the consideration of the adoption of Dewey's classification-coding system.[5]

Mark's connections were central to Herbert Field's career. Mark was a powerful force in academia and an influential player in sciences' emerging professional infrastructure. He had worldwide connections among the newly formed national and international science organizations, and he had an international reputation for training only the best-and-brightest graduate students. Among the many future luminaries in science and politics that Mark educated were the psychologist and intelligence-tester Robert M. Yerkes, the geneticist Charles B. Davenport, and, none other than the future nature-appreciating president, Theodore Roosevelt.

Importantly for Herbert, Mark knew of a growing demand and hopes for international science bibliographic services, and of the intensifying frustrations of those attempting to create them. While his international colleagues were pessimistic and felt the best they could do was to encourage the many small national science bibliographic services to begin to cooperate on a voluntary basis, Mark remained optimistic about a much stronger organization. He passed that vision onto his students.[6]

Edward Mark provided more than optimism and professional connections to his students. He demanded intense efforts, and Herbert and his friends responded. But that did not mean Herbert became a narrowly specialized scientist or a person without wider involvements. He took the time to go with his family on expensive European sojourns while he was at Harvard. He used those trips to reinforce ties to the French Havilands and to meet some of the scientists who had become aware of the accomplishments of his mentor. His family's frequent summering at fashionable Nantucket was used to conduct experiments under the tutelage of Mark's colleagues as well as to learn of the doings of New England's high society.

In the Family or Not—Mark Fosters Another Great Connection

Herbert Field had joined a high-powered group when he entered Mark's program. One of Mark's students had an especially great influence on Herbert's life and the lives of his children. George H. Parker, who was four years older than Herbert, but only a year ahead of him at Harvard, was also a favored Mark student. Parker soon became Mark's colleague and a world-famous zoologist and philosopher of evolution. George became very close to Herbert, partly because Parker was courting Herbert's charming cousin and Brooklyn Heights neighbor, Louisa "Lulu" Stabler. The two first met at a grand party Aaron Field held at Harvard to celebrate Herbert's accomplishments. After a romantic courtship, including weekly letters while George was studying in Europe, Parker became part of the extended Field clan. He married Lulu in a Quaker ceremony at the family's Brooklyn Heights mansion. After that, Herbert and Parker remained close friends throughout their lives, with Herbert frequently staying at the Parker home when he visited Cambridge.[7]

Herbert Becomes a Scientific Insider and Crusader at the Beginnings of the University System

Herbert had quickly won a master's and then a doctorate in zoology. As well, he had become one of Mark's favorites, nearly guaranteeing a major role in academic science. An indication of the respect Mark had for Herbert appeared during the early twentieth century's impressive celebration of Mark's career, when Herbert was seated next to his illustrious teacher. Mark also arranged to have Herbert awarded another honor, the status of an Associate of the Harvard Museum, and he helped Herbert gain recognition as an official member of the Smithsonian Institution's delegation to international information meetings.[8]

By the time Herbert completed Mark's program, he had proven himself a phenom and had connections that any young academic would envy. At age twenty-three he belonged to a new and tiny American intellectual elite of professionalized PhD-holding academics. An American with a doctorate was uncommon. Harvard, for example, had awarded less than 250 of those degrees in the preceding thirty years. There were reasons for the lack of an American or even European-trained American scientific professoriate: the scarcity of American graduate schools; the cost of enrolling in higher education in the new learning, even in Europe with its favorable exchange rates; and the fact that few professors with advanced degrees could envision making a generous living in academia through their scientific skills.

Herbert had decided not to spend time pursuing any limited professorial opportunities. Despite Mark's encouragement, he did not seek a teaching position. Even if good teaching jobs were available, Herbert knew he would have an uphill battle as a professor because he feared he had not fully eliminated his stammer. Fortunately, he did not have to go into teaching. He did not even have to seek a job in 1891. The indulgent Lydia and Aaron again came to the rescue. Herbert began to lead a professional life more like the older type of "gentlemanly" scientists, not one like that of the new professors such as his mentor Edward Mark.

The French Connection, Multilingual Herbert, and the Higher Circles of Science Information

Herbert's parents determined to maintain him while he continued his research and writing. He traveled to Europe to see the Havilands and to pursue advanced study, gaining three more doctoral degrees in Germany and France. Also, in a period of just two years, he produced a string of articles that made him a world-recognized zoologist. He became an established young academic and was no longer regarded as just a favored alumnus of Mark's program.[9]

Herbert was also quickly becoming a man-on-the-make in the institutions and politics of the professionalizing sciences. His shyness receded when he dealt with technical matters. In 1891, he was gaining notice by voicing his own special concerns about what came to be called a "science information problem." By 1893, the young Herbert had made sure he had met or become known to leaders in all the biological fields. He attended professional meetings in Europe and America where he built a reputation as an advocate for scientific bibliography. His integration into the higher circles of European and North American (even South American) science was speeded because of his financial ability to make trips to all the major zoological centers and to participate in many international congresses.[10]

It was not just money that propelled his career. Herbert had talent and the ability to read, write, and speak all the major science languages of the era. His amazing multilingual skills were important to the next stage of his career, and to his coming contributions to science information.

Aaron and Lydia's Surprise: Hamilton's Desires, Herbert's Calling

Herbert was a rising scientific star, and he was a young man without practical life responsibilities. He remained a bachelor and there were indications he might never have to find regular employment. That was not a great worry to his parents.

Their youngest son's ambitions were also tolerated. Hamilton, after experiencing, as had Herbert, some health difficulties while at Poly, entered, then quickly left, Columbia University's School of Mines where he had begun the study of architecture. After that, he moved to the more liberal arts–oriented Harvard—but only for a brief stay. Lydia and Aaron forgave him for that. Then, to fulfill Hamilton's "potential," Lydia decided he could continue his education abroad. She was determined that Hamilton Easter Field would realize all his dreams. Lydia sent Hamilton to Paris to study architecture.

She felt confident because her influential Haviland relatives, she knew, would always provide advice. As well, Hamilton already spoke and wrote in several languages, including Greek.

However, just as Hamilton went to France, Herbert (figure 3.1) made a life-changing decision, one important to the history of science information but one that did raise Aaron and Lydia's concerns.

Carrying On the Field Family's Commitment to Social Liberalism and International Peace

In 1893, the twenty-five-year-old Herbert had informed his family that he was not actively seeking a job, even a research post. More startling, he announced he was going to postpone his career as a zoologist. He told them he had experienced what the old New England Puritans and Congregationalists might have termed a Calling (an Inner Light for Quakers). He had decided to pursue a new interest, at least for a time. Perhaps the change in life goals was triggered by what he later told a few friends: worry about failing eyesight, something that threatened his microscopy work. Or, the decision might have come because he could not find a pure research position in a university, American or European.

Aaron and Lydia were apprehensive but did not protest—although the new life path Herbert described seemed unlikely to provide a livelihood. Herbert announced to his parents and to his professional colleagues (in an article in the important international science publication *Nature*) that he had found his own way of continuing the family's commitment to social liberalism and international peace through a dedication to the rational transformation of science information systems. While it meant putting aside zoological research, for which he had been expensively trained, Herbert declared he

Figure 3.1
Herbert Field, from Gotlieb Archive

felt a deep and unshakable obligation to his new causes. He promised he was going to apply a revolutionary new tool, Efficiency, to bring order to the world's scientific literature.[11]

Herbert's choice of such a vocation was unusual. He had decided to become what the nineteenth century might have called a bibliographer or a librarian—mid-twentieth-century people would say that he had become an aspiring information scientist. He told his family and colleagues that his new ambitions ran deep. He was going to be much more than an ordinary bibliographer; and, he said, he was not going to concentrate on just zoological literature, although at the time that area of study was near all encompassing. Nor, he said, would he use the old methods of managing information or build a system just for America's scientists. Probably unaware of the many failed attempts since the sixteenth century to create universal bibliographic systems, Herbert was optimistic and energized.[12]

He began telling his peers that he looked forward to creating new techniques and tools to serve science around the world once he gained approval, time, and resources. To obtain them, his first step was to persuade his family to continue to support him. That was not difficult. Then, for the needed endorsement and aid of his colleagues, he began to contribute additional short articles to scientific journals announcing his determination to find groundbreaking ways to solve the sciences' information problems.

Addressing the Researcher's Bibliographic Challenge

Herbert soon detailed the reasons for his radical change of life goals, both to his family and to his scientific friends. He had, he said, first become of aware of a great information problem as a graduate student under the information-minded Mark. He detailed how he and his fellow Harvard post-grads had experienced the anxiety-arousing threat of missing a critical paper when committing to a topic for their doctoral research. Repeating what had been done previously, or not including the findings of another scientist, could have led, he said, to the rejection of theses and make-or-break academic papers vital to building a career. Worse, if a scholar overlooked an earlier publication, there might be damaging accusations of plagiarism.

Herbert explained that being aware of all relevant research in even a narrow field had become time consuming and costly, prohibitively so, even for those who had finished their scientific apprenticeships. What scholar could know of the research in every nation of the world, he rhetorically asked? The best American universities held less than one-third of the world's science periodicals—few European libraries did much better. There were many national bibliographies in Europe that aspired to list all the major publications of a nation, but they were just national, usually incomplete, for library books only, and always late in appearing. The many calls for international bibliographies since the mid-1800s had usually gone unheeded, he emphasized. So, Field said, even if he and his fellow researchers had the time to leaf through all available publications and bibliographies, they would miss at least two-thirds of the relevant journals and articles.[13]

Herbert described how older guides to the zoological literature, such as the one created by Harvard's Agassiz, and the newer ones published in Europe, were dangerously incomplete. He emphasized that despite the current attempts to provide up-to-date information, all the bibliographic efforts failed to provide adequate coverage of significant articles, books, vital professional proceedings, and, importantly, other less-than-formal scientific communications that were becoming essential for those trying to keep at the edge of research. Herbert went so far as to criticize England's famous *Zoological Record*. Begun in 1864, it was one of the world's first bibliographic journals. Herbert claimed it was inadequate for modern needs.[14] That shortfall had led him, Herbert reported, to begin a search for alternatives as early as 1890. He noted that he was not alone in sensing a dire need for bibliographic help. His European travels had shown him that keeping up was not just a problem of young, struggling American academics and their money-starved universities.

He detailed how he had taken the time and spent the funds needed to visit scientists in all European nations, except Portugal and those in the Balkans. He had found the same needs and complaints everywhere. Science was expanding over all of Western civilization. It had become international and multilingual, but it remained without the methods to allow its workers to remain current on even a fraction of any field's literature. Herbert's writings emphasized that although scientists in many subjects had been urging their international congresses to explore the possibilities for international indexing journals, they had been unsuccessful.[15]

The Problem of Information Plenty

Herbert and his contemporaries were correct about the growing information challenge and its consequences. Even half the estimated increase in the scientific literature in the nineteenth century presented a daunting bibliographic challenge. The number of new books published each year in just the United States had increased three-fold in the twenty years after 1800. As the number of the world's scientists increased from a thousand in 1800, to ten thousand in 1850, then to a hundred thousand in 1900, there was a flood of scientific articles. The number of science periodicals had tripled in a generation. Some estimates are as high as ten thousand scientific, medical, and technical journals and other publications being in print by 1900.[16]

There were many reasons for the explosion of science publication: the nineteenth century's amazing outpouring of scientific knowledge; increased specialization; the belief of every Western nation that it had to have its own scientific societies and publications; the conviction that each specialty "had" to have its own journal; and, as science began to sustain professional careers, scientists were expected to publish on a regular basis. There was an additional and expensive complication: editorial imperialism. A journal attracted more buyers if it carried items on a broad range of subjects. That meant a scientist or a bibliographer had to search through seemingly unrelated publications to ensure full coverage of a subject matter. (The young Herbert Field believed that a complete survey for zoology alone called for a search through, at minimum, a thousand journals—it was worse in other specialties, although not too much worse).[17]

Information Shortcuts, Plus the Index and Abstracts Challenge

Because of the abundance of publications, scientists needed to develop shortcuts for accessing the articles within them. They sought more than improved versions of older aids containing just titles and very short summaries culled only from major publications in their fields. They now desired full "indexes" and formal "abstracts" of the publications from around the modern world—not just from a few of the more popular journals. But the cost of obtaining, surveying, and reporting on the contents of the many new publications seemed unbearable. According to observers like Herbert Field, even if the literature burden was reduced through the expediency of establishing additional traditional indexing and abstracting journals, the current information methods could not eliminate the "overload" problem.

If only twenty items per issue of each of the world's scientific journals of the late nineteenth century were indexed yearly, each annual index's volume would contain fifteen hundred pages, Herbert predicted. Users interested in a decade's developments in a topic would find searching through fifteen thousand pages more than an unwelcome challenge. It would be unbearable. If the index followed the traditional pattern of organization by only author name, not by subject, a decade's search through a topic would be impractical. However, Field and other interested scientists never gave much thought to reducing the problem to one of examining a subset of only the most important publications.

No One, or Government, to the Rescue

The publications and bibliographic problems were grave, so much so that few scientific societies could finance even a continuing survey of publications in their own narrow subject areas. For example, a leading nineteenth-century annual bibliography of zoology, England's *Zoological Record,* could not attempt to cover all of the field's literature, and even then it had to be published at great financial loss. Germany's zoological contribution survived because of the devoted efforts of a professor who had been confined to his sick-room and could no longer do research; and, the Italian's famed Naples zoological bibliographic publication had to be limited to one aspect of the field. Typically, its compilers could not afford to search all the relevant journals.[18]

Money was one of the problems, but only one. Organizational weaknesses had to be overcome. Many governments were seeking new relationships with science in the late nineteenth century, but none had found an adequate solution to the information problem. Every field had small national journals, each unable to fulfill the needs of comprehensive bibliography and each worried about competition. Governments had tried to solve at least their own nation's science information problem with subsidies to scientific societies; but the increasing subdivision of all scientific fields and the multiplication of journals had overwhelmed the traditional systems of government support.[19]

There had been some large initiatives. Germany, whose government heavily subsidized scientific development, was making a valiant effort to further science bibliography, especially for its own scientists and industries. The ambitious German program included publishing useful summaries (abstracts) of papers, but many of those publications, some eighty by the year 1900, were in narrow fields and Germany resisted attempts at the creation of international bibliographic and cataloging standards. France had a similar national science bibliographic effort and various projects in England and the United States emphasized government's role in fostering cooperative efforts among science societies.[20]

A Few Steps Forward

Signs of progress were appearing, but not many. There were promising national attempts at comprehensive indexes in medical and chemical literature before the late 1890s. Pressing military and industrial needs spurred those advances. But work such as that by America's army medical librarian John Shaw Billings, and by the chemists at the Massachusetts Institute of Technology, did not mature until the twentieth century. Some applied fields, such as engineering, whose organizations could provide liberal support in the 1880s, did begin to establish successful indexes that kept their readers up to date. But such success was a rarity.

Ambitious attempts at transnational and general, not specialized, science periodical indexes were also emerging. But achievements were limited, even though their owners were driven by a for-profit motive, and although they tailored their literature coverage to a very wide audience. Despite reporting on only a few hundred of the most popular magazines, so that indexing could be affordable, just a handful of such indexes were self-sustaining. Practicalities such as reasonable subscription fees, mailing costs, and locating trustworthy translators had always been a block to attempts at comprehensive international bibliography.[21]

Beyond National Failings, the Plight of Information Internationalism

The condition of "international" bibliographies, even for single disciplines, was much less satisfactory than for national indexes and abstracts. The state of international multidiscipline catalogs was far, far worse. There had been a few midcentury attempts at all-nation catalogs for particular disciplines but each, like the path-breaking 1860s *Bibliographica Zoologica* and *Zoological Record,* fell short of timely and comprehensive coverage. They had to resort to quick indexing by only author name or by near ad-hoc subject classifications based on a document's usually inadequate titles. The results of using titles or short abstracts as the basis for classification and indexing were frustrating.

Attempts to minimize the cost of classification and indexing by using an early version of "keywords" (single words) had led to chaos because no one could agree which words from the titles would tell readers if an item was of interest. In addition, to meet economic constraints, a publication was typically indexed with terms relevant to only one or two scientific areas. There was neither the time nor money to support reading the full text of articles and then classifying them through a comprehensive and informative theoretical structure.[22]

Of course, no one had overcome the great challenge of finding terms that would inform scientists from many different nations accustomed to using different words for the same concept. And no one found an affordable solution for providing indexes and abstracts in all the major national languages.

Even the Great Royal Society Came up Short

The shortcomings in international bibliography had motivated a few grand attempts at a solution. In the 1850s, just a few years after the Smithsonian Institution was established in the United States as a trust (from the bequest of a British citizen named Joseph Smithson intended to increase and diffuse knowledge), a Smithsonian initiative headed by Joseph Henry called for an international bibliographic project for all the sciences. The failure of the United States' government to provide the funding motivated a plea for help from abroad. Help was needed because there were few resources in America. There was no powerful unifying science organization, just small and weak professional societies; the Library of Congress was only a roomful of books; and America's university system remained tiny.

Joseph Henry pleaded for international cooperation. To take a first step toward his bibliographic goals he called on England's science societies, which were then the world's leaders in scientific information, to finance what he knew was a monumentally expensive and politically sensitive mission to achieve international consensus on the standardization of scientific terms. Henry's plea went unfulfilled, partially because of national scientific pride.[23]

But within a decade, the British decided to develop an empire-wide, if not fully international bibliography. That led to the mighty British Empire's *Catalogue of Scientific Papers* project of the 1860s. Its goal was to index all the science literature of the past then catch up to contemporary publications. The society's ambitious plan remained the only major attempt at an "international" and on-going science bibliography project for a generation. Unfortunately, it was unable to generate even timely yearly volumes, or ones with vital subject indexes to their contents. By 1896, only ten volumes were produced, and they covered just the years 1800–1883. That was after thirty years of work.

As well, a scientist had to plod through thousands of essentially unordered citations in the *Catalogue* to compile a useful bibliography. Very disappointing, the *Catalogue* effort certainly did not solve the international classification standards problem. Although useful, the *Catalogue*'s bound volumes did not satisfy scientists, especially because their organization, as only author indexed, meant that notice was given to just the works by authors with a well-established reputation. Criticisms of the *Catalogue* in its later years emphasized how scholars missed important papers, such as Gregor Mendel's now-classic works on heredity.

There was an additional and significant point that Herbert Field repeatedly stressed. Few if any indexes or bibliographies used a viable subject classification system. Without such a system, he constantly warned, even the best bibliographies were useless. He called for the creation of such a subject-organized system. Herbert would soon declare he would do the seemingly impossible, build an international classification system.

4 Challenging the British "Lion" of Science Information

Complaints about the Royal Society's *Catalogue* had been evident as Herbert Field began to notice the bibliographic debates of the 1880s. Oft-repeated criticisms he encountered during his student years included the *Catalogue*'s failure to cover all science publications, its lack of an index, its late arrival, its incomplete historical coverage, and its failure to report on recent scientific journals and proceedings. He also heard rumors the society, a veritable "Lion" of science information, had found the project so expensive that it was about to terminate it. However, Herbert then heard there was to an attempt to begin a new British endeavor.

Too Much Hubris?

Herbert was aware of the possible new British initiative when he made his youthful 1893 bibliographic boast that some form of a new international information institution had to be established and that he was the one with the original and sound vision needed for it. He was unaware of the dangers posed by the implication that he could outclass the work of the British Empire's greatest bibliographic experts.[1]

Herbert felt he was uniquely well prepared to fulfill his new Calling. He had attended especially to conditions in his own domain, zoology. Although that subject had several bibliographic efforts in operation, many zoologists pleaded for major reforms. They argued for the creation of some type of international bureau that could channel all publications to bibliographers and would register all the new discoveries of creatures.[2]

Critically, Herbert believed that any system would fail if it depended on voluntary cooperation or funding by national governments and their scientific societies. He not only declared he had viable alternatives to all existing methods, but also suggested that his plans would work better on organizational and financial levels.

Initially, Herbert considered himself the only viable contender on the international information scene. He soon recognized he was not the only one conscious of the need for radical international information reform in the 1890s. A growing number of scientific congresses were readdressing the issue. Herbert attended most of these meetings, where calls were made for effective international cooperation. Some groups had already begun to employ innovative methods; the mathematical association, for example, was building an ongoing bibliography at its French center. Herbert had also heard of a group in Belgium led by Paul Otlet and the Nobel Peace Prize winner Henri La Fontaine that was exploring effective ways to allow cooperative assembling of all types of catalogs. Their work was especially attractive to Herbert as, like him, they were hoping to create international associations and information systems to counter nationalistic isolation and competition.[3]

Herbert quickly sought out the Belgians. He also followed some hints their government was willing to provide him the economic support he needed to begin turning his ideas into an operating system. He remained convinced that he, not Otlet or the Royal Society, would be the one to find the cures for the science information malaise.

Toward Centralized Solutions—Finding Dewey but Not Money

Herbert fervently believed in his cause in the early 1890s, but he knew he had many obstructions to overcome. His first challenge was to develop an acceptable way to intellectually organize and distribute scientific information—one that could serve a community that used many languages. It is not certain when he placed his faith in Melvil Dewey's then-revolutionary numeric system, but by the 1890s Herbert thought it had the potential to meet the challenge posed by multiple languages and to satisfy the critical need for orderly subject indexing. He seems to have learned of "the Decimals" while at Harvard, and he and his fellow students may have experimented with it during their graduate studies.[4]

Whatever the origins of his interest, Herbert published the first announcement of his "numeric" intentions in 1893. He declared he would create a unified zoological classification-retrieval service based on something like the Decimal system. However, he soon faced several hurdles. The first arose when his initial advertisements for his proposed but not-yet-developed service did not lead to a rush of money-generating subscriptions. The lack of response from America deeply disappointed him because he had personally outlined the nature of his system to his North American friends.

Herbert had thought one of his ideas would immediately attract contributions from around the world. After redrafting his original idea of creating a type of clearinghouse that would send items to selected scholars for classification, he decided to create a central site where the entire world's bibliographic literature would be scientifically classified and indexed. That arrangement would be more efficient, he claimed, than the older international cooperative projects. A central bureau would free individual countries from the burdens they had experienced while participating in projects such as the one run by the Royal Society.[5]

Besides the financing problems, Herbert faced intellectual challenges. Could a structure constructed by a college's student librarian, and which divided all knowledge into just ten general categories (and then into subcategories) by succeeding decimal numbers), be altered to accommodate the complexities of the many realms of specialized scientific knowledge? Could a relatively fixed set of categories ever be useful to constantly changing fields of study? Could a hierarchical system remain viable if scientists kept shifting the placement of a creature or subject within it? Could the international language of numbers provide the codes required for efficient and complete retrieval of requested science information?

Herbert bet that Dewey's systems could satisfactorily answer all those questions. Moreover, he believed his own work would help prevent the creation of ad hoc systems for each scientific field.

There were more challenges. One was psychological and touched on professional pride: Would scientists and their societies accept a system that referenced publications by what many regarded as crude and demeaning numbers? Many scientists wanted to continue with elegant books and the sophisticated terms of their fields of study. Another involved maintenance: Herbert knew the importance of finding feasible ways for his subscribers to update copies of his bibliographies in their own libraries and offices. Like his future mentor-collaborator, Paul Otlet, Herbert Field may initially have investigated the possibilities of photographic reproduction and microfilm storage. They, however, were information technologies of the distant future, not the 1890s.[6]

Postcards over Paper—Addressing New Problems of Distribution, Storage, Retrieval, and Access

On a trip to Russia, Herbert was impressed by the huge paper-slip files used by the government's bureaucracy, and he knew of the loose-leaf paper binders that some libraries were using. His first bibliographic plans had centered on paper slips. But within a few years he believed only one techno-solution for distribution and information searching was practical. Field concluded that over the previous decade the postcards, turned Dewey library cards, had proven themselves durable and economic. American libraries, some museums, and a few small bibliographic centers were already employing them. Even the United States Department of Agriculture was experimenting with a card distribution system to help fulfill its mandate to keep farmers up to date on the latest scientific and economic information.[7]

Cards seemed the only robust and practical nineteenth-century technology for immediate updating of files and for distribution to distant customers. Paper sheets, even in the new-fangled folders, were costly to search and maintain, and improvements in paper manufacture would, Herbert felt, continue to lower card prices. He also believed innovations in printing techniques would soon lead to cost-effective reproduction of sophisticated versions of bibliographic cards.

However, while a Dewey-like card system seemed a good choice for the distribution, storage, updating, and retrieval challenges, it had some economic and logistical drawbacks. The cabinets required to turn the cards into what was a very primitive (but still a best-choice) "random access" memory device were expensive and hard to find. Herbert also worried that if his system depended on the subscribers maintaining the files he provided them, they might be unwilling to pay for the sorting and filing of the cards. Would they carry the cost of immediate filing when they received updates? The system would be near useless unless the users could file the cards promptly and correctly.

Herbert also faced an access problem. How could his classification center afford to acquire the entire world's journals, books, notes from professional congresses, and erratic publications? Could he ever gain the cooperation needed from all the nations in the modern world to have their publications sent to his center at, hopefully, no cost to him? Critical and worrisome was the question of whether national scientific societies would feel that any sort of cooperation with his center endangered their own bibliographic work. He also knew he faced a significant cultural problem, one that is now termed "diversity." Would he be able to build a classification system that did not alienate or slight scientists and their national societies in countries outside Western Europe?[8]

Step-by-Step: Assessing Competitors, Consulting Colleagues, and Concerns about Funds

Herbert had growing concerns about competitors. He heard some vague rumors about possible rivals when he began discussing his plans with colleagues. Because he concluded the new British catalog was to be in book form and only indexed by author name, it might not be a challenge. However, would a more commercial-like bibliographic company such as that of the American H. W. Wilson decide to tackle the scientific information problem? He'd also heard that a card bibliography for all the natural sciences was going to be proposed at a prestigious 1896 international congress in London. And, just as worrisome, were indications that his alma mater, Harvard, might launch a major science bibliographic effort.

Despite his anxieties, Herbert continued his quest. He gained help and advice from Raphael Blanchard, an influential mentor at the Paris natural history museum, and by 1894 he was ready with an initial model for a centralized bibliographic institute for the world's zoological literature. Significantly, he had decided that he was not going to wait for aid from any government. He began making new requests for approval and money directly to family, friends, and fellow scientists. Then, he followed another round of articles and letters in scientific journals with something of a grand fundraising tour. He contacted scientific leaders in Europe and America asking for advice on classification and for money.

Herbert was thorough. He even consulted with Alphonse Bertillon of Paris. Bertillon was an information expert and had developed complex methods for criminal identification and indexing. His system fit the use of card files. Oddly, Field did not seem to have consulted two others in France who were attempting to develop card-based scientific information systems, Marcel Baudouin and the socialist Charles Limousin.

Unfortunately, most of the stops on Herbert's tour, like the one to the powerful Smithsonian Institution in Washington, resulted in a "pat on the head" and a pledge of moral support, but no money for his project. That was not the only disappointment. European traditionalists voiced opposition to any inelegant Dewey-like number system. Sometimes, such as during a presentation at an 1895 British meeting, Herbert and his proposed methods were subject to rather sharp, disheartening, and hurtful rejections. He even had to fend off accusations that he was trying to redo the intellectual structure of zoology. The objections in England worried Herbert so much that he pledged to turn his work over to the British if the Royal Society ever began to do a comparable bibliography.[9]

Idealism to the Fore

Herbert kept on going—although he backed away from promising the immediate launch of a full-blown bibliographic service. In late 1894, he admitted that he had only the outlines of the initial parts of a viable program. However, he suddenly had a few indications of possible contributions of support for what he had begun calling an "experimental information research" center. His research, he was saying, could lay the foundations of an advanced service, one that could satisfy the needs of all nations' zoologists.

Herbert developed a new theme in his writings and pleas. It showed that he had not abandoned his Quaker reform tradition. His broadsides began to include an important and obvious "subtext." He stated that he desired to establish an institution that would contribute to a higher role for science—international peace. His centralized bibliographic service would erase national intellectual boundaries and equalize access to scientific discoveries. He soon proved his commitment to those loftier ideals.

Money, Money, Money, at the Price of a Scientific Career

Herbert Field's initial fund-raising successes in the scientific community came with an unexpected demand. It was one that had the potential to change his life. When he first began soliciting professional approval and contributions, he conceived his role as only a designer and organizer, not as a full-time hands-on operator of a bibliographic service. But two likely contributors, a duo of small national zoological organizations, asked him to take both intellectual and operational responsibility for all aspects of his project, especially the indexing work. And, they advised him to return to his goal of immediately establishing a service, not just an information research center. With that, he gave up any lingering visions of returning to zoological research until he fulfilled his new Calling.

Herbert immediately began work on the details of his "center." He had to, because the small amounts of money from those French and Italian zoological societies were to come, Herbert was informed, only with his development of a complete and acceptable plan. Assurances that his project would not interfere with the work of others were also demanded. There were special worries about one project in Leipzig, run by Julius Victor Carus, and the *Zoological Record* in London.

Herbert rushed to consult and gain the approval of Carus, the famed zoological bibliographer in Leipzig. Carus, however, had not begun as an information man. After an illness terminated his brilliant scientific career, he spent the rest of his life in a sick-room organizing the literature of zoology, partly through his editing of the *Bibliographica Zoologica*. An assistant brought materials and Carus turned them into sometimes monthly, but usually yearly, bibliographic/abstracts volumes. ("Abstracts," the term used for short summaries of articles, would become one of the most hotly debated elements regarding Herbert Field's work at the Concilium.) Given that Carus had become the most recognized zoological bibliographer of his time, Herbert needed his blessings. If he won Carus's approval it would be easier to convince the influential English not to block his dream, as Herbert feared they would.

Carus was more than hesitant about Herbert's ideas. He had bad memories of large-scale bibliographic efforts and continued to have faith in the older classifying and indexing methods he used for his work. He also favored sparse abstracts over classification. In addition, Carus's contacts with the Royal Society's decades-long and failed attempt to devise a subject classification for its early *Catalogue of Scientific Papers* had made him skeptical of international projects. Herbert's plan to use a new-fangled classification system, using numbers rather than words, made Carus especially wary of Herbert's ambitions.[10]

As a first step to overcome Carus's concerns, Herbert assured him that the special nature of his proposed service meant that it could not endanger Carus's abstracts volumes or England's *Zoological Record*. Herbert also argued that his yet unnamed operation would help Carus obtain and classify more items, as well as aid the distribution of his works. Herbert pledged that his operation was going to focus only on sending out subject-classified cards, and he promised to provide Carus with the cards free of charge so Carus could use them for his traditional volume. That pledge dispelled some of Carus's fears, but he continued to withhold his approval.

Herbert described other aspects of his plan, hoping to finally convince Carus to give his approval. The older scientist knew about American libraries using separate cards for each book but was not convinced they would work for scientific publications. He doubted that "numbers" could ever inform a user of the contents of a publication or that numbers would be an "international" language. Greater resistance came when Herbert described details of his most radical alterations in classifying. Herbert stated that some parts of the old zoological taxonomies might have to be abandoned.

Field worried that Carus would kill his project, so he pressed on. Herbert spent much time stressing the organizing powers of well-ordered hierarchical subject classifications. He put forward an especially persuasive point: a freestanding bibliographic service was not going to depend on gaining the cooperation of hundreds of voluntary indexers, and it would not be hampered by picky ongoing evaluations of methods and procedures by contentious academics. Such interference, Herbert emphasized, was something that had marred the Royal Society's projects.

Carus eventually gave his endorsement. However, he warned Herbert that financial disaster marked the history of all bibliographic efforts. Herbert's description of a vast market of libraries and individual scientists willing and able to support another science information service had not fully convinced the older bibliographer.[11]

Making Career and Financial Commitments

Herbert then tackled the next part of his challenge—turning ideas into an institution. He was optimistic although most of his limited financial backing was still coming from his own and his father's and the French Havilands' pockets. The two small zoological societies had given him only a promise of minimal support for a five-year trial period and other donations remained just pledges. Yet, he began to fill in the outlines of his operational strategy and, importantly, the business plan for his centralized facility.[12]

His irreversible commitments were to base his service on the cards; on some way of numerically coding items; and on a focus on zoological literature until his service was well established. Of great significance was his plan to send his patrons card packs of bibliographies every two weeks. His customers could arrange items by subject, and perhaps by author or title, to build a cumulative file.

As he again searched for supporters, Herbert added more to his arguments. His stated that his potential market was composed of more than institutions or rich scientists. If an individual scientist could not afford the full file, he could subscribe to a smaller one devoted to a subsection of zoology. Herbert hinted he might offer as many as 230 variations of the smaller files. Furthermore, individual scientists without access to any of the files might pay to have bibliographic searches done by professionals at his central bibliographic office.[13]

Making an Indispensable Retrieval System

There was another attractive feature that Herbert soon publicized. His preliminary scientific analyses led him to the firm belief that the relative cost of retrieving information from his type of card file made his product indispensable. He calculated in his mind how long it took to go through the old individually bound bibliographies to compile lists of relevant items. Then, he compared that to the results using his well-organized, cumulative, subject-indexed card file. There was, he claimed, no contest.

Critical to his plan was the assurance that his professional staff would base their indexing on reading the full text of every item, not on a supposedly cost-saving use of just a title or abstract, as had most other efforts. Herbert was convinced that valuable indexing resulted only from a full-text reading provided by a specialist in a subject's field. He told potential subscribers that even if a biologist had indexed an article, a zoologist's needs required another reading to find significant items. There was another feature he later emphasized. Full-text analysis allowed the indexing of ideas, not just a document.

Importantly, Herbert restated that numbered classifications would avoid offending the sensitivities of the many nationalities he would serve, and thus the markets of all nations would be open to him. An Indian, a Lithuanian, or a Croat could accept and use his cards without feeling insulted, or having to call upon translators to use the system.

Next, he guaranteed that his bibliography would survey all of the literature in a field—it would be the only complete bibliography in the world. He also promised to create what the later twentieth century called a nonprofit organization, with a board of overseers monitoring its finances. He pledged that his service would be affordable because he had ways of minimizing costs of production.

Then, with the innocence of youth, he guaranteed that if he did continue as the organization's director, he would ask for just a minimal salary and nominal money for basic expenses. Herbert soon offered another pledge to customers and donors—one that came to haunt him. He confidently predicted that while some negligible startup subsidies might be needed, his service would quickly become self-supporting.

The Economics of Information

Early on, Herbert had explored the economic challenges he and his customers would face. He was confident that he had a solution to the expense-of-distribution problem. He stated the International Postal Union had twenty-five years of practice and had proven reliable, and the postal systems of most nations had also developed enough, so that mailing costs for delivery of his fortnightly updates would be bearable. Herbert emphasized other economic advantages of his system. Using some sort of numeric classifications would save huge amounts in translation and other production expenses, he said. Also, card systems did not require expensive binding.

He argued that cards and numbers reduced many other costs. Sorting and filing hierarchically numbered items were much easier than with any system using alphabetic names—especially when dealing with items from dozens of languages. He also touted some cost savings through his chosen technology. As cards were more durable than book pages or slips of paper, there would be fewer replacement expenses.

As he fleshed out his plan and appealed for more startup funding, Field believed financing was only a short-term difficulty. He declared he could quickly develop a constant income stream and he told his prospective supporters that sales of subscriptions would cover all expenses. He promised to put all contributions to immediate use and he emphasized that donors would never again be bothered.

At the core of his financial plan was the income from long-term subscriptions. Herbert made some exuberant claims. He predicted that he needed only a few hundred subscribers to meet all expenses. At this early stage of his planning he foresaw significant income from journal publishers. He predicted they would all provide their publications at no charge and make small contributions to have their works gain notice in his bibliographies. He also predicted that publishers would regularly pay for advertising. Unfortunately, he found those hopes to be unfounded. While bibliographic services for rich users, such as those in the chemical industry, could rely on advertising revenue, most academic bibliographies could not. In addition, Herbert soon developed qualms about any payments from publishers. Customers and professional societies might interpret them as signs of favoritism, he concluded.[14]

That and other early miscalculations had not discouraged Herbert. His latest research, he claimed, proved there was a greater market for his services than he had thought. He made additional travels and began a new round of correspondence with scientific leaders. He reported there were more than enough large libraries and established scholars ready and able to pay his proposed moderate fees, ones that, he promised, would be below those for bound-volume services. To assure future customers and sponsors they would not be investing in a soon-to-disappear operation, Herbert wrote that men of honor had given their word to purchase the files once his center began producing cards. Soon, his refined calculations led him to declare that he needed only a few full subscribers and less than five hundred partial subscribers to make his service self-supporting.

His sales promotions continued. His newest formal prospectus included an important promise. He would be able to market segments of his files to libraries that served the general public. He announced he had ideas for new products for such markets. He planned to create, for example, mini-databases for each of the American states. Local public libraries could have small files on hand for their patrons to look up all the references to the insects and other creatures in their area.

Compromises to Reduce Cost yet Maintain Quality

Then, under renewed pressures from the donors who were demanding guarantees that he would maximize his markets, Herbert began to think of being flexible in respect to his commitment to cards. He explored offering both a paper bulletin and card packets for each issue of his planned bimonthly bibliography. A user might choose either or both. Herbert would print the bulletin on one side of a page so that entries could be clipped and pasted. Herbert then proposed another moneymaking service he thought would solve a critical problem for the fast-professionalizing sciences. His bureau could be the world's central registry for establishing priority of publication and scientific discovery. He did not mention that he might be alienating the editors of the abstracts journal, the *Zoological Record*. They had carved out a niche as the international registry of biological names.

Herbert sought additional ways to minimize costs. He held direct conversations with several zoological societies to convince them to send journals and publications to his center at no charge. In exchange, he promised to supply them with bibliographies to be printed in their publications, at no cost to them. (That seemingly logical offer would prove to be one of Herbert's great business mistakes.)

Unaware of the future consequences of those agreements, he then made inquires about printing techniques he hoped might keep his overhead and wages to a minimum. He rejected some good suggestions. He turned away from using the cost-saving stereotype printing technique that publishers, such as W. F. Poole, had introduced. That process would not lead, Herbert concluded, to the type of high-quality image that his customers would demand. Traditional and labor-intensive repeated typesetting was the only possibility for his operation. Field stood with that early decision long after library publishers such as H. W. Wilson showed that the stereotyped plates were creating works widely accepted in libraries. Stereotyping would have solved one of Field's greatest economic problems, printers' wages.[15]

It took Herbert additional time and much correspondence, but his persistence led the French and Italian zoological societies to reaffirm their pledges of five-years of trial subsidies. Although the amounts were minimal, they were vital. But he needed more. Luckily, Lydia's connections came into play. Herbert's French relatives seem to have been the ones who raised an additional special startup subscription to help cover his first year's expenses.[16]

The Politics of Information

A major political campaign came next. Herbert needed to gain the endorsement of the world's most important zoological organization, the International Zoological Congress. It had sponsored Victor Julius Carus's indexing/abstracting journal since the 1860s, so Herbert worried the organization would see his new center as a threat. Without its approval, Herbert stood little chance of success; with it, his credibility would increase. He planned well for his meetings with the congress leaders.[17]

He used his Harvard connections and managed to obtain the status of the United States' official representative to the 1895 International Zoological Congress convention. With such credentials, he was able to lobby the meeting's delegates and officials.

His first approach was one of reassurance and deference. He informed his colleagues that he had no thought of altering zoology's intellectual order or endangering Carus's bibliography. However, there was much to overcome because of the worries about what many thought was really an American, not international, bibliographic project.[18]

Herbert's earlier visit to Julius Victor Carus proved to be critical. Citing Carus's approval, Herbert was able to convince the congress that his service would not be a threat to established science or to its member countries' own publications. His arguments and lobbying succeeded. He received the congress's blessing.

After that, Carus's growing fondness for Herbert continued to be important. Carus was not just a respected bibliographer; he was a highly admired and influential international scientist. He was a close friend of the greats of nineteenth-century science including Charles Darwin. Darwin, in fact, had chosen Carus as a translator of his epoch works. So, Carus lightened Herbert's burdens when he became an intellectual mentor and a long-term political advocate for the young American. Herbert soon gained more supporters. A few American scientific societies gave approval and some hints of additional financial contributions. Happily, there were indications that influential European universities and libraries might join in, at least with subscriptions.[19]

Devils in the Details—In Paris and in Dewey's Decimals

With that, Field felt secure enough to begin to work on final details. He spent more than a year in Paris exploring methodological and technical alternatives. His ability to concentrate on a problem and his Quaker discipline helped him avoid the allures of the city. He began his work with a focus on the zoological literature since 1896. He planned to leave the coverage of earlier publications to others. He then turned to his primary technical-intellectual challenge, to see if a classification system designed for general library books could, as he had predicted, be altered to fit all types of scientific publications.

Although Herbert's goals were limited, and although he gave no thought to a radical redoing of the concepts and taxonomies of the established sciences, his task was not as easy as he first envisioned. He knew of C. W. Woodward's advice that Dewey's scheme fit science classification's needs, but it took Herbert months of frustrating work, and correspondence with Melvil Dewey in America, to decide that Dewey's system truly was the best of the new classification systems.

Unfortunately, immediate and practical problems confronted Herbert as he labored over modifications to Dewey's system. His weak health caught up with him. He had a bout of exhaustion that fed a significant infection, which disabled him for some critical weeks. Herbert had to deal with other difficulties. His initial hopes about access to the entire world's zoological literature proved to be naive. He concluded that it was impractical to rely on authors to voluntarily, and without prompting, supply notifications of their publications.

Likewise, his initial canvass of publishers had attracted only a few responses. He also had to put on hold one innovative idea, knowing it would be very costly. He put-aside the possibility of repeatedly sending out postcards to publishers and authors asking they return the cards with notification of their pending publications. In the meantime, the only possible solution he could think of was to again go to additional professional societies, asking them to provide free copies of their journals and papers.

Out with the Old "Abstract" Ways, Says Herbert

Herbert's continued investigation of the history of classification reinforced his earlier conclusion that no matter how one might coordinate one's work with the international scientific community, the traditional method of having classification and abstracting done by a huge number of volunteers was unworkable. In response to questions by those who thought he should use volunteers, Herbert took the chance of insulting his colleagues.

He risked much by telling his sometimes overly sensitive supporters that being a good scientist did not mean being a good bibliographer. He emphasized what a difficult job classification was, even for short, topic-specific articles. He reasserted his belief that there was no alternative but to have classifications done by a professional staff and in a central place. Importantly, he said, the old volunteer process led to classifications that were not standardized, and editors invariably had to redo them.

Herbert soon made additional critical commitments. One was critical to his center's fate. His further exploration of operational costs led him to decide that his service would avoid providing abstracts of papers. The expense of composing and printing even a typically minimal synopsis would be beyond the reach of his center. Herbert never emphasized it, but a serious flaw in abstracts was the language problem. There was no universal language and the cost of translating abstracts into at least the three major languages of the era would be too costly.

Herbert emphasized that with a powerful numeric classification system, abstracts were not required. Only in cases in which a document's title and its classification failed to provide what he thought was adequate indication of content would any abstracts be attempted—and they would be very short. He also decided that additional information, such as "subject terms" that were appearing on book catalog cards, were unnecessary.[20]

Herbert soon abandoned a feature suggested as a possible moneymaker. He had thought of providing the full text of an article, a service that might be attractive to researchers who were far away from a major library. But he soon realized such a service faced too many economic and legal hurdles. There could be copyright problems, and reproducing a text was too expensive given the technology of the time.

Classification, Dewey Style: The Library of Congress Challenge and Herbert's Allegiance

Meanwhile, Herbert worked on the complex problem of a classification scheme for zoology's literature. Creating a hierarchical and numeric system for the entire field was an intimidating task, even though zoology's focus had traditionally been on creating a hierarchy of "systems." Herbert had to consider intellectual, user, and economic aspects. He wanted to devise a system that could serve as both a means of data organization and as a substitute for the traditional abstracts, and he required a system that was logically pleasing, practical, and affordable. Above all, it had to be one that fulfilled his central promise to his future customers: creating the first easy-to-use system for the indexing and, importantly, the retrieval of science information by subjects, not authors or titles. That was a formidable challenge. Field again consulted Julius Carus. The two bibliographers had faith that they could make Dewey's system correspond to the current conceptual structures of the zoological sciences.

Herbert then intensified his work on modifying the Decimals to fit with zoology's concepts. There were few models available of how to subject-index articles and few examples of how to alter any general classification system to make it useful to scientific specialists. Herbert knew about and rejected a somewhat different method the U.S. Library of Congress was beginning to develop from Charles Ammi Cutter's very sophisticated system. The decision by the library to create its own system had consequences for the grand dream of a single classification system for the world and all knowledge.[21]

While the system emerging in the U.S. Library of Congress was a bit simpler than Cutter's, it appeared suited only for very large collections oriented to books with a single subject classification per volume. It was also oriented to the English language and to the particular needs of the Congressional library and its special type of user. And, it used letters in its identifiers. Despite that, and Melvil Dewey's decision to remain aloof from the effort, the library foresaw its classification as a national standard for all information. In fact, just a few years after Herbert launched his crusade, the library began to establish its role as a national classifier for new books. The library's administrators knew they were in competition with Dewey's classification scheme, which was already dominant in local public libraries, but they thought their more complex system would become the "gold standard" of classification. Soon, the library recognized that users needed more information than the classification number, title, and author to find new items. The library began adding as many as three additional alphabetical headings (Subject Headings) at the bottom of its cards to indicate additional subjects covered by an item.[22]

The Library of Congress (LC) staff pressed its advantage. They sent classification cards for each publication they received to subscribing libraries across the nation. As the LC became the nation's central cataloger, it did bend to the demands of small libraries. By the 1930s, its cards also contained Dewey classifications. As a result, within a few decades the LC classifications became the standard for books in the same large American research and university libraries that had become Herbert's best customers. Many were quite willing to make the expensive shift from Dewey to the LC codes in later years, especially during the library budget–boom era of the 1950s and 1960s. While European research librarians remained committed to versions of Dewey's Decimals, such as the Universal Decimal, using any system based on a non-LC classification became less and less common for large American research collections.

However, in the mid-1890s, Herbert did not foresee the emergence of serious competitors to Dewey. Of importance, he believed that Dewey's system had more than purely intellectual advantages. It facilitated easy filing and information recovery by subjects, something not easily done when using the partly alphabetic codes of a Cutter-like classification. In addition, by 1895, Herbert's experiments reaffirmed his belief that file updating and searching by long and complex scientific names or titles would always be too costly and error-filled. Importantly, Herbert became convinced that with the use of short and purely numeric identifiers, unskilled and low-paid workers could easily store and retrieve information.

5 New Information Ideas in Zurich, Not Brooklyn or Paris

The man his peers regarded as the king of zoological classification, Julius Victor Carus, had become a devoted admirer of Herbert Field's innovative ideas. He even approved of Herbert's radical plan to classify not just a document or a book, but major ideas within them. With Carus's blessings, during the winter of 1896–1897, Herbert made a public commitment that became of great importance to his information service. His experiments, he said, had reconfirmed his belief that a good information system could not be based on indexing just the traditional units. He found, despite the increase in specialization and rigorous style mandates in science publication, that documents usually contained more than one idea or fact that deserved recognition.[1]

Herbert then made what he held as an unbreakable pledge. He declared he would follow through on the plan of indexing major ideas by maintaining a file on facts and thoughts, not just on articles or volumes or even paragraphs. However, he did not discuss the economic ramifications of that promise.

Not Just Books or Articles but *Ideas*

Indexing major ideas meant that all words in a document had to be read and all distinct and significant facts or concepts indexed. With this primitive "full text" approach, many cards might have to be created for each publication, not just a single one as Melvil Dewey's original system foresaw. The dedication to ideas entailed more than increased classification expenses and larger files. It called for a more detailed reexamination of the Decimal system and related procedures before Dewey's creation could become the basis of Herbert Field's center.

Herbert admired the simple elegance of the Decimal system and believed it could be adapted to specialists' needs. However, his first experiments revealed that applying it to scientific domains required some major reworking of the Dewey methods of classification. A first priority was a change in procedures. Herbert had concluded that subject specialists, not general librarians or low-paid clerks, had to do the classifications. There were too many technical terms and concepts in the literature, and too many expert judgments to be made when dealing with the ideas in scientific materials.

Experts attuned to the latest scholarship were needed, for example, to bridge the subject matter divisions in Dewey's system. Only a working zoologist could know when to merge a fact or idea from a biology journal's article on embryology or ecology into a zoology classification.

Such expertise in applying Dewey classifications was required to make them acceptable substitutes for abstracts. Although Herbert had announced in 1894 that his bibliographies might include short abstracts for each item, he now declared that skillfully assigned numbers would, by themselves, point to all items of importance. Only when a paper's title was inadequate or misleading would his classifiers need to write a short "telegraphic" clarification using abbreviations and simplified text. Those who wanted the expensive full abstracts, he intimated, could use his service and then wait for the appearance of the established publications, such as London's *Zoological Record*.

Herbert then made more rather intemperate comments aimed at the demand for abstracts. He stated that truly professional scientists did not take shortcuts such as reading abstracts rather than an entire publication.[2]

Large-File Logistics and a Link to the Universal Decimal Classification

Herbert soon began to worry about problems with "full text" reading, precise classifications, and the need to satisfy a wide range of specialists. Their solution called for a larger, more complex and expensive file system than he had anticipated. It would have more entries, partly because it had to allow multiple entry points for retrieval. His conversations with zoologists showed they might demand multiple classifications for each idea so they would be able to find items by several aspects. Each article might require, for example, having a separate card for the type of creature, its structure, time period, and geographical location. A single journal article might call for a handful of cards if it referred to many topics that needed such multiple classifications.

The large-file challenge was critical. Although Herbert cooperated with Paul Otlet and his partner Henri La Fontaine in their grand attempt to standardize information classification, he had a different vision than theirs for his services. That difference was important. Unlike the center that soon flourished in Brussels under Otlet and La Fontaine, Herbert was emphasizing a distributed system. His intent was to have users around the world possess their own files, ones updated by cards mailed from his center. Herbert feared that customers would not accept a system with huge files. Otlet and La Fontaine, however, worried little about file size because their system centered on one central file in their offices. It would be maintained and searched by their staff to answer specific bibliographic requests submitted by customers. Later, Otlet attempted to create similar centers in other countries, even the United States. Predictably, logistic and economic problems defeated those overly ambitious efforts.

Otlet was and continued to be important to Herbert, however. Otlet had already decided that he could turn Dewey's Decimals into a system for legal and humanities information, and he began to muster scientists in all major fields to create Dewey lists (schedules) for their subjects. He soon looked forward to indexing all types of information through a single system for all knowledge. That, and arrangements Otlet was making with a bibliographic center in France, the Bibliographic Bureau of Paris, reinforced Herbert's faith that some version of Dewey's system would become the single international standard for all knowledge. Such an international system would bridge the harmful diversities of language, culture, and national loyalties that seemed to impede scientific progress and lead to conflicts.

The name for Otlet's very sophisticated version of Dewey's classification system, the Universal Decimal Classification (UDC), was appropriate to academic aspirations and to the dream of internationalism and a harmonious world culture. However, it was a very complex scheme that was labor intensive and that required people highly skilled in both subject matter and the system. It soon had three times the number of main categories as in Dewey's system, and dozens of codes and relationship symbols. However, after his latest interchanges with Otlet, and learning of Dewey's granting Otlet permission to publish the Universal Decimal Classification at least in French, Herbert felt comfortable in cooperating with the Belgian and in helping to build the UDC.[3]

Determining Customers and Limiting Markets

Now intending to work in harmony with Otlet, Herbert Field proceeded to finalize his own zoological classification system. As he prepared to become operational, his changes reflected the need to balance the quality of his work, the market appeal of his service, and the hard truths of the economics of information. The first decision was who to serve. He knew he might lose potential customers if he oriented his center only to specialists. But he had been promising a service for the benefit of natural science professionals. He decided to give priority to the needs of experts, knowing that could mean general libraries might not subscribe to his service.

There were several advantages to targeting specialists. One was that it led to savings in translation of terms. Herbert knew that most specialists used the Latin terms of their fields, unlike people in the general public. Therefore, his commitment meant that he would not have to translate the names of creatures and eras into the colloquial languages of the many nations he hoped to serve.

Market concerns almost made him bend from one early commitment. Despite his need to minimize expenses, he almost abandoned the idea of using Arabic numbers after he received some very negative comments from established bibliographers and publishers who stood by the more esthetically pleasing Roman numerals.[4]

Fine-Tuning the Decimals and Minimizing Costs

Time was running out and Herbert's anxieties deepened as his operational deadline neared. What parts of the Dewey system could he eliminate to reduce expenses? One decision was easy. Because each of his files were to be organized around one major Dewey number, such as "590" for zoology, there was no need to have that number in each item's classification code. Furthermore there was no need for the last number (the "0"). So, to save time and money Field decided to print the general code for each of his series as a constant at the bottom of each bibliographic card. That was a wise decision; it saved his center a great deal of expensive typesetting, and it fit with the emerging standards for Otlet's system. The next cost-saving decision avoided using additional numbers. Librarians amplified the Dewey classifications through lines of appended information to give a unique identification to each book and volume. Because the Concilium Bibliographicum's system was not to be for shelving or inventory purposes, dropping the use of the extra numbers would save hours of the classifiers' time and costly printing.

In contrast, Herbert did not try to save money by using short titles or to substitute numbers for title words or authors' names. He thought full title-lines were necessary. A customer who wanted to order or cite an item needed them. There was also no need for his classifiers to turn, for example, to a table of Cutter-like conversions and change an author's name from Scanlon to, perhaps, "S29." Nor would Herbert's systems use, in more than a few special instances, the numerical publication-type codes sometimes employed to augment Dewey numbers. Such codes told what kind of a work was indexed: an encyclopedia, a society publication, an essay, and so forth.

A few decisions were painful. Herbert studied Otlet's very sophisticated approach to classification, which included a host of special symbols that allowed a classification's codes to indicate such things as links from one item to others and to behave like what became known as "faceted" classification. Otlet developed many other advanced features, including the indicators for language used, the place where an item was published, and the physical characteristics of a work. Herbert, with some anguish, decided the best way to help his intended audience was to try to be a classification minimalist, partly because he wanted to keep the cost of his service as low as possible. He determined it was wise to avoid expensive features such as links or subject terms. He predicted his users could create them if they desired. He believed his multiple classifications per publication and the well-organized subject files would allow that.[5]

Herbert realized that information minimalism had its drawbacks and that his systems might need to be tailored to the demands of different audiences. He knew that zoologists and their professional kin usually wanted precise information about the location of any creature or plant. So Herbert vastly expanded Dewey's list of countries and their subdivisions. He went through gazetteers and zoological journals and created a long list of numerically coded locations for the entire world. All the American states had a code. Africa had nearly forty different areas coded, and Herbert subdivided each European nation into regions.

The geographic modification was a minor challenge compared to the task of expanding the Dewey classifications so the Decimals could allow scientists to quickly locate an item. Field had to create a classification system that went far beyond Dewey's. For example, Herbert added seventeen classes of worms. And, to match the ever-widening interests of zoologists, he extended the boundaries of the field. Items from biology and even paleontology were included in the zoological lists and coded in a manner that fit into the useful and hierarchal format scientists expected.[6]

Organizing Files in the Pre-Computer Era

Herbert also had to determine the best approach to the organization of his files. He decided the simple ascending Dewey numeric order (with author and date of publication as secondary sorting items) was the best choice for his major files. Herbert may have explored, but rejected, the use of a version of the index to books as a basis for retrieval. Later, the method was called "inverted files." With inverted files, items are assigned a serial number based on their order in the processing chain. Documents are filed by that number, but retrieved through a table. In the table, each subject classification is followed by a list of the serial numbers of the articles relevant to the subject. The inverted-file methods had the potential of vastly reducing the number of cards, could allow more multiple subject classifications, and had the advantage of quick, easy, and inexpensive filing. But the method did not facilitate casual and direct browsing within subjects and it was impractical to attempt to keep the index tables of hundreds of distant users correctly updated.

After the decision about file organization, Herbert returned to the intellectual problem. With the help of Otlet, he obtained permission from Dewey to proceed to a final version of the "schedule." His efforts received plaudits from Dewey. In his revisions of the Decimal's handbooks, Dewey included his thanks to Herbert.[7]

Letter from Paris: "To My Dearest Friend" in America

Herbert had settled in Paris to compose his schedule of codes. At first, he lived with Auntie Haviland in her large home. Then, in the mid-1890s, he rented a small and dreary unfurnished apartment in the city. It was there, in a setting much like that of a struggling artist, that Herbert worked out the general scheme for his zoological classifications. When he thought he achieved enough, he wrote it all out in careful, tiny handwriting and sent off a double-sided twelve-page letter to his best friend in America, the zoologist Charles B. Davenport. Herbert hoped it would gain the approval, and financial support, of American zoologists.[8]

He knew he could count on his old friend and classmate who was becoming a rising star in American science to spread the word. Davenport had already helped convince the Americans to give their approval to Herbert's initial plans. He also aided Herbert's 1895 campaign that resulted in the International Zoological Congress giving its blessing to the future zoological service.

Expanding Classifications, Honing Search Categories

Herbert had become a near recluse, something hard to be when living in Paris in the 1890s, especially while being in contact with the Havilands and their exciting artistic and political friends. Sitting at his desk in his mid-city apartment, Herbert read hundreds of professional journals and reference works to create a satisfactory yet practical list of subjects, terms, and codes that would conform to the new international standards for zoology. In his 1898 guidebook, the alphabetical list contained over 1,600 zoological terms and their new Dewey codes. He had extended the classification for his zoological system to eight major and some two hundred subcategories. For example, embryology was increased from one category to twenty, and reptiles from one to a dozen. Herbert could thus tell potential subscribers who were interested in "Embryologia Ovum, Segmantatio" to go directly to section 18.1 in the zoological file to locate all the articles on that subject published after 1895–1896. Or, if they had a general interest in the topic, they could locate the first card in section 18 then browse through the cards until they reached section 19.[9]

Herbert was proud of his "expansions." However, he was never able to eliminate all the inefficiencies in his system. He had sought schema that would place all related items together so that a user had to make only a single entry into his files when looking for a particular item. To his disappointment, in some cases he was unable to avoid having to ask users to search more than one category when exploring a subject. He could find no way to avoid, for example, placing cards relevant to cellular histology in three different file areas. He told his customers that besides looking in the main section 18.1 they should check the entries in 13.1 (Embryologia ovum) and 14.63.1 (Sperma).

Marketing the Service: Herbert Worries about Delays and Compromise

With the progress on the expansion, Herbert sharpened his marketing. In 1897, although he remained unsure of how he could access all the journals he needed, he guaranteed his possible customers that his service would be comprehensive, without mistakes, and timely. He even went so far as to say that his service trumped the Royal Society's offering which would appear only every few years. Herbert began to advertise that he was designing a service that could beat any competitor, even the established ones subsidized by governments.

Herbert worried that any delay might prevent his center's opening, or that its integrity might be undermined if he followed the usual procedure of having proposals such as his reviewed in detail by numerous peer organizations. Herbert especially feared waiting for the hordes of national and subject-field scientific committees to revise and then give approval to his new version of the Dewey classifications. He believed that if he allowed such reviews, his scheme would be a jumble of unworkable compromises, ones reflecting what he was growing to fervently hate: nationalism.

Herbert had reason to worry. One of the problems of the Royal Society's international classification project-in-the-making in England was that it depended on "cooperative" support from hundreds of national groups. Battles over what and how to classify, whether or not to publish both a card and book catalog, what languages to use, and how to secure financial support from hesitant national governments bogged down England's project. Especially because England was basing its plan on decentralized classifying by voluntary indexers, Herbert again let his colleagues know that he was sure the society's effort was bound to fail. He and his supporters in America and Belgium also feared the Royal Society's project would not adopt the Universal Decimal Classification system.[10]

Retaliating to a Message from America, Herbert Twists the "Lion's" Tail (Again)

Surprising and unsettling to Herbert was his American friends' sudden recommendation that he now adopt a "cooperative" approach. They advised him to wait for the Royal Society to contact him. Herbert rejected the suggestion. He failed to appreciate the importance of the Royal Society's diplomatic links to the governments that were to participate in its international project, and he continued to be unrestrained in his criticisms of the British effort. He wrote that the society's leaders would reject any technological innovations and would resort to book-form bibliographies because of the pervasive influence of "old-fashioned" thinking. He was correct about the cards. Although there was an early tentative agreement to employ them, as well as a periodic distribution system, later investigations of the cost of providing cards led the society to abandon their use.

Herbert was not alone in criticizing the Royal Society's and other cooperative projects. Among their most energetic critics were Otlet and La Fontaine who were startled by London's rejection of their classification ideas.[11]

Retreating from Information Allies, Advancing the Center to Its Zurich Locale

Herbert was convinced that he had the solution to the science information problem and he wanted to begin his service before the Royal Society's project endangered it. To save time, he decided to discontinue consulting others. On the list of those excluded were Paul Otlet and many of Herbert's other old information allies. He wanted to be free of interference—in all possible ways. He had first thought of utilizing standing advisory committees in every nation to provide practical aid in the form of donations, subscriptions, and publications. In 1895, he had established such committees and organizations of associates in France, hoping they would supply journals and ensure that every publisher provided publication information before a book or periodical became available to the public. Now he viewed such advisory committees as burdens that would not provide much help until his system was in operation and had proven its worth.

He put aside any idea of establishing what librarians came to call "cataloging in publication" because it called for too much cooperation. And he retreated from his early hopes of integration or direct cooperation with all the existing major zoological-related bibliographies. There was too much political baggage involved in that, he decided. At most, he would supply the other publications with his cards after his service had independently produced them. In addition, seeking quality and promptness, he decided on an expensive alternative to the usual procedures in science publishing. He wanted printing, cutting, binding, sorting, and shipping to be in-house operations. He would not bother with unreliable contractors, as did other science publications.[12]

Other decisions shaped Herbert Field's family's life, even the history of the Cold War. The most significant of these decisions he made in late 1895, when he settled on the location of his new center, the details of its unique organizational form, and its name. The selection of the location for his center was a great surprise to his American supporters. He declared it would not be in New York City and close to his Quaker community. Nor would it be near the supportive Harvard museum or anywhere in the United States. Nor would it be in France, England, Italy, or Belgium—although those countries had great libraries that facilitated easy access to journals. Instead, the twenty-seven-year-old bachelor gave up his Paris apartment in late 1895 and leased a large villa overlooking a lake in Zurich, Switzerland.[13]

Although Herbert loved Switzerland's scenery and its sports opportunities, and although he had already found it a quiet setting for tackling tough intellectual problems, the reasons for choosing the country as his permanent base were other than personal. Idealism as well as business and political factors determined the location. Switzerland's readily available collections of scientific books and journals from around the world; the country's role as a crossroads of Europe; its multilingual population; its skilled academics, who could be trusted to make the difficult judgments needed for classifying; the Swiss government's hints that it might subsidize the effort; and a guarantee of loans of materials from all its libraries were among the reasons. There was another critical factor: Switzerland's reputation as a neutral and being above the suspicions that defeated international cooperation were of extreme important to Herbert Field.

The Fear of Information Nationalism

Herbert knew from his many European contacts that locating in the United States would be a political blunder. Some Europeans were already voicing objections to American library methods, as well as to the nation's arrogance. In addition, Herbert sensed that France, England, Belgium, and Germany had their own information visions, ones that were less than friendly to a service directed by an American.

Herbert had good reason to fear nationalism's impact. He had spent time in Naples, Berlin, London, and Paris. All of them had zoological centers and the necessaries for his project, but they were in nation-states that had become entangled in patriotic conflicts and prideful competition. London, to Herbert, was a city with people who were hostile to his work, as well as to any "internationalism" they could not manipulate. The new German nation also seemed a center of political hostility to international operations. The German government's policies had quickly turned the country into a world-class center of science publication; one that other nations feared might develop into a monopoly. Herbert correctly predicted that it would try to ban his service.

However, an important practical consideration that led to the selection of Zurich was its proximity to Germany. Herbert calculated that its libraries would be a valuable resource for his work. Zurich was also close to zoological centers and libraries in Leipzig and Naples that already had pledged to aid him. Herbert might need their help because he now estimated that he required access to as many as a thousand different publications a year to cover the zoological literature.

As soon as he leased the Zurich residence, Herbert asked his brother Hamilton, who was viewing art in Europe, to paint the outside wall of his villa with a map of the world and, then to overlay that with a logo for Herbert's international operation. Herbert also made his final choice for a name. Seeking a shorter and sparkling version than ones he had previously contemplated—the International Bibliographical for Zoology and Comparative Anatomy, or, the Central Bureau of Zoology and Comparative Anatomy—he opted for Concilium Bibliographicum. The name had the ring of prestige and permanency, it left room for covering other subjects, and it signaled that the spirit of cooperation and internationalism ruled.[14]

Herbert then made a very dangerous financial commitment. He signed documents making the Concilium a strange legal entity with only loose oversight by representatives of the International Zoological Congress that had approved his project and the Swiss government. Significantly, all the financial responsibilities and obligations were put in his name. Herbert did not seem to realize the consequences of that arrangement. One reason for that blunder was his failure to examine the legalities of publication in Europe. He unexpectedly had to register as a bookseller and to set up a legally responsible entity in order to be allowed to market his services.

After making those arrangements in November 1895, using his own funds Herbert rented two rooms near Zurich's university, opened an office, began to build the files that would be the heart of his service, and announced that he had begun the "experimental phase" of his work. He immediately became a busy, but unpaid and lonely, classifier. With only a secretary to help, he read through article after article writing out the Dewey codes. With the later aid of a part-time and temporary assistant, Herbert Field managed to classify some six thousand zoological articles, books, and societal proceedings in just two years, even though he had again taken to fever soon after he began work in Zurich.

Reconnecting with Friends and Taking On the "Lion"

With demonstration cards in hand, Herbert sought additional approvals, donors, and finally subscribers. The first step was to recontact Paul Otlet. After a bit of confusion, Herbert reached new agreements about methods, and about how Otlet would monitor and coordinate classifications with his larger bibliographic effort in Brussels. Otlet soon provided help—including advertising that Herbert's service was associated with his. As well, at least in the 1890s, Herbert was willing to have his publications marked as part of Otlet's series.[15]

Reaching this stage in the Concilium's history had been exhausting, and Herbert had never been strong. He had spent much time traveling to acquire and survey several hundred publications. However, he needed to take on some additional responsibilities to ensure a flow of journals and future cooperation. To please his colleagues, he accepted partial editorial responsibility for two other bibliographic publications, Julius Victor Carus's in Belgium and Paul Richet's of France. But all his demanding work had a payoff. His zoological allies stated that his modified Dewey classifications worked, and that they were the only acceptable ones.[16]

Herbert was physically drained, but the beauty of the surroundings in Switzerland compensated for his sacrifices in the Paris apartment and in the rather cramped office in Zurich. He had found it difficult to classify the zoological articles, and completing the instruction manual, *The Conspectus,* was a major chore. That guide for classifiers and users contained fifteen-hundred entries for just zoology. To ease his customers in locating the code numbers for the classification terms, the manuals had to include an alphabetically arranged appendix and his own "relativ index." To save money, he gambled that relying upon Latin for all its subject terms and translating only the introduction and instruction pages into national languages would be accepted.

The Conspectus guidebook was critical to the marketability of the Concilium. Unfortunately, Herbert could not make it the simple exposition he desired. It, and the system it described, were a bit forbidding. As well, Herbert had less than a reader-friendly writing style and he assumed that his customers already knew the basics of his numeric classification system. Fortunately, most of Herbert's clientele would be sophisticated, scholarly, and accustomed to a very formal style.[17]

With the classification tables and the booklet now available, Herbert began to move his project to the next stages, production and distribution. Herbert could tell his supporters their faith was well placed—he was about to begin sending cards to subscribers.

But Herbert's future was not assured. Just as he was about to offer his cards to the public, he faced a great, near defeating, threat.

London's Royal Society posed it. It seemed determined to prevent anyone from endangering its latest bibliographic ambitions. There were rumors that London was on the verge of introducing its own version of a numbered subject classification system. That led to another burst of extreme pressure on Herbert to, at minimum, yield his independence. England's bibliographers had so much power that he bowed to the request by his initial sponsors that at the end of five years he merge with the society's bibliographic project.[18]

The danger from England did not end. Hoping to eliminate a competitor, it soon demanded that England's universities terminate any Concilium subscriptions. Herbert was not the only casualty. London's great *Catalogue* planning meetings had been conducted in such a manner that even Otlet and La Fontaine decided to again cease any further contact. They determined to continue to build their great system despite the threat of future competition if England decided to expand into bibliographies for the social sciences and to create its own subject classification system.[19]

The Royal Society soon made some conciliatory overtures to Herbert, however. It asked for an immediate merger, citing, for example, how England's *Zoological Record*'s manager had been cooperative. Herbert rejected the offer on professional grounds. Then, the society tried to hire him away from the Concilium. Its leader went so far as to offer Herbert the role of director of a significant part of its international projects. Herbert refused the financially generous offer when the society demanded that he abandon the Dewey system and perhaps use Latin words, not numbers, for indexing.[20] The society did not stop. It soon let the academic world know that it held Herbert Field's methods and technologies, as well as Paul Otlet's, in less than esteem. It then hinted that any organization using the Concilium's services might endanger its relations with the society and its projects.

Herbert Whimpers, then Roars Back

In response to the Royal Society's challenges, and to the International Zoological Congress's desire for a merger with the British effort, Herbert explored a compromise. The formal announcement of the society's monumental new worldwide cooperative *International Catalogue of Scientific Literature* project led him to reexamine its methods with hopes of reaching some acceptable agreement. However, when he tested the society's classification schema for zoology against three thousand zoological articles, he found London's methods woefully inadequate. He thought its division of scientific knowledge into just seventeen major categories, and one thousand subcategories within each, was misguided.

Then Herbert examined the economics of the grand London project and concluded that its startup cost of $4 million, its dependency on rather uncontrolled classification centers around the world, its use of a yearly book-form publication, its translation into several major languages, and the high cost of subscriptions made the project impractical. Herbert also had negative thoughts about the suggestions to save expenses by indexing and classifying using only words from article titles.[21] (Unless specified, all dollar amounts are in 2006 U.S. dollars.)

Herbert then became aggressive. In his writings in professional journals, he stressed that if a library did subscribe to the British *International Catalogue* its high subscription price would drive out all other scientific bibliographic services. His predictions about the competitive impact of the *Catalogue* proved close to the mark. The number of subscriptions to the *Zoological Record* dropped by half when the *Catalogue* finally appeared. In order to survive, the *Record* had to merge with the *Catalogue* for a few years.

Herbert's Political Coup

Fortunately, Herbert convinced his Swiss oversight board that he should remain independent. That was a true political accomplishment because the Swiss continued to have obligations to the Royal Society's project and because of the agreement to merge with the *Catalogue* after his first five years of operation. In addition, despite London's threats, the Concilium gained friends. Herbert attracted supporters, even a few in England, as some scientists noted flaws in the Royal Society's project.

During the period when the society seemed unable to arrive at a sensible classification scheme, and when it appeared powerless to create a functional international organization, others began to criticize the *Catalogue*. Detractors pointed out, for example, that the various subject-field professional associations the society was relying upon could not even agree on a standard set of codes for geographic regions and nations. It was also noted that when subject matters overlapped, there were different and confusing terms.[22]

Herbert even found an unexpected American friend whom he originally thought was determined to undermine the Concilium. Herbert had been receiving encouragement from his Harvard contacts, but the initial correspondence with the University of Minnesota's zoologist-bibliographer and eugenicist Henry F. Nachtrieb was disheartening. In 1895, Nachtrieb notified Herbert that the university had already started its own zoological card file. There was an interchange of letters, some of which were heated. Nachtrieb warned Herbert that American academics in the West distrusted Harvard University projects and wanted to have any bibliographic center in a western city such as Minneapolis. Herbert replied that his system was not to be Harvard based and its faculty already knew that it was looked upon as elitist. But he also warned that a place like Minneapolis was inappropriate for a world center. There were some additional tense moments when Nachtrieb complained about the overly complex nature of Herbert's indexing system.

Much to Herbert's relief, Nachtrieb eventually told him that all of the Minnesota work would be turned over to the Concilium and that the college's staff gave him their full support. There was an added bonus to that. His university's library was building a reputation as an innovator in card cataloging and Nachtrieb was on his way to fame within the academic community.[23]

More Political Power: The Grand Documentalists' Dream

Of growing importance was Field's refreshed and deepening relationship with the two idealistic Belgians, Paul Otlet and Henri La Fontaine. Both were becoming influential international figures. Herbert had first sought them out when he was exploring the possibility of gaining support from Belgium's scientific societies; now they were becoming a means for Herbert to connect to the international community. La Fontaine had been, and continued to be, a prominent legal scholar, an expert on modern art, and a socialist politician who had sparked Belgium's interest not only in scientific information but also, significantly, organizational internationalism. Otlet was also a devoted internationalist and was the information idealist of the pair. Younger than La Fontaine, he was born the same year as Herbert Field, 1868, to a father of high political and economic position. Otlet's family had an international outlook, even more so than the Fields, with business interests across Europe, South America, and Africa. Like Herbert and Melvil Dewey, Otlet was a dreamer, but with some important twists. As lavishly documented in W. Boyd Rayward's works, during his teens Otlet drifted from his Catholic Jesuit education to a search for a grand secular "sociological" answer to life's fundamental puzzles.

So, unlike Herbert Field, Otlet's commitments to social welfare were not based on religion. However, like Herbert, Otlet had experienced a Calling and decided to change careers in his mid-twenties. He became determined to solve the information problems that prevented international cooperation. As a consequence, he and La Fontaine began their Brussels bibliographic center with a focus on legal and sociological information. Their database was to serve the needs of governments and reformers as they attempted to reduce nationalistic tensions. The Brussels database fit with their successful efforts to create the Union of International Associations, which they envisioned as a basis for cooperation and social reform throughout the world.[24]

Otlet had taken a great chance when he abandoned a legal career for "information." But he did well in his new pursuit. While Herbert's work never received much notice, and while Herbert was not a truly prescient information theorist, Otlet has been hailed as one of the most visionary of all Documentalists. The term "Documentalist" appeared when people such as Otlet sought a new and broader name for those who were concerned with information management. In its original meaning, the term pointed to those who wished to deal with more than books. They looked to all types of documents for the many facts and ideas they contained. In the terms of twentieth-century information scientists, they saw the documents as "works" containing ideas, not just as physical objects.

In addition to running a more grand information service than Field's, Otlet established himself as a philosopher of information and international cooperation. He seems to have been aware of the long history of bibliographic thought, and he was sensitive to technological opportunities. In his forty years of writing on information, Otlet predicted the use of microfilm, automated search engines, photograph-retrieval systems, and new types of libraries that dealt with separate ideas, not just books or articles.

Otlet's achievements and recognition did not come quickly or without sacrifice, however. He shifted from his boring law career just as his family experienced a financial downturn in the early 1890s. Then, with no guarantees of subsidies from his family, he decided to join with La Fontaine to establish their universal bibliographic service. Like Herbert, Otlet and La Fontaine had circled in on the Dewey methods. In 1894, they became aware of the Decimal numerical system, and they were soon convinced it was the starting point for their centralized files. By 1895, they had done enough work on amplifying Dewey's categories to hold an international congress in Belgium. At that and later congresses, Otlet and La Fontaine worked for more international cooperation in all aspects of life. Their long-term efforts in behalf of world peace and international organizations became linked to the 1920s League of Nations and, then, to the 1940s United Nations.

Otlet and La Fontaine's political connections helped them to obtain the Belgian government's financial support and, later, other significant financing. The funding helped them to establish the International Institute of Bibliography (IIB), and then a sister organization to build a card-based international database, the Repertoire Bibliographique Universal. Otlet hoped the members of the IIB would help to supply his center with the bibliographic items he needed. During the 1920s, Otlet worked to establish a neutral city-state in Belgium to house all world organizations. The center was never completed, but it had his Mondaneum, a library that at its height in the 1920s held over a hundred thousand books and twelve million bibliographic cards on what Otlet called "facts."[25]

For unknown reasons, Herbert sometimes shied away from highlighting his link to the great Brussels center, but he gave Otlet some important aid. He convinced him to employ Dewey's standard-size card so that users could buy the mass-produced catalog cabinets sold by Dewey's Library Bureau in America.

6 Starting an Information Revolution and Business, the Hard Way

By 1896 Herbert Field had many friends, but the Concilium's Belgian, Swiss, Italian, and French connections did not mean immediate success. None of those tie-ins provided adequate income or enough customers. Herbert had to continue his search for recognition and financing. That involved additional efforts to gain acceptance of the classifications for zoology's subfields. He could no longer avoid contacting many small scientific societies and engage in much frustrating negotiating. Herbert held his temper while he fought off adding new codes because he was determined to preserve his system's logical consistency against the demands of what he saw as nitpicking specialists.

Herbert could not evade confronting one very heated issue. It hurt, but he had to break with his commitment to numbers only. He realized that fixed codes for families of creatures were impractical. New ones were constantly being added and their numbers were so great that a codebook for them would be unmanageable. So, he decided to have their alphabetic names appended to his Dewey/UDC classifications. Also, he soon appended personal names to the codes for biographies and obituaries. Although he refused to use the expedient of letter-names for geographic indicators, he gave way to the demands for more precision. He completed a modification that had number codes specifying locations down to the category levels of rivers, deserts, and even lakes. A researcher could even know the ecological type, the region, the country, and the state location of an item. Herbert knew that meant increased cost for classification and printing, but he saw no alternative. Pressures also led him to make yet another visit to his earlier decision about his cards-only policy. He decided to implement the older idea that he had recently abandoned. His service was to send out short biweekly book-form bulletins as well as cards.[1]

The changes gained Herbert additional allies. Yet, before he could begin operations, he needed to overcome other hurdles. He had to solve many technical and economic problems. Printing techniques, even card-cutting machinery, took development time and hard-to-get money. Herbert was very disappointed when he realized that a newly invented automatic typesetting press was not suited to his project. That meant increased labor costs. He also had to launch additional surveys of continental libraries to ensure that, at the worst, he could travel to the various cities to use their journals.

To his credit, Herbert was able to fulfill his full-text pledges, at least in the early years of the Concilium. By 1904, the center based only 2 percent of its file entries on titles alone, but that was costly. As well, to meet the promises of worldwide coverage the Concilium found it necessary to gain access to and index over 2,500 different journals during its first decade. Many times, when he discovered issues were not being forwarded to the Concilium, Herbert had to make hurried and anxiety-filled trips to German and Italian libraries to attempt to fulfill what later information scientists, such as S. C. Bradford and Eugene Garfield, thought was an unattainable (and unnecessary) goal of surveying every journal.[2]

Labor Problems, Critics, and Herbert's Deep Pockets

Herbert continued to face practical ramifications of his commitment to a high-quality centralized service. The donations and the promised subscription income required for a skilled professional workforce were insufficient, and he knew he could not continue to be the Concilium's only professional indexer/classifier. He also needed more than a secretary to help with logistics and finances

He contacted the Swiss zoological and biological societies asking them to advertise his need for classifiers. He also had to recruit workers to maintain the database. When he recalculated the cost to sort, package, and ship his cards, and the wages of a bookkeeper and in-house printers, he was shocked. Nevertheless, he was not deterred. He decided that he would devote more of his own money to his cause although he was worried about renewed resistance to his ideas by influential old-school bibliographers and scientists, even those he once thought were his allies.[3]

Herbert encountered greater disappointments. His mentor Edward Mark signed off on a new and threatening Harvard report on scientific bibliography. It did not include a recommendation for the use of any numeric classifications system. It recommended short abstracts by authors. Even the forward-looking American medical librarian, John Shaw Billings, was taking swipes at card systems and expensively classified bibliographies. As disheartening was news that the influential Billings was in favor of the Royal Society's project. He was one of the American representatives at its London meetings, and he voiced misgivings directly aimed at the worth of Decimal classification in science in general.[4]

Additionally, criticisms of the particulars of Herbert's own classifications surfaced and spurred him to write rejoinder after rejoinder to heated denunciations of the Decimal approach (and his and Otlet's use of it), the cards, and the specifics of his classification structure. At one time, he was so frustrated that he penned an emotional public letter that took on the scientific establishment by declaring he was proud that his revised Dewey system did not employ all of the standard classifications.

Continuing complaints about a fundamental aspect of his system infuriated Herbert. He repeatedly had to point out that Arabic numbers were a universal language that eliminated the need to translate terms into a multitude of tongues. He also had to repeat that cards were the only efficient way to keep bibliographic files up to date. There were times, however, when he had to be a bit apologetic. For example, in reaction to complaints that the Concilium was not keeping up with the many changes in physiology, he soon created a new set of codes for the category. That meant leaving some older cards in their previous place while placing additional entries into a new subcategory. He did not admit that such duplicate placements created both filing and retrieval difficulties. He did, however, find it appropriate to defend what later began to creep into his lists, English language terms such as "weight of the body" as he tried to bend to the predispositions of new types of users.[5]

Meanwhile, adequate financing remained uncertain. Herbert had to continue to spend his own money to sustain his dream. Then, a tragedy eased the economic burden. His father passed away in 1897 leaving small fortunes to his children. Aaron's death gave Herbert the funds to continue the Concilium during what he thought would be its short formative period.

With over one-half million dollars in investments Herbert was not worried about finances, but in its first five years the center would spend 50 percent more than it received. In 1897, Herbert was not overly alarmed about any such temporary shortfall, nor was he deeply concerned about how to make a living once he turned the Concilium over to a permanent director. In contrast, his friends and sponsors remained anxious. When pressed, Herbert was unable to tell them when his emerging center would reach financial stability.

Herbert worried about immediate problems. As 1897 was ending, he wondered if he could guarantee distributing his bulletin and a packet of new cards every two weeks by his promised deadline of 1898.[6]

Professional Setbacks and Poor Health, but Optimism, 1897–1898

There was one encouraging breakthrough. The Swiss governments finally consented to provide seemingly permanent subsidies. Herbert (incorrectly) told his supporters in early 1898 that he did not have to use significant amounts of his own money, and he declared the work in Zurich was going beyond the purely intellectual task of expanding the Dewey system and writing manuals for use by the staff. However, there were some additional minor, but worrisome, setbacks. They took an emotional toll.

Unfortunately for Herbert's ego, the Nobel-honored Professor Charles Richet of Paris prepared a two hundred–page book of Dewey/UDC classified items on physiology before the Concilium could send out a regular card pack.[7] Because of that, Herbert believed his chances of gaining further support were endangered. Fortunately, the impact was minimal. Richet was in a cooperative relationship with both Paul Otlet and Herbert, and he had no thoughts of competing with the Concilium. Soon, he turned his work over to Herbert.

Although Richet's achievement did not prove a major problem, the project in Zurich had almost died. Because Herbert had fallen ill in the last half of 1896 (a frequent occurrence during his adult life) he had not met his initial goal of having his classification codes and methods in place by early 1897. Without them, the time-consuming classifications and card production could not begin.

Herbert had turned the further revisions of the Dewey codes over to a Swiss colleague while he took several months to recover. Then, he was stunned. The reworking proved unacceptable. His replacement had not followed directions. Herbert rushed to correct the problems while fearing that he might never find good men for the Concilium's major task, classifying articles. He was especially frustrated by the refusal of local printers to guarantee cooperation. He had rushed to them in desperation when he thought that time would not allow setting up his own press.[8]

Herbert feared he might have another breakdown. Fortunately, at the end of 1897, he finished redoing the classification scheme and processed enough documents to tell his sponsors that he had finalized his methods. Yet, Herbert was not elated. He had not attracted a significant number of subscribers to be able to predict when he could fulfill his earlier promise of establishing a self-supporting operation. He wrote his close friends that his greatest disappointment was that there been no immediate responses to his announcements in the major American scientific journals. As of mid-1896 he had not gained a single large American subscriber. A year later, only a few had signed on.

Herbert was not deterred. Once more drawing on his own funds, in 1898 he went from experiment to full production. Herbert purchased an expensive printing press and a top-of-the line card-cutting machine. Then, he rented a house near his office for the printing operation. He hired several typesetters, each with multiple-language skills, and a printer—along with a new skilled multilingual zoologist assistant to help turn document content into Dewey/UDC numbers. Also, a time-consuming complex file was completed. It contained all the mandatory abbreviations for any word or term used at the Concilium.[9]

There were more difficulties. Herbert had to reverse an earlier decision and retry a new version of a frequently advocated method for speeding the acquisition of publication information. Rather desperate, the first task he assigned to his printer was to create double-fold postcards. On a side of one of the detachable cards was the address of a publisher or author. On the other were instructions on how to fill in the card so that the Concilium could be quickly notified of a coming publication. The return portion of the card was to contain the usual bibliographic information, as well as a short abstract. In the absence of a full text, the abstract might be used to classify an item.[10]

Herbert again put on an optimistic public face. He announced that his unique system was ready for more subscribers. His renewed energy and buoyancy paid off. In fact, just as he was ready to begin sending out the first packets of cards, a greater number of subscribers than he expected came forward. Success seemed at hand. In six months, some three hundred full and partial subscribers received an amazing 1.6 million cards. The Concilium seemed to have a chance to attain the economic foundation necessary to survive, if and when Herbert decided to pursue other interests. Herbert was proud. Importantly to him, he had, or so he thought, constructed an efficient system, one that might carry on without him. He was considering returning to scientific research.[11]

The Hard-Won Achievement of User-Friendly Efficiency

The center was becoming a marvel of the application of the new learning and methods. Herbert's days at Brooklyn Poly made him sensitive to the value of the application of rational processes in the workplace. The Concilium began with well-ordered procedures in place, and Herbert continued to improve them. He, or a skilled assistant, read a document to see if it was at all relevant. Then, to determine how to code items, professional judgment was applied to each significant section of the work. Ideally, there was to be a class code assigned to each major fact or idea in a document so that users with different needs could recover all the relevant cards for each query.

Herbert had designed an intense indexing/classification scheme that went far toward providing a complete summary of an item. For example, a typical card concerned with one type of creature that built coral-like skeletons contained two general class codes, an era code, the alphabetic name of the specific variety, and a geographic code. The codes for other items in a document might be listed below the article's title in a small font. Separate cards for each of those multiple classifications could be created if customers desired them. They could place them in different sections of their file. However, to reduce costs, Herbert soon reserved card replication to cases in which a classification made sense only if it required codes from two separate classification hierarchies—an Otlet-like "colon" classification (figure 6.1).

Filing and Retrieval

Correct filing and timely updating were critical to the Concilium's offerings, and Herbert did his best to avoid problems. He knew that if his customers misfiled too many cards they might well discontinue their subscriptions. Herbert thought that he could balance the needs for efficiency in filing and retrieval with user friendliness by using special symbols and fonts, and by permuting the order of the classification entries in versions of a card.

Despite providing such aids, the Concilium's system was not burden-free. Preparing to search the files took some effort. A user had to look up the various codes for an

Hall, T. S. **47.13 Adeona** 04.99 (118 : 94.5)
1897. On the occurrence of the Anchoring Tubes of *Adeona* in the Older Tertiaries of Victoria with an account of their structure. **Proc. R. Soc. Victoria N. S. Vol.** 9, p. 1- 4, 1 Pl.

Figure 6.1
Concilium card, from Hoyle, "Concilium Bibliographicum at Zurich and Its Work," 1899

Streeter, George L. .602.858
1907. Some Factors in the Development of the Amphibian Ear Vesicle and Further Experiments on Equilibration. **Journ. exper. Zool. Vol. 4** p. 431—445, 6 figg. [Ear Vesicle transplanted to opposit side, assumes normal posture. Primitive left ear vesicle transplanted to right side retains left-sidedness. (Rana).]

Figure 6.2
"Telegraphic" Concilium card, from Concilium Bibliographicum, *Conspectus*

item in the guidebook's alphabetical "relativ" index. For example, a user interested in a single geologic period had to consult the table of "period" codes in the Palaeontologia section of the conspectus and memorize the time period codes in the Stratigraphia (11) subsection. Those numbers ran from 111 (Archiacum) to 119 (Quartaericum). For someone seeking articles on cellular histology related to the Herault region of France, the alphabetic index had to be consulted to find 18.1 for cellular histology and the geographic code 44.84. The demands on a user's time and memory increased in cases of broad-scale browsing. A searcher had to remember many codes if he was making a general survey of the cards.

Herbert, however, thought the other advantages of his system more than counterbalanced the code memorization difficulty, and he made some additional concessions to possible user demands. If the scientist-classifier judged that an item's title was not meaningful, he wrote a short explanatory text in a "telegraphic" format (i.e., one that uses abbreviations and simplified text, figure 6.2).

Error-Proof Coding, Printing Safeguards, and Sorting

Herbert devised many safeguards against the kinds of errors that marred other bibliographies. To avoid transcription errors, after codes had been assigned the document itself was sent to the typesetters who copied exactly the item's title and the telegraphic text, if any. Then, using the Concilium's official abbreviations, the printer set the other information. Those abbreviations saved type and labor costs. The typesetters' next step was to set up separate cards (with the same text) but with a different card format for each of the permuted codes. For that step, all he had to do was switch the classification codes' order. To reduce costs and errors, Herbert designed special equipment so that the codes could be moved while the type-frame remained in the press.

Herbert established other procedural safeguards. Once the type was in place, typesets were sent to the printing room. After a first trial run, the printer forwarded card sheets (each holding thirty-five cards) to the editor's desk for a final proofreading and correction. Before a full printing, another sample was sent for proofreading and editing. The final sheets were sent to the cutting room and then to the room where the machine for punching the card holes was located. Those holes allowed for integrity of the card files. Before the Concilium's staff placed any cards in the master file's cabinets, they transferred them to the sorting room.

The sorters assembled the card packs for the main file and for each subscriber. Those were complex tasks with much potential for error. So, there were several checks made to ensure that each customer's shipment included the correct type and number of cards. That was not a simple step; Herbert allowed some subscribers to take only parts of the master run because there were not enough patrons for the entire bibliography. To ensure correct billing, a supervisor attached a "totals" card to each pack. Then, the shipping clerk wrapped the rather large and heavy packages (of as many as two- or three-hundred cards) and mailed them. The remaining cards went to a special storage area to await future patrons' requests, with one set to the master file.

The Inventory Dilemma

One aspect of Herbert's system was very critical, but impossible to fully control—especially because he rejected "stereotyped" printing. After the proofreading, the center's manager had to determine how many copies were to be run before the printer broke up the type block. Ill-judged decisions about quantity were prone to costly errors. One of the major selling points of the Concilium was the assurance of providing a complete cumulating bibliography, one dating from the mid-1890s. If new subscribers were to be satisfied, the Zurich center had to have copies of all the cards on hand. Too few copies meant much pricey recopying of items from the master file. But printing more copies than were being sold meant added expenses, including the cost of a storage room to warehouse the extra cards.

There was no sure method of calculating how many cards to keep on hand. At first, Herbert typically made runs of slightly more than the number of subscribers, but even that meant buying space in another building to house the possibly back-ordered cards. The storage problem grew over time as the Concilium began producing several thousand primary cards a year. By World War I, just its central file had hundreds of thousands of cards. Warehousing multiple copies (figure 6.3) was becoming financially prohibitive.

Herbert was unable to devise a solution to the inventory problem, but he was pleased with the other procedures and the overall effectiveness of the Concilium. He felt that

Figure 6.3
A card storage area, from Doolittle, "A History of the Concilium Bibliographicum, Zurich," 1929

his service was more than price competitive with other bibliographies. He proudly told his friends and his overseers that his cost per item was a small percentage of the amount predicted for a Royal Society *Catalogue* entry. Herbert's comparison must have increased the tension between the two organizations. It was difficult for England to acknowledge that he could produce thirty-three items at the anticipated cost of one of their entries, and that his system allowed more entry points for retrieval than did theirs. His "random access" feature was also an irritant.

Success and Failure in the World of (Information) Plenty: Making Dollars, Sense, and Promises

By the end of 1898, the Concilium seemed to be on the way to success. Herbert felt he was near to fulfilling his Calling. But, some new problems were already appearing. Costs began to rise, but not the number of subscribers. The cost problem was another result of information-plenty, not of a lack of efficiency. Herbert had expected that his zoology work would call for some six thousand entries a year selected from about one thousand different publications. That was an underestimate—a considerable one. Just keeping track of the journals became a job in itself. It took constant correspondence and much non-too-diplomatic pressure to convince the world's scientific societies to send their publications to Zurich for review, and such efforts were not always successful.[12]

More difficulties arose. Just as he began to mail his cards in 1898, and just as the Concilium's review committees were meeting to decide whether to continue to approve

Zurich's work, Herbert was persuaded to take on the task of helping to edit two additional traditional bibliographies—and to do it with little or no compensation. With that, he overreached himself and endangered the Concilium. Herbert and his tiny staff did their best to carry out their promises, but they soon had to notify the scientific community they could not complete the work. Herbert felt humiliated. He worried that the Concilium's reputation was permanently damaged.[13]

However, Herbert went on to assume a greater editing and publishing responsibility, one that continued for years. Taking up Carus's *Bibliographica Zoologica* was both a blessing and a curse. The annual volume gave the Concilium prestige, but it was a competitor to the card file. Herbert did not stop assuming such burdensome but prestigious editorial chores.

Additional pressures for Herbert centered on a classic business dilemma, the tug-of-war between rising cost of production and the need to keep the prices of products low in order to attract customers. As Herbert discovered additional relevant journals, and as more cards per document seemed to be required, his processing costs escalated. By the end of 1898, zoology alone called for fourteen thousand primary cards per year. Soon Herbert had to hire additional science professionals to do the coding. That called for cutbacks of other activities. For example, he abandoned the expensive attempt to use those double-fold cards for prepublication notification by publishers and authors.

The Concilium's production costs were increasing. That meant a need for more subscribers. Herbert once more became a traveling salesman, in Europe and North America. He also became involved in a costly attempt to introduce new information products designed to expand his markets tand income. At the same time, he had to eliminate parts of his service. As a first step, he and his classifiers became highly selective as to which items would receive multiple cards. There was one expense he could not reduce, however. He had decided that titles were to be kept in their original language to avoid translations costs and errors, but there were downsides to that decision. He had to buy and maintain an increasing stock of expensive and exotic type; and, his printers had to devote much frustrating time to typesetting in unfamiliar languages (figure 6.4).[14]

To gain new subscribers, Herbert created several custom-tailored card bibliographies. He and his classifiers built subsets of items as they read the regular journals, and then searched for additional items in specialized publications. The following example (figure 6.5) is from a later attempt at a bibliography for evolution, the subject that would cause so many political problems for his most generous American supporters. To keep the cost of the new bibliographies at a minimum, he used one general subject code.

Juggling Supply and Demand, Plus a Few Bells and Whistles

Herbert's new information offerings were producing some income, but they led to added expenses that, in many instances, were greater than the returns. An ambitious file containing a list of all the new discoveries in zoology was central to the hope that

Godlewski, E. **4.38 Helix : 14.63.1**

1897. Wielokrotna karyokineza w gruczole obojnaczym ślimaka Helix pomatia. **Rozpr. wydz. matem. przyrod. Akad. Umiejętn. Kraków Bd. 33 p. 171—208,** 2 Tab. — Ueber mehrfache bipolare Mitose bei der Spermatogenese von Helix pomatia. **Anz. Akad. Wiss. Krakau 1897 p. 68—81,** 2 Taf.

Figure 6.4

Concilium card, from Concilium Bibliographicum, *Conspectus*

Bumpus, Hermon C. **575.4**

1899. The Elimination of the Unfit as illustrated by the Introduced Sparrow, Passer domesticus (a fourth Contribution to the Study of Variation). **Biol. Lect. mar. biol. Labor. Wood's Holl 1897/98 p. 209—226.**

Figure 6.5

Evolution card, from Concilium Bibliographicum, *Conspectus*

the Concilium would become the world's authority on the priority of discovery and names. Herbert invested heavily in the file, thinking hundreds of customers were willing to pay to have their claims validated. He even imagined that major universities and libraries would purchase their own copies of those cards. A "discovery" card took hours of experts' time and as many as fifty lines of text. Despite Herbert's faith that it could be a moneymaker, the service remained only a huge in-house file. Even a stripped-down version for new discoveries only was costly to compile. As a result, the file was often left unattended.[15]

The behind-the-scenes overhead at the Concilium was also growing. Large files that brought no direct income had to be maintained. Two of those, lists of all journals and publishers, were innovative and reflected Field's bibliographic thoroughness. They held full information on addresses and cost of each publication. Another file recorded all items that the center had indexed by journal and issue. That large file on previously classified items prevented embarrassing duplicate requests to publishers.

A surprising, perhaps disappointing, customer demand added to the workload and financial challenges. Herbert thought that all modern scientists desired retrieval not by author name but by subject. The demand of some users for the old-fashioned file type, however, arranged by author name, soon became an additional burden. At the same time, Herbert had to accept another financial and professionally distasteful obligation. He tried to avoid it, but customer demand soon prompted him to make available more and longer abstracts of items, not just their classifications and telegraphic description.

Abstracts continued to be disliked at the Concilium. An important point for Herbert was that no one abstract of an article could satisfy all users. He argued, for example, that an abstract written by an entomologist might overlook an aspect of the article that would crucial to an evolutionist (and vice versa). That implied criticism of attempts by any one abstract journal to cover many fields quickly earned Herbert more enemies. But supporters of later general publications, such as the *International Catalogue* and *Biological Abstracts,* became quite unhappy when they later had to admit to the validity of Herbert's critique of short summaries done by a single abstractor.[16]

Meanwhile, Herbert searched for additional ways to attract customers. He created sets of "guide" cards. They were a version of the tab cards patented by a YMCA administrator, James Gunn, which Dewey's company marketed at the turn of the century. The special cards had tabs that stood above the file cards and indicated in words major subject divisions and subdivisions. Herbert had found that such cards greatly speeded retrievals because they partitioned the Concilium files into subsets and allowed users to sidestep the numeric codes. In modern terms, they made his files more like "random-access" mechanisms. By World War I, Field offered over twelve hundred of those cards for the zoology main file, hoping that more and more divisions of the file would speed both updating and retrieval. Commonplace today, they were at-the-edge of information techniques at the time. Those bells and whistles for card files also had the potential for profit. Their price was three times that of a bibliographic-entry card.[17]

By 1903, Herbert added other possible revenue sources. One was selling cabinets. His European customers had found it difficult to obtain such furniture from Dewey's Library Bureau in the United States. In response, Herbert devised compatible designs and arranged to have them manufactured in Europe. Although he advertised the cabinets in America, his design reflected European concerns about the space required by card files. Herbert gave that precedence over ease of use. His first model was six-and-a-half-feet tall but only two-feet wide. That saved precious room but made card access a bit difficult. The top and bottom drawers were hard to reach. Like the cards, the Concilium cabinets (figure 6.6) were of high quality, but heavy and expensive. A cabinet with enough drawers to hold seventy thousand zoology cards cost almost $5,000. At the usual rate of publication of cards, in 1940 a full subscriber would have had to invest at least $50,000 in cabinets to hold the collection.[18]

Herbert also hoped to gain income by selling blank custom-made cards. His in-house printer produced different color combinations. Herbert priced those cards to make a tidy profit. Other income-generating activities soon included ordering hard-to-get books for subscribers and acting as an agent for all of Paul Otlet's Brussels publications.

New Publications and Services Amid Ties to the Rich and Powerful

Another activity stemmed from more than a business motive. Herbert had not shed his social-idealist background. He began selling the bulletins of an organization, the

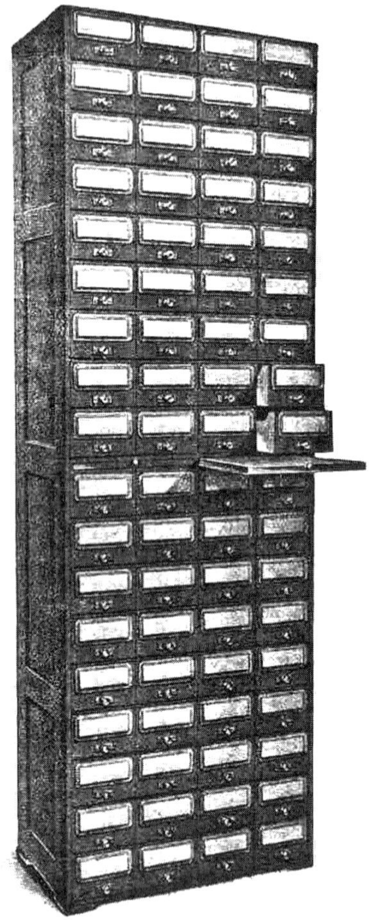

Figure 6.6
Concilium cabinet, from *Annotationes*

International Peace Institute of Monaco, which supported his Quaker pacifism. Monaco's prince, an acquaintance of Herbert and the Havilands, had recently established the institute. Herbert published peace books on his own. For example, in 1916 and 1917, his center published two works by Howard Fowler McCormick, the heir of the American farm machinery magnate Cyrus McCormick Sr. Howard was the husband of Edith Rockefeller (of the oil family) and trustee of the ultra-rich Rockefeller Foundation that came to play a critical role later in the Concilium's career—and in the lives of Herbert's children.[19]

Another hoped-for source of additional income was an on-demand service. Herbert worked out the details of how to do more than cater to long-term customers who subscribed to either the full or the partial sets of cards. For extra fees, a request could be sent to Zurich for the creation of a one-time bibliography, even one that might call for searching through several different categories and files. An inquiry could be placed through the high-tech long-distance telephone in Herbert's office, Zurich #4418, or through the "bibliography.zurich" address on the telegraph system.

Herbert also began designing an ongoing Concilium newsletter, the *Annotationes*. It was to contain announcements of publications by and about the Zurich center, promotional pieces about the Concilium's services, and articles on scientific bibliography. It took some years to begin its publication but in 1905 Herbert believed it would promote sales and bring in additional income.

One Pesky Question—Can a Nonprofit Make a Profit?

In continuing to satisfy users' demands, Herbert Field was nevertheless undermining the Concilium's financial position. For example, by 1902 he was receiving complaints that his service was missing much vital information because it was indexing only the materials in the major body of a text. In response, he created a new and very costly file for "minor" information found in notes or small entries on a page. Identifying and classifying such bits of information proved unprofitable because few of those who had demanded the service subscribed to the new Supplemental Bibliography.[20]

Then, Herbert found himself burdened with another costly, unpleasant, and eventually embarrassing task. It was to fulfill the Swiss portion of the Royal Society's great *International Catalogue of Scientific Literature* project. Herbert could not avoid the responsibility because of the obligations that accompanied the Swiss government's Concilium subsidy. But, he found an excuse to send the work out to others. That meant little income for the center and some damage to its reputation. The quality of the product that came from the Swiss work pleased neither those in charge of the catalog nor Herbert. Another new project was less discomforting and lasted much longer, perhaps through to the 1940s.

As an income generator, however, it was a disappointing and costly failure. In 1901, the International Zoological Congress asked that Herbert begin yet another version of his earlier "discoveries" project. They wanted all new taxonomic names registered in a central file. Thinking the demand of the Congress indicated that customers were finally ready to pay for such an effort, Herbert and his staff began work. In the first year, they recorded almost twelve thousand card entries in manuscript form. By the late 1920s, there were five hundred thousand. That achievement may have received accolades from zoological societies, but it never brought funds. Only a few "names" cards were ever printed and only a handful of inquiries were processed.[21]

There was another economic threat. It came close to exhausting Herbert's patience. His assistant for physiological classification demanded a raise, one so high that his income would have exceeded Herbert's. The employee became abusive when Herbert refused. Herbert fired him. The man then filed a lawsuit. Fighting the suit endangered the Concilium's funds and its ability to attract skilled indexers. Fortunately, Herbert persuaded the angry professor to drop his legal action.[22]

7 Sustaining the Concilium: Big Debts, Big Gamble, Big Building, Big Friends—and a Special Librarian

By the early 1900s, the subsidies from the zoological societies and the Swiss government proved inadequate solutions to bridging the gap between the Concilium's costs and the income it derived from subscription and services. As psychologically depressing for Herbert Field were the required but increasingly difficult-to-meet payments on growing debts. He had to keep pumping his personal funds into the Concilium and began to show signs of bitterness, especially because he was regretting his decision to remain at the Concilium rather than to return to zoological research.

Herbert launched another quest for remedies but discovered few. He explored a new pricing policy, which varied subscription prices according to the size of an institution (and which seemed to be working for commercial publishers), but he did not want to alienate his best customers by adopting it. He was especially troubled by an unexpected financial challenge. Many important subscribers were not paying their bills on time. To Herbert's surprise, "gentlemen scientists" were not being gentlemanly. While trying to avoid losing any customers, Herbert had to engage in some very delicate negotiations with major institutions.[1]

At the same time, Herbert was still unable to control his expenses. In hopes of attracting additional subscribers he started new bibliographic series for anatomy, physiology, paleontology, botany, microscopy, and general biology. Some of those specialties required their own guidebooks with their own codes and alphabetical listings. By the 1920s, the offerings even included museum methods such as cataloguing; unfortunately, some of the efforts in that field were short-lived.[2]

Herbert soon went beyond those subjects. Just before World War I, the center was moving into applied fields such as electro-chemistry and had increased its coverage of the hot-button topic of the period, evolution. Meanwhile, Herbert kept his loftiest information goal in sight, a bibliographic service for all the natural sciences. That far-reaching ambition included the vision of services for those interested in medical research, mineralogy, pharmacology, and anthropology.

He hoped that he could market such products through cooperative arrangements. He made, for example, attempts to tap the medical researchers' market through the

Wistar Institute of Philadelphia. That proposed arrangement proved to be a mistake. Herbert also thought he could increase his customer base by reemphasizing the Concilium's possibilities as an information bureau—a true research center that could act as more than a provider of bibliographies and might also be able to answer many substantive research questions.

A Needed Boost and a Gamble: Building an Information Palace

Such hopes were not realized. None of his new initiatives were paying off. The Concilium was barely surviving financially. Fortunately, in 1900, the center had received some help. The Swiss government increased its yearly subsidy five-fold. It came at a price, however: more outside supervision. To monitor the contribution, the government appointed a more powerful committee of respected Swiss scientists to audit the Concilium. They provided yearly reports to the nation's Society of Naturalists.[3]

Soon, when Herbert had a surge of optimism, he pledged his own investments to secure a very large bank loan for a new and imposing building. Herbert wanted to make a statement. The Concilium was here to stay and subscribers would never be abandoned! He had concluded that if the Concilium looked secure, it would become a permanent international information center. He purchased a lot on the hill near the university for his new bibliographic palace and then hired an expensive architect. The two spent months working out the details of an Efficient as well as impressive building that would make the Concilium and Zurich (as well as Frederick Taylor) proud. Herbert announced to the world that the building would be in full operation by 1908. It would be large enough, he wrote, to give each classifier an individual room. There were to be reinforced floors to support the huge cabinets, and a large in-house printing plant on the ground floor.[4]

The handsome four-story building at 49 Hofsta would be expensive for more reasons than the bank's requirement that it be designed so that it could be converted into apartments in case Herbert defaulted on his loans. The architects had to make its interior an efficient space for the passage of people and materials. There were special staircases and lifts. Herbert had an elegant office that he could use as a meeting room. There were both gas and electrical service. A telephone system with a long-distance connection ran throughout the building. Herbert had special versions of the card cabinets built into the walls to allow access to the cards from two workrooms at once. Extra rooms were included to accommodate the projected growth of the workforce and to provide housing for the full-time janitor and his family. In addition, the new Concilium had a lovely view of Zurich and had the added bonus of being close to the university, a tramline, and a post office.[5]

Even though the service was not paying for itself, Herbert continued to expand the Concilium as he awaited the completion of the new building. He leased more offices and increased his staff to fourteen. Production grew. By 1904, the Concilium had sent out nearly thirteen million cards and the main file had only twenty thousand fewer entries than in the U.S. Library of Congress's general collection catalog. Other files had 150,000 more. In less than a decade, an impressive 150,000 articles had been read and classified.

Herbert continued to do much of the classifying, although he had to spend more and more time as a manager and as a version of traveling salesman. He was again rushing to professional meetings, such as that of the International Association of Botanists. Despite that and his debts, he bought land on a hill outside Zurich and constructed a new home.[6]

Herbert's efforts did increase the number of customers. There was a growing list of individual and institutional subscribers. Also, several scientists were making requests for special bibliographies, although that service was expensive. Those on-demand users were willing to pay as much as $.20 per retrieved subject card. Regular customers paid much less. A full subscriber was paying $600 for all the cards and perhaps another $1,400 for local filing and maintenance per year. A subscriber to a complete set of the main files (some fifteen thousand cards per year) paid less per citation than a partial subscriber, but still an appreciable amount. The yearly cost of a full subscription and maintenance was approximately two-thirds of an undergraduate's tuition at an eastern U.S. college.[7]

Yet, in long-term perspective the Concilium seems a scientific information bargain. Future competitors, such as *Biological Abstracts,* which began its service in the 1920s charging less than a fourth of what the Concilium was asking for its cards, also became expensive. In 2006, in an era of automated production, a subscriber to *Biological Abstracts* was paying its publisher more than $9,800 a year. The Concilium's oldest competitor, London's *Zoological Record,* was asking close to $5,000. That was eight times what Herbert's subscription price would have been.[8]

Herbert was selling his card-file products at a comparatively low price, but that did not allay his fear of competition. Realizing that a subscriber's maintenance cost of older book-form science publications was near zero, because only shelf space was required, Herbert always worried that his subscribers would abandon his unique service for a lower total-cost traditional offering.

Despite the fears about customers returning to book-form bibliographies, Herbert had to take a dangerous step. He raised his prices. He had no other solution to escalating production expenses. Under pressure from his advisory board, he took the chance of losing patrons by raising fees by one-third. Then, just after World War I, he raised them to double his original rates. His increases far exceeded the general inflation rates. Fortunately, in the early 1900s, Herbert's unique subject-classified "random access" service had little direct competition. Herbert had subscribers in all major European nations. In America, he had customers in every region except the South. Even Minnesota, California, and Hawaii had subscribers.

Herbert did try to serve the science-information "needy." He hoped that some new products would silence critics who were complaining that his system served only large and rich institutions, providing nothing for the typically isolated and poor individual scholar. Besides giving discounts for large orders, he offered a lower-cost but thinner and much less durable card intended for those who did not plan to frequently search their files. In combination with his special series that provided cards for narrowed subjects, those brown cards provided a "desktop" version of his service. However, they attracted few individual subscribers.[9]

The Era of "Big" Science: Important Contacts and "Bigger" Ideas

Although the Concilium had new and more popular offerings, and although it was international, its survival depended on its large-scale customers, especially those in the United States. Europe remained the center of modern science, and continental governments were subsidizing science and technology, but North America was more hospitable to new information initiatives. As a result, although located in Switzerland, the Concilium's fate rested in New York and Boston, and on those Americans who were guiding the emergence of the United States as a significant force in world science.

Central to the Concilium's cause in America were Herbert's former Harvard zoological colleagues. They had always provided Herbert important contacts with wealthy customers and sponsors. However, more factors led to America's central role in the Concilium's destiny than the ties to Edward Mark's students. The attractiveness of Herbert's service to cutting-edge researchers in America was a result of zoology's link to the most critical scientific issues of the era, and to the struggle to professionalize, define, and modernize American academia.

There was another critical factor. Just as the Concilium was born, so was America's new industrial wealth. Industrialization gave America the money to build expensive science museums and to change more and more once-small colleges into research universities. With the new well-financed institutions, and a flood of their eager young professors rushing to make their way, came an acceptance of fresh techniques.

Ideas also played a role in the American support of the Concilium. One was central. The Darwinian evolutionary hypothesis had torn apart old grand explanations of the origins and value of life, and there was a frantic search for new answers. At the same time, new understandings in biology, and even agriculture and medicine, were making scientific natural history relevant to the practical world. Investments in science seemed ready to yield economic benefits well beyond those imagined by the founders of the American state universities. American zoologists and their institutions were no longer just collectors of curiosities. They were beginning to contribute to economic, medical, and social progress.[10]

That energized many young scientists and provided them with arguments for the ever-greater funding needed for their research projects. America's new generation of scholars was intent on establishing a place in more than traditional science. Many justified their costly work as vital to industry, modern agriculture, and to the determination of sound social policy. Some men trained as zoologists even went on to become founders of a new branch of psychology, intelligence testing. Herbert's zoological ally, Robert M. Yerkes, for example, was a creator of mass IQ testing, a tool increasingly utilized as an objective, efficient way to sort people into or out of jobs, military service, or colleges.[11]

Given zoology's ties to such ambitions, it was no wonder that it was one of America's first almost "Big" Sciences. It was becoming costly, relatively well funded, and the producer of famous academics and renowned explorers. America's leading zoologists could afford to be the Concilium's best customers, and they made sure Herbert Field's file was at all the major zoological research centers and universities in the East, Midwest, and West. Yale, Harvard, Princeton, Columbia, Cornell, and Michigan, among many others, were early subscribers. The Smithsonian and the Library of Congress in Washington took full sets, and even the new form of a public library for science information, the John Crerar Library in Chicago, supported the Concilium.[12]

The Smithsonian's subscription was a very pleasant surprise to Herbert because it had been the official American member of the Royal Society's catalog project. As encouraging, the Concilium's North American customers were not just zoologists. There were American ties to institutions and practitioners in medicine, entomology, veterinary science and horticulture. The United States Department of Agriculture gained access to the entire Bibliographicum through its station at Ohio State University. Other entomologists, anatomists, and physicians across the country took partial subscriptions.

Very Special American Friends

Of special importance to Herbert Field's cause were two American institutions and two science leaders. They were also crucial to America's search for answers to the questions raised by Darwin. They would even become important to Herbert's sons.

A former neighbor and classmate of Herbert's, "Charlie" Davenport, was one of the men. The Brooklyn Museum, founded in 1897, was one of the institutions. The museum, only three miles from the homes on Columbia Heights, was the product of the surge of progressive reform in Brooklyn that had created a cultural complex next to the museum. That complex included the city's magnificent public library and its exquisite Botanic Garden.

Charles B. Davenport was a "Brooklyn Heights boy." He was born just two years after Herbert, growing up a few blocks away from the Fields at 11 Garden Place. Like Herbert, he came from a wealthy family. But, unlike Herbert's father Aaron, Davenport's father was very strict and a conservative Congregationalist, not a liberal Quaker, whose views on child rearing meant that his son's life was far different from Herbert's, at least during his formative period. After years of barebones home schooling and serving as unpaid office help for his father, Davenport's parents begrudgingly allowed him to enter liberal Poly. He stayed seven years, eventually receiving a bachelor of science degree in civil engineering. Like Aaron Field, Davenport's father wanted his son to be a businessman, and he resisted any idea of college for him. Charles Davenport had to go to work after leaving Poly, but he did not obey all of his father's dictates.[13]

While working as an engineer on railroad projects, Davenport schooled himself in Latin and eventually gained admittance to college, soon securing a career-saving scholarship. He became, like Herbert Field, a Harvard man and a natural science major under Edward Mark. Davenport did well and continued to shine in Harvard's graduate zoology program. Like all American graduate students of the time he wanted to continue his studies in Europe, but he did not have Herbert's financial resources. So, in 1892, "Charlie," now a newly minted PhD, had to take a position as a junior instructor at Harvard. At first, that appeared to be a dead-end job. In the new high-status American universities, a European degree seemed essential to promotion. However, Davenport quickly climbed the ranks in Cambridge and in America's expanding scientific professional organizations. One thing that aided him was the mathematical and statistical training he received while he studied for his engineering degree. Davenport wrote a path-breaking text that introduced American scholars to the use of statistical methods in the study of biological variation.[14]

After spending some years as a professor in Cambridge, Davenport was enticed into moving to a shining star among new American universities, the Rockefeller-supported University of Chicago. A sabbatical leave in 1899 then reshaped his professional life. He sailed to England to study with the geneticist-psychologist-eugenicist (and relative of Charles Darwin) Francis Galton, and with the statistics wizard Karl Pearson. Davenport returned to become a dynamic leader in American genetics. His focus on genetics was what led to Davenport's appointment to a soon-to-be prestigious institution founded and sponsored by the Brooklyn Museum. Davenport took charge of the museum's Cold Spring Harbor research station.[15]

The "Harbor" was located at the tip of Long Island, not far from the museum and Columbia University, which made it a favored site for New York's academics. Its zoological, plant, and animal genetics research (and later human genetics and eugenics work), made it and Davenport world famous and to some, infamous. The Harbor's evolution-genetic effort was what led the Brooklyn Museum to subscribe to the Concilium's files and to spread the word around the nation that it was an indispensable tool—any respectable scholar had to have access, Davenport asserted. In 1905, the museum, because it had subscribed to all file variations, had over 260,000 Concilium cards. And it was an active supporter of Herbert's system, publically stating that complaints about it were unfounded. The museum's leaders were also declaring that the value of its contents outweighed its expense and bulkiness.[16]

The second American academic central to the Concilium's fate, Henry Fairfield Osborn, was also one of the new professionals, but one different from men such as Davenport. His family's wealth allowed him to be as much an old-fashioned gentleman scientist as a new researcher. This famous and influential American contributor to Herbert's cause was a New Yorker who grew up in Manhattan attending its best private schools.

The second institution critical to Herbert's dreams was being reborn with the infusion of money from America's new upper classes, and by the policies and skills of Osborn. The important institution was New York City's competitor to the Brooklyn Museum, the American Museum of Natural History. The inclusion of "American" in its name reflected its founders' ambitions. The museum was created just after the Civil War as a means of educating the public and as a way of establishing New York's standing as a city of cultural importance. But, a generation later, its leaders decided to begin to turn it into a world-class scientific research center. To accomplish that, they linked its programs to Columbia University's goal of ascending into the ranks of America's leading educational institutions.[17]

Osborn and his family's connections had much to do with the change in the museum's position in American science, and much to do with Columbia's rise to becoming a powerhouse among American universities. Henry was born into even greater wealth and position than Herbert.

Although Osborn was ten years older than Herbert Field, he chose a life devoted to academic science. Initially Osborn was a Princeton and New York medical-school-trained biologist, then a student of the evolutionist-zoologist Thomas Henry Huxley in England, and eventually a vertebrate paleontologist. Among many other distinctions, he became known as one of America's great museum "dinosaur men." Osborn also became one of America's most respected and widely known experts on evolution and, then, eugenics. At the time, eugenics seemed a natural outgrowth of emerging biological and genetic research, not an expression of a belief in a hierarchy of human racial capabilities. It was then a liberal and forward-looking movement. Even the progressive Alexander Graham Bell and futurist information guru H. G. Wells sponsored eugenics research. Initially, Osborn took a rather liberal public stance on the questions of human evolution and genetics, and he was able to accommodate evolution and Christianity. Importantly, it was not until the turn of the century that he turned from being a Lamarckian to a supporter of a more severe Darwinian interpretation of evolution and the distribution of human capabilities.[18]

In 1891, as Herbert Field was finishing his theses in Europe, Columbia University and the American Museum of Natural History offered Osborn a joint appointment. By the early 1890s, Osborn was already in the higher circles of America's science organizations, as significant a boon for him as for Herbert. And, through his elite family, he was integrated into the new generation of American millionaires who were beginning to use their fortunes to underwrite universities. Osborn was one of America's most gifted early university fundraisers. He brought in an enormous amount of money for the museum, $6 million in two decades. That was what allowed the museum to finance such fame-generating efforts as the thrilling and widely reported expeditions to Central Asia to uncover the secrets of human and animal evolution. The museum became the home of a score of noted zoologist-explorers and culture theorist–social anthropologists such as Franz Boas.[19]

Osborn also became a central force in science organizations including the elite National Academy of Sciences. Fortunately for Herbert, Osborn developed into a Concilium supporter. The two met at zoological meetings and became correspondents. In the 1890s and the early twentieth century, Osborn was a coveted friend of and advocate for the Concilium. He convinced the "American" museum's board to allow its staff to act as unpaid distributors of the Bibliographicum's cards, and he had much to do with the aid Herbert received from financially hard-pressed organizations such as the American Association for the Advancement of Science.[20]

Despite Friends (Even in England), Trouble Breaking Even

Herbert's important connections did not end with Davenport and Osborn. His friends ranged across Europe and North America. Even England, the nation that was home to many influential scientists who did not want the Concilium to prosper, had a few university subscribers and supporters. The faculty at England's Manchester University, and William Evans Hoyle, the head of its famed natural history museum, became staunch allies. In addition, vital cost-free marketing for the Concilium's offerings appeared. The family's anchor, Herbert's older half-brother Edward, served as a New York agent for the Concilium, as well as being the financial advisor for the entire family. He sent out the Concilium's bills and collected payments. That saved American customers the burden of changing dollars into Swiss francs and simplified Herbert's operation.[21]

Besides those in New York, Herbert had other American friends. The director of the Brooklyn Museum was attempting to establish an American board of directors for the Concilium. In addition, Herbert became excited when, in 1900, someone convinced the public library in Springfield, Massachusetts, to take a subscription of forty thousand cards. That held out some hope the Concilium could enter a vast new American market.[22]

Despite Herbert's admirers, the Concilium remained in financial difficulties. But there were indications that by 1905, for the first time, there would be slightly more receipts than operating expenditures (i.e., counting only the costs for the main card file). That was not enough to make the center self-sustaining. Its income was not covering the capital debts, the cost of its book-form publications, or maintenance and general administrative costs. Herbert had to launch another sales and fundraising campaign.

During his new quest, in his enthusiasm, Herbert tended to bend the facts about the Concilium's financial situation. To placate his advisors, he failed to emphasize, for example, that a full 10 percent of his best customers were in tiny Switzerland, and a majority in far-off America. Also, Herbert did not mention that he was obliged to provide Swiss libraries with sets of his cards at great discount.

Herbert also avoided stressing that he had to break with his pledge to have a fully centralized operation. He found it impossible to obtain all the necessary journals, or to hire scientists with the required language skills, to make a complete survey of Central and Eastern European science. After much consideration (and travel) and with the aid of some Russian subsidies, Herbert set up three classifying subcenters. One was in that fragile entity called Poland. Another was in the heart of Russia. The third was in the area that became the land of chaos soon to be called the Ukraine. Although an always worrisome and not financially rewarding commitment, the Eastern operations meant better journal coverage and important personal connections.[23]

A Mixed Bag of Sales Tactics, but No Great Returns

The continuing economic pressures forced Herbert to hone his skills as an entrepreneurial-like executive. He achieved some early but minor victories. For example, his pleadings led to twelve years of small but important support from the prestigious American Academy for the Advancement of Science. More was needed, however. Herbert put much energy into composing buoyant reports on Zurich's activities and accomplishments. At times, Herbert was less than humble when promoting his center. He even tried to use guilt as a sales tactic. He once admonished American ornithologists for being the only ones not officially supporting his great scientific work.

He also tried to connect with possible supporters outside of academia. He began an attempt to convince the growing numbers of industrial and corporate research libraries to take subscriptions. At the same time, he had to pacify customers unhappy with the logistics of Concilium service and to respond to those complaining about a European science bias. At one point, he announced he was laying plans to create a fully independent secondary Concilium center in the United States. It would, he stated, be able to concentrate on surveying and quickly processing American scientific literature.[24]

Unfortunately, Herbert made some bad decisions during his early-1900s money-raising campaigns. And, in some instances, he created his own competition. He hoped to gain additional income through the Concilium's provision of book-form bibliographies for several additional scientific societies and journals, and yearly pamphlet versions of some of his card series. The work yielded little.

Herbert's energetic efforts were barely keeping the Concilium alive. In contrast, in the early 1900s, Paul Otlet's center flourished. It had many users and a database that was growing to millions of cards. Otlet's staff wrote, or pasted on cards, some four hundred thousand entries during his center's first years. In 1903, there were three million entries, and the Belgian operation was processing more than five search-and-retrieval requests a day. Its name was recognized worldwide. In addition, Brazil purchased a sizeable portion of the Brussels cards and the U.S. Library of Congress received several thousand of them in exchange for copies of its cards. France also tried to maintain a copy of Otlet's bibliographic cards. In a few years, there were an astounding eleven million cards in Brussels' rows of Dewey-like cabinets, and sixteen million by 1930. Otlet had even bent to outside demands and created an author- as well as subject-indexed file.[25]

A Wannabe "Special Librarian": The Concilium Takes on a New Man

The Concilium was not doing as well, although its financial situation at times seemed to be on the verge of improvement. Herbert once again thought he should expand to gain additional income. To do that, in late 1903, he brought a young and ambitious American relative, Adolph Law Voge, to Zurich. Voge's addition to Herbert's staff would signal a change of the Concilium's orientation and send a message that Herbert was ready, despite his mandate to focus on the natural sciences, to move into new fields. He was now willing to gamble the Concilium's future on a foray into the world of applied science and technology information.

The timing seemed right to Herbert; Voge had contacted him just as Herbert was preparing to address a major gathering of librarians at the grand World's Fair in St. Louis. The librarians' meeting was a historic event. Its goal was to establish the foundation for international library and bibliographic cooperation.[26]

Adolph Voge (who had spelled his name "Voege" while in college), was an 1899 graduate in electrical engineering from Cornell University, a school that was an extreme example of the attempts by American universities to become relevant to an industrial and technological era. Cornell put less into the liberal arts than Herbert's alma mater, Harvard. Located in upstate and isolated Ithaca, New York, Cornell's goals were to provide occupational training for the common man (not for elites) and to create the knowledge required for modernizing the American economy. The school did not focus on creating gentlemen scientists or humanists, but its growing reputation led the most advanced engineering and industrial corporations to seek out its graduates. Cornell trained students in mathematics, chemistry, and the new field of Efficiency, and it provided a much greater dose of that statistically oriented movement than Herbert had encountered while at Brooklyn's Poly.[27]

That new orientation, Efficiency, was identified with a Philadelphia Quaker, Frederick Winslow Taylor. He went to the famed Philips Exeter Academy but had to forsake a Harvard degree when his eyesight began to fail. Although he was from a wealthy family, Taylor went to work as an industrial apprentice. Then, he gained an engineering degree by attending New Jersey's Stevens Institute of Technology—at night. Taylor was inspired by Melvil Dewey's writings on efficiency while he was a young student. Adolph

Voge may not have known of Dewey's role in the movement, but Cornell's professors made him aware of Taylor and attuned him to what he and the other fledgling engineers would encounter in the workplace, Taylorism applied to shop-floor activities.[28]

Although they were Brooklynites, the Voge family was not from the Fields' upscale Heights area. Adolph Voge was a distant cousin of Herbert Field—one from the wrong side of Brooklyn and of Quakerism. His American Quaker mother had broken a Quaker code by marrying outside the faith to a Prussian-born man lacking an established career. Consequently, Voge grew up in his maternal grandmother's home where his father and mother raised several children under very difficult conditions. He had a stressful childhood as his parents and a nurse had to devote their attention to caring for ill members of the household.[29]

Adolph Voge seems to have wanted to forget or redraw much of his past, even telling passport officials that his father was of Swiss descent and born in the United States. Voge's American maternal line, the Law family, dated to the 1600s when the Quakers first landed in America. Like the Fields, they settled on Long Island, intermarried among the faithful, and prospered. The family centered on Great Neck, as did the Fields, and some of the family migrated to Brooklyn. While keeping in touch with the Fields and others of the extended family, the "Voeges" had less and less contact with other relatives, especially as father Augustus became a Brooklyn politician tied to the city's ethnic-based Democratic political machine. Serving as Brooklyn City Auditor, he fought against such Progressive things as civil service reform and competitive contracts for government work. Augustus Voege did not last long in office, and he did not prosper after the 1880s. After his wife died, he became a salesman, one not accomplished enough to be able to afford a house.

Because of his father's behavior Adolph Voge had a very different life than did his cousin Herbert. His family's difficulties prevented him from entering Cornell until his twenties and he always found it hard to match his social ambitions to his achievements. In addition, like Herbert, he had some speech difficulties in his youth and he always spoke with a German accent. His tendency to put others at a distance and his flaring temper made personal relations difficult.

Those deficiencies made it hard for him to settle into a career. Voge could not afford to make the usual post-graduation trip to Europe for culture or for graduate school. As a result, he went directly from Cornell to work in a shop as a beginning electrician. However, that first job was at the huge General Electric facility in Schenectady, New York. It was an industrial marvel and a great example of science applied to industry. Significantly, it had one of the first extensive company library systems. Its librarians were even developing a classification scheme for electrical items. By the turn of the century, the firm had its own Dewey-like "relativ index" and it was aware of efforts to use Decimal systems in other engineering libraries. Working at the plant was an education for Voge. But he soon took another position in a new industrial center on the Canadian border, left the shop floor, and became directly involved in resolving the growing information problems of industry.[30]

Voge became interested in bibliography when he secured a job as an assistant engineer at one of the large chemical factories utilizing the awesome power supply from the engineering wonder at Niagara Falls, New York. As soon as he arrived, his employers assigned him to the task of quickly finding all the literature related to a critical chemical production problem. The assignment frustrated him. He decided to find ways to overcome the bibliographic hurdles to the solution of practical (not theoretical) applied-science problems.

Ambition (and a New Title) in a Different Information Culture

Adolph Voge's bibliographic motivations, however, were unlike Herbert's. Voge was not an academic scientist, an idealistic internationalist, or a philanthropist. He was a young engineer-businessman on the make, and he aimed at becoming one of the first Special Librarians who could make a good living serving the needs of business customers. His bibliographic task was not, as was Herbert's, to survey and serve a single academic field, or to meet the needs of those working in "pure" science. Voge's tasks demanded searching over many disciplines to find items to quickly solve immediate problems.

During the era, Special Librarians performed that new kind of librarianship. They were a novel breed that had to wait for formal recognition for a decade after Voge began his bibliographic work. It was not until the 1920s and 1930s that viable, professional organizations, such as the Special Libraries Association in America, and the Association of Special Libraries and Information Bureaux (ASLIB) in England, brought them some degree of recognition.[31]

In Voge's lifetime (and, later, in the lifetime of bibliographer Eugene Garfield), one American sector, the chemical industry, became a leader in establishing special libraries and in developing new information methods and systems. However, in the early 1900s the "Specials" had few developed tools. There were some blossoming theoreticians, such as Julius O. Kaiser, who explored methods to maximize card files' potentials. He developed new "keyword" type searches using individual terms from the vocabularies of his users, rather than the specialized hierarchy of a Dewey-like system. Unfortunately, he and his theoretical work on indexing remained relatively unknown, and there were few other Special Library theoreticians or training programs.[32]

So, Voge needed to devise his own ways of organizing and retrieving useful information for technology and business. He examined the major classification systems of the era, including those initially created by his old employers, as well as those of the Stone and Webster engineering firm, London's Royal Society, and even Paul Otlet's emerging version of Dewey's system. He found all of them insufficient. However, Voge knew there was a growing demand in industry for up-to-date information in specialized fields. He believed that huge gaps remained in science-technology (sci-tech) bibliography, despite the appearance of England's indexing/abstracting service for engineers (later to become the Institution of Engineering and Technology's INSPEC), and hints that others were planning to begin publishing indexes suited to the engineering field.

Voge wanted to fill the void before others so he could establish a rewarding career as an information entrepreneur. But he did not have the resources of a Herbert Field or Paul Otlet. He had to find different paths to bibliographic success. One was through family connections.[33]

Voge had known of Herbert Field through the extended family, and he had heard of the Concilium while working in industry. He saw potential in Herbert's idea for a central and universal card-based bibliographic service and he admired Herbert's modifications of the Dewey classification system. When he and Herbert came into direct contact, Voge was already convinced that Concilium had a bright future. In turn, Herbert came to see Voge as a man able to move the Concilium into a new organizational era—and into new markets that had the potential to attract wealthy corporate customers.

Herbert brought Voge to Zurich in late 1903 and assigned him a position as the Concilium's "Technologist." The job title required him to cover varied "information science terrains." First, Voge was in charge of an initiative to develop the products and markets for specialized bibliographies in the applied sciences. His initial mandates were to expand the Dewey classifications to accommodate electrochemistry and to create a bibliographic file for the subject. Then, with Herbert's consent, he formulated a plan to begin to make Zurich the center for every branch of technology—each with its own Concilium subcenter and independent board of directors.

Herbert seemed to accept the idea of franchising his center's name and Voge wasted little time in launching his plans. After calculating the cost in materials and the staff time required for the technology files, and after establishing precisely how long it currently took to search through technology's book-form bibliographies, Voge advertised that the Concilium could offer special libraries a system with huge cost-saving potentials.

There was a financing problem to overcome, however. Herbert could not afford the startup investment. That did not stop Voge or Herbert. They worked out an imaginative funding plan. Voge soon made an offer to the public. The Concilium would invest $20,000 in each technology file if any reputable group would also invest $20,000 and guarantee seventy ongoing full subscriptions for a decade at an annual rate of $600 each. Voge did not explain how the Concilium could find the money for such initial investments. Nor did he demonstrate how he had estimated that enough organizations would be willing to spend $600 a year over ten years for bibliographic service in one limited field.

Voge was so confident in his plan that in 1908, while on one of his many trips to the United States to market the future services, he approached Melvil Dewey. As he had told others in America, he informed Dewey that his databases would pay for themselves because of the search time saved by using his versions of Herbert's "random access" system.[34]

More (and More) Efficiency

Voge was doing more than establishing a new Concilium business. There was an important second meaning to Voge's title of Technologist. It meant Efficiency expert. Herbert Field and his three top-level, university-trained, science-classification assistants knew they did not have the required skills to modernize the Concilium. They turned to Voge because of this engineering training. Herbert asked him to bring his center up to the standards emerging from the era's scientific-management movement, and to develop cutting-edge, modern database techniques. Voge quickly got to work. Formal cost analyses came first. His statistical time studies showed that with good work habits a classifier could complete a reference in just eight minutes. If an abstract was needed, he concluded, a well-trained worker could classify and write one in fifteen minutes. At those rates, a single classifier should add almost fifteen thousand cards a year to the great files.

Hearing the claim that one truly efficient man could do as much as three or four current employees was unsettling to the Concilium's staff. The term was unknown in Swiss academic circles, but the Concilium's classifiers must have thought that Voge was an advocate of what laborites called the "speed-up."[35]

After that initial efficiency exercise, Voge applied himself to the data-management problems. One vital question for Zurich's customers was the cost of updating the card files as the number of entries grew. Voge put one of the three young teenage sorter-updaters through a month-long training regimen and concluded that filing of new cards, even with large files, was quite affordable. When the boys followed the procedures Voge developed, it was determined that a year's additions (fifteen thousand cards) could be correctly updated in a large existing file at the cost of a small percentage of a librarian's salary. Adolph calculated equally low costs for file searches, showing how retrievals from Concilium files required a tiny fraction of the time needed to search through bound bibliographies.

Voge believed future customers would be attracted to his and Herbert's services by knowing that all the maintenance activities for a large subject file took less than 5 percent of a librarian's time per year. He never seemed to have considered that librarians might protest that his methods and calculations were both unrealistic and inhumane. Significantly, he tended to see workers as cogs in a system.

Pricing, Payment Collecting, Sorting, and Coding

Voge also performed a thorough statistical study of pricing policies. Analyzing the labor cost for compiling different sizes and types of orders, he discovered the center had not been maximizing its income. Herbert had recently been using a simple linear price scale that charged more per card for small orders, less as the size increased. Voge discovered the best scale was a parabolic one. It reflected the high-per-unit labor cost of small orders, the lower cost for moderate orders, and, then, the rising costs for large orders with cards widely scattered over the files. Voge's analysis showed that Herbert had been charging too little, even for the cost-saving order sizes.[36]

Voge's analyses brought another old problem to the forefront. Herbert had thought that his earlier remedies had solved the late-payment problems. Out of professional courtesy and compassion, Herbert had allowed some customers to continue with balances due because they had, they said, forgotten to send in their payments. Voge called for a different approach, a strictly business one. Herbert took the advice. He announced that remedies were in the making: a 10 percent price increase for delinquents and a new policy to immediately cut off any shipments to those who were lagging too much in their payments!

Voge increased his efforts for solving data-management problems that had been leading to complaints. First, he devised new methods for the validation of the card packs sent to users. There had been protests that important cards were missing from packets. In response, Voge trained the Concilium's boys so that one silently read the outgoing card while another read off the identifier from the master pack. To ensure that items were not lost in shipment, Voge insisted that after the content double-check, a sturdy cord be placed through the cards' holes. Then he had its ends secured with a strong hot-lead seal.

Voge also regularized the sorting procedures. He devised minutely defined sorting techniques for the complex arrangement of the cards by Dewey number, date of publication, then author. During the sorting and order-building process, a brown card was placed after every tenth item. When the next step was performed an out-of-place brown card flagged a problem.

Voge also worked on the integrity of files. He devised a new type of rod for the card cabinets so that a librarian would know if a "hacker" had removed an entry. The new rod had sharp notches. If an unskilled person withdrew a card, the rod sliced off a portion of the card and a sliver remained in the file. That indicated which card the thief had removed.

Voge performed additional studies on one of the most sensitive Concilium issues. He claimed he had proven that Dewey/UDC codes allowed unskilled workers to perform file updating without any errors. He then conducted more numerical experiments on the human and financial cost of the inflexibility of book-form bibliographies and demonstrated that updating them was prohibitive just in terms of typesetting. He then produced new statistical data that demonstrated the labor cost of paging through dozens of annual or semiannual volumes far exceeded that for card-file searches.

Before the "Micro" Era—Problems of Journal Size and Journal Acquisition

There were some issues that Voge had to sidestep. Of most importance were the enormous growth of the physical size of the Concilium's card files, the space they required, and the time needed for filing. Despite Voge's studies, finding the correct place to insert a card was becoming, some users complained, overly time consuming, even for series which contained just one card per item.

There was no easy solution given the technology of the era. The special tab cards Herbert was selling were only an expedient. By the late 1930s, one American subscriber to the zoological series, the Department of Agriculture's (DOA's) Ohio experimental station, found itself with over 412 drawers of cards. Its administrators protested they were taking too much room. More threatening to the Concilium was the department's assertion that keeping the files in sorted order was too costly. As a result, new card packs were often left unattended. Then the Department of Agriculture found a grave and more fundamental difficulty to overcome: they discovered too many filing errors in their own files.[37]

When Herbert could not fulfill the promise of sending fortnightly update packets, other subscribers became irritated as well. That left them to deal with large packs of new cards that arrived, at best, every other month or just three times a year. As a result, the customers just haphazardly placed the new cards into the drawers by their general category. Such difficulties led, for example, to many of the Department of Agriculture's scientists turning to the less bulky yearly book versions of the file although that meant searching through the separate volumes. Voge also failed to examine one irritating aspect of the Concilium's card-file system. Because Herbert's classifications and technology seemed the most efficient of the time, few realized they were not ideal for fast and inexpensive retrieval of information through multiple terms. However, neither Herbert nor Voge could be faulted for that weakness: it would be decades before ideas and technologies appeared to ease such combinational ("and") searching.

Voge tried to help with another recurring difficulty: a solution was essential. By 1908, Herbert had to again confront a long-standing challenge: the inability of the Concilium to gain access to all publications in its fields. The expense and logistics of obtaining copies of the journals was becoming unbearable. Herbert and Voge sought new ways to keep the promise to index all the relevant publications. They had to keep that pledge because a severe credibility problem had developed.

For example, of the four hundred known relevant American journals, the Concilium had access to only 175 on a regular basis. As a result, Herbert was handling an increasing number of complaints from users who found that the Zurich center had not indexed a significant percentage of items in their specialty. Herbert again turned to the scientific and publishing communities for help: they did not respond. Then, he made yet another attempt at having publishers and authors send in those double-fold postcards with prepublication bibliographic information and a short content description. That failed, as before.[38]

Herbert could not give in. He once more published appeals and made visits to scientific congresses. Frustrated and worried, he and his allies composed a series of articles explaining why all authors and publishers should regularly submit their works to the Concilium. The newest version of the argument emphasized self-interest. The Bibliographicum was now the primary database for other index journals, they claimed, so not having an item classified by the Concilium meant a loss of income and professional recognition. Herbert expected that he would solve much of his credibility problem through such pleas, as well as through a proposed new system. The newest one would maintain direct contact, he said, with the publishers of the twenty-eight hundred journals the Concilium had on its lists. In addition, he planned to launch another round of visits to the offices of all the major publishers in Europe and America.[39]

A Collaborative Effort, an Eventual Parting of Ways

Despite the pressures, Voge and Herbert maintained warm relations in the early 1900s. Both had great hopes for the Concilium's products, especially the ones that might tap the expanding lucrative corporate markets. They had become somewhat of a team. They worked together for almost a decade and Herbert foresaw many more years of collaboration. He even planned the great new Concilium building so that Voge could have an elegant apartment within it. In turn, Voge gave his first child a name with the initials "H. H." and began serving as a Concilium emissary—and not just for the technological series.

In 1907, Voge traveled to one of the Concilium's most important allies, the U.S. Library of Congress. Arriving in Washington to help with the use of the Concilium files, Voge found that what he had heard about the library's director, Herbert Putnam, were understatements of his abilities and accomplishments. Putnam, an embodiment of library progressivism,[40] came from a famous New England publishing family. He was a Harvard, then a Columbia, man who carried out both universities' liberal social agendas. He was a driving force in bringing the Boston Public Library into the modern era. After that, he was the first librarian to head the Library of Congress. Within a decade of his appointment in 1899 he changed it from a rather old-fashioned institution into a library for the entire nation.

Putnam helped to establish the new classification method at the library, one based on the Cutter, not Dewey, system. Under his guidance, a card catalog was developed, and the library was becoming the nation's cataloging center. Putnam was also at the top of the new profession, librarianship—he led the American Library Association.

Putnam impressed Voge. In turn, Voge's ideas, his publications on chemical classification, and his methodological skills impressed Putnam, so much so that Voge received assurances that the library would find a high-level position for him if he ever desired to leave the Concilium. Voge did not take the offer in 1907, but he kept it in mind. Significantly, Putnam's proposal set him thinking about career alternatives, especially because, after four years of development, his Concilium technology services failed to attract customers. Voge was depressed, but a letter that arrived a few months later lifted his spirits.[41]

In 1908, an American industrial firm that wanted a Special Librarian offered Voge a salary so attractive that he was willing to cut his ties to the Concilium. He informed Herbert that he was leaving Zurich. Then, an economic downturn led the firm to rescind the offer. Adolph quickly turned to Melvil Dewey, asking for work preparing the new Dewey categories for electricity in industry. Dewey agreed. Furthermore, he approved of the idea of Voge doing the work abroad. Voge contacted Herbert, dealt with his hurt feelings, and received notice that he was welcome to return to Zurich. Voge packed to sail with a year's salary in hand, and something else.[42]

Adolph Voge returned to Switzerland with a surprise: the confirmed thirty-year-old bachelor had a twenty-year-old American bride, Lillian. They quickly had two children, but Voge did not let up on his information work. He introduced his file for electrochemistry, worked on a major project for inorganic chemistry, and then published a grand manifesto for the technical section of the Concilium.[43]

When problems with Voge's work began to appear, Herbert decided to overlook them. The contract work that Voge was doing for Dewey was rejected by experts in Europe and America as being too much of a departure from standard practice. As well, his electrochemical effort was criticized as too complex, and too expensive; he was asking $1,000 a year for ten thousand cards. Ironically, his moderately priced index to the huge collection of patents at Zurich's technical university was ignored.[44]

Creating a hierarchical code for a complex subject such as electrochemistry, which was tied to so many fields and processes, was difficult. The field lacked a stable vocabulary, called for index numbers much longer than those for zoology (as many as ten or more digits), depended on the use of additional special codes for such things as "point of view" of an article, and required what Herbert detested, short abstracts (usually in English) for every entry.

Voge had also departed from Herbert's internationalism. Because of his expected customer base, every title was translated into English. Electrochemistry also required many more of the secondary ("see also") classifications that were listed after an item's title. In addition, the field was so interlaced with other subjects that almost every item had to have a colon classification. Colons were used when a classifier judged that additional important classification from another series was required to make sense of the central topic of an article. Herbert's favorite example of a "colon classification" was "81.2: 14.31.4" (the code for an article about the teeth of snakes, in which 81.2 came from the list for systems while 14.31.4 came from the classification list for anatomy). A colon usually led to a card for each part of the code.[45]

Despite Voge's threat to the Concilium's pure science–internationalist reputation, he seemed a permanent and important part of Herbert's team. In 1912, in another attempt to secure the financial success of his applied-science systems, Herbert had Voge's family sail to the United States on a marketing tour.

Voge never returned. Perhaps that was because his plan to make the Concilium an information provider for the new applied subjects failed; perhaps it was because he saw the Concilium itself on the decline; and perhaps his lonely, discontented, and pregnant wife had enough of Zurich. And there was a hard economic fact, competition. When Voge first laid out his chemical information plans only Germany and England had chemical bibliographic services and their short summaries of articles were not up to standard. Then in 1907, to Voge's dismay, American chemists launched their own abstract-indexing service using volunteers. Under the flag of the American Chemical Society, and with the editorship of the head of the National Bureau of Standards, *Chemical Abstracts* began a meteoric rise to leadership in chemical information—one that was helped by having its subscription bundled into the cost of society membership.[46]

Such a membership-subscription option, which guaranteed a market, was unavailable to information entrepreneurs like Voge. And, few other professional organizations thought it politically wise to conduct what some viewed as coercive marketing. *Chemical*'s financial success was also due to something that Herbert mistrusted: the use of low-paid but essentially volunteer abstractors who worked at their own institutions. Both bundling and the volunteer system performed well for a time, but in the 1930s *Chemical Abstracts* was forced to abandon bundling because of protests over high membership dues. As well, as Herbert had predicted about volunteers, by the 1960s the journal had to turn to a professional staff in order to maintain quality and timeliness.[47]

Disappointment and Change: Family Dynamics and Career Paths

Voge may have decided not to return to Zurich because of Herbert's Quaker moral commitments. Herbert's alienation had become evident after he found that Voge was having dalliances that endangered his marriage and his work which, by 1912, Herbert thought was becoming less and less worthy. In addition, there may have been a less than perfect relationship between Lillian Voge and Herbert's new wife.[48]

What is certain is Voge's acceptance of a job at the Library of Congress in 1912. He expected that he was to play a role as an expert in chemical and technical classification. He was hired just as the library was extending its own classification systems. The library was creating a bibliographic national standard. That meant that it needed experts in all subject areas.

Voge, as well as Herbert, then became involved with another potential threat to any Zurich science initiative. The new challenge was the grand and idealistic 1911 proposal by the famed German chemist Wilhelm Ostwald to establish his "Bridge." Part of his larger, but unrealized, goal of organizing international science through cooperation and Esperanto—(an auxiliary language constructed in the late nineteenth century by L. L. Zamenhoff to foster peace and transcend nationality)—was an information component. Based on a standard format for scientific reports and paper sizes, the Bridge was envisioned as becoming a clearinghouse for all scientific information.[49]

Voge's job at the Library of Congress did not mean that he had a settled life. His situation in Washington quickly became a source of discontent. He had expected to become a department head but found himself working under a man he detested. His "slow-witted" boss, Voge wrote in a heated letter to Putnam, was even preventing him from completing his library science research and his sociological work on the genealogy of his American Quaker family. As much of a cause for his unhappiness was Voge's salary, which was not sufficient to cover the expenses of raising three children, with one on the way, and would not allow him to purchase a home.

Within two years after moving to Washington, the frustrations at work and his marriage difficulties caused Voge, with his pregnant wife Lillian and his children in tow, to move across the country to take a position as the reference librarian at San Francisco's Mechanics Library. In 1914, the couple was renting a small apartment in a working-class neighborhood. Lillian was soon employed as a clippings cutter and typist for a printing/information firm. It was a bad family situation, and the always-emotional Voge's continuing professional frustrations intensified Lillian's alienation. A divorce seemed inevitable.[50]

Despite his family's traumas, Voge continued on his job while he kept trying to build a professional reputation as a true industry-based Special Librarian. He did make some professional progress. Then, America's entry into World War I, obligations dating from his years at a student-soldier at Cornell, and his expertise in chemistry changed his life. In 1917, Voge was called to serve as a captain in the army. Because he was a chemical engineer, he was considered valuable as the war became "chemical" after the first poison gas attacks. With the correct background to manage and deploy chemical weapons, he expected to rise quickly to a position of importance.

At first, it seemed he would. He spent a month in Washington doing research with leading chemists, and then he received intensive training from the French. But after only a few months in the "gas" service in Europe he was transferred to boring jobs as a billeting officer, an escort for German prisoners, and as the American liaison to the library at the Sorbonne. He was so frustrated that he wrote to his friend Henri La Fontaine asking him, as a good fellow socialist, to intercede and find him an appropriate scientific slot in the American Army.

There was more to Voge's wartime stay in Europe. While there, he became involved in the library-related issues during the formation of the League of Nations. That led to a none-too-happy reunion with Herbert Field. Voge also took a new wife, decided to explore the Amazon jungles, and followed those escapades with a conflict-filled career as one of the new independent Special Librarians, and as a researcher for the American military during World War II.

8 On the Eve of World War I: Two Brothers, Two Callings—and Tests of Faith to Quaker Ideals, Business Plans, Family, and Friends

Adolph Voge was not the only family member who traveled to Zurich. Another visitor had a great influence on Herbert Field's life. It was Hamilton, Lydia Field's "baby boy." Hamilton Easter Field was a late-life baby, born when Lydia was nearing forty years old. As she was with Herbert, Lydia had been indulgent as Hamilton grew up and as he decided to leave college and go to Europe to learn to be an artist rather than an architect. As a result, as soon as Herbert occupied his first home in Zurich, Hamilton began visiting.

His later visits, after Herbert had finally married in 1903, made Hamilton a part of the new family. He formed emotional bonds with Herbert's new wife, Nina, and their children. They held him as a caring and mentoring uncle. He was freewheeling, strikingly handsome, energetic, carefree, and inspiring as he sat beside the villa's lake to sketch and paint. He could tell great tales about the arts and artists in Europe and play the most up-to-date tunes on the mansion's piano. Although he seemed a bit eccentric, he was an exciting, romantic model for the children. Hamilton had another appeal. By the 1910s, he was rather famous. He was frequently in the news, and regarded a major although controversial figure in the American art world.[1]

Two Quaker Brothers, Two Different Callings

Hamilton's Calling was different from Herbert's. He seemed to devote his life to being a nonconformist. He was building a reputation as a crusader for reform—reform of what his friends saw as the backward and stodgy styles, and crusty institutions, that dominated American painting and literature. There was more to his revolt. He took a different ideological direction than Herbert's. Hamilton crusaded against middle-class lifestyles and culture. Moreover, he was involved with leaders in America's version of radical socialism, if not communism. Hamilton's unquiet confrontations with the standing artistic and social orders could not be hidden from his family, and he passed an early version of the 1960s' distaste for anything "established" to Herbert's children.

Hamilton's youth had paralleled Herbert's. But there were important variations. He successfully went through the Friends school and Brooklyn Polytechnic. Then, Lydia had to tolerate his early withdrawal from the architectural program at Columbia's School of Mines and his unexpectedly short stay at Harvard. She did not disown him. Rather, she excused his lack of persistence as the result of health problems and gave him money in the mid-1890s to travel to Paris to become an architect.

As he came under the influence of the European branch of the Haviland family, he pushed architecture and a moneymaking occupation into the background. The Havilands introduced Hamilton to a new lifestyle. The Haviland factory in the suburbs of Paris and the firm's distribution center in America had brought the family fame, wealth, and comfort. They had led the way in bringing modern art into the design and manufacture of ceramics and tableware. Importantly for Hamilton's future, the family's experimental center in Paris employed Europe's leading Impressionist artists, and a Haviland had married France's leading art critic.

Consequently, Hamilton was introduced to the best art teachers in the French capital. He began a full-time, nearly life-long career as a student of art and cared little about the worries of earning a living. His bohemianism was not confined to art. Hamilton's ideas about society and religion changed as he socialized in Paris' cafés with young artists who held revolutionary views about politics, as well as painting and literature. He became fascinated with Picasso and Matisse, and with the social views of their colleagues.

It was not long before Hamilton informed Lydia that he, too, had found his life's Calling. Hamilton declared he would become a great artist and a patron of the arts. With Aaron's death in 1897, Hamilton had an inheritance and Lydia gained control of the family's estate. She gave Hamilton more latitude than she had given Herbert, much more. She allowed Hamilton to be a full-time student and art collector for as long as he wanted. That allowed him to spend the next eight years in Europe traveling, learning, and collecting—not working. Hamilton was not even like his cousin Paul Haviland who was able to become a famous art photographer, art collector, and the financial angel for Alfred Stieglitz's celebrated New York City gallery while running the American side of the family business.[2]

Hamilton returned home to Brooklyn in 1902. He arrived with Robert Laurent, a twelve-year-old protégé and native of Brittany whom he had met while at a European arts colony. Hamilton, Laurent, and Laurent's parents stayed with Lydia in the ancestral townhome for a time, and then Hamilton decided Rome was the place for "true" artists. Another eight years of touring Europe followed, during which he traveled to Rome with Laurent. As before, his mother usually accompanied him. Hamilton finally resettled in Brooklyn in 1910, guided by a more specific goal. He told Lydia he was now determined to be the man to lead American art into a new and liberating era. He would reshape the art world and, to a degree, American literature.

Lydia again cooperated with Hamilton. She and Hamilton turned the Columbia Heights house into an arts studio and gallery. They soon used family resources to buy the adjacent townhomes to allow for the gallery's expansion and a hoped-for arts school. They filled the homes, from cellars to attics, with the art Hamilton collected in Europe and America, as well as his famous collection of Japanese art. Next, farms in Maine, and one in the White Mountains of New Hampshire, were purchased to provide inspirational rural settings for summer art excursions. Then, Hamilton began publishing an arts journal.[3]

Hamilton painted, taught, wrote, and tweaked the American arts establishment. He declared he would bring art directly "to the people" and would found galleries whose exhibits were too cutting-edge for the recognized ones. Representatives of the "people," such as he, were to determine whose works would be shown. Hamilton introduced America to folk art through his own paintings; he made America aware of Impressionism and Abstractionism through his commissioning of many works by major artists such as Picasso; and he became one of the nation's greatest experts on modern and Asian art.

Of much importance for Herbert's children, Hamilton trained and mentored some of the most important men and women of the first generation of American modernist artists. Significantly, they typically favored the politics of the left, usually socialism and communism.

His friends also convinced Hamilton—the Quaker ex-student of a Friends school—that Freud and others who emphasized the irrational and un-peaceful had to be recognized as contributors to the new theory of art. Although he continued to support some Quaker charitable efforts, Hamilton rushed toward the secular, thus giving his contribution to the Field's tradition of social services a new twist. His "social service" did not always fit the older family patterns. For example, he followed a rather self-serving version of Quaker pacifism. He did not put his life or welfare on the line to fight the German menace in World War I. He did not directly help, as other Quakers were doing, World War I's victims. He did not join any relief organization, and he did not volunteer to serve in such government approved organizations as the Quakers' American Friends Service Committee that had been specifically created to allow pacifist Friends to aid Europe's victims. He managed to escape the draft, although several thousand American Quakers complied through draft-based noncombatant service and many others volunteered for such duty.[4]

Hamilton's definition of Quaker service was unusual. It was at a distance, only intellectual, and confined to the artistic. He did publish, however, two heart-wrenching letters from a young French acquaintance about his firsthand experience of trench warfare in 1916. Hamilton also published a broadside suggesting that he was a martyr because of his opposition to the war and his support of the Russian Reds. Meanwhile, he continued his comfortable life in New York focusing on his art and artistic friends.[5]

That did not mean Hamilton was a conscious hypocrite. He thought he was sacrificing himself to the cause of social and economic reform. The writings in Hamilton's famed magazine, *The Arts,* reflected his artistic populism and his commitment to use the arts to help ameliorate the social problems created by capitalism. However, his contribution remained indirect. With perhaps the last major chunk of the family money, he founded a school and colony in Maine that became a center for radical thought, as well as radical art.

Hamilton had moved far left of liberalism. The weekly discussion groups Hamilton led at his Ogunquit, Maine, arts colony have been described as cultural versions of a John Reed Club meeting. For example, an attendee was the writer John Dos Passos. Dos Passos was the son of a rich lawyer, but in his teen years he developed more than liberal predilections. He had left Harvard and volunteered as an ambulance driver in World War I. He visited post-revolutionary Russia to see firsthand the conditions in the "workers' paradise," then he returned to America where he later wrote a popular trilogy that was a near materialist's bible for those seeking an ideal social-economic order in America.[6]

Herbert, Married with Children, Puts Quakerism to the Test

While Hamilton was traveling in Europe and while Voge's Zurich projects were getting underway, Herbert entered into a new phase of his life. By 1903, he finally had some time to himself. The first critical five-year mark for the Concilium was reached and the zoological societies had concluded the Concilium's work was valuable. So, at age thirty-five, he could think of marriage and family.

He met an English Quaker woman of German descent, Nina Seiten Eschwege Foote, while she was on a ski trip to Switzerland. Herbert was enchanted. She had a sparkle and was intelligent. She had an independent streak and had worked as a journalist. Like Lydia, she had delayed marriage in an era when even educated women wed in their early twenties.

The courtship took some time, and much of it was conducted through the mail. Finally, Nina and Herbert's wedding was held in London in 1903, just as Nina was approaching thirty.[7] The couple moved to Zurich rather than staying in England or returning to the Quaker community in Brooklyn. That decision had an important impact on Field's children. Although Herbert still bowed to Quaker fundamentals, he could not raise sons and daughters in a closed Friends religious community while in Zurich.

The marriage was good and activity filled, despite the pressures caused by the fragile status of the Concilium, and by Herbert's frequent fund-raising trips. Nina was content, although she was frequently alone while carrying increasing family responsibilities. She had four children over a decade: two boys, two girls. Noel came first in 1904, the second boy, Hermann, in 1910. Elsie was born two years after Noel, and the last child, "dear" Letitia arrived in 1913 as Nina was nearing forty.[8]

For her first child's birth, Nina and Herbert traveled to London to be with family and to ensure the American citizenship rights of their baby. Two days before Noel's birth, Herbert rushed to the American consul and made out a will that reaffirmed Herbert's citizenship and attempted to secure Nina's and his future children's financial interests. There was an important coda to the will. It stipulated something that would prove a burden to Herbert's family and friends. The Concilium, it stated with emphasis, must be continued after his death.[9]

Back in Zurich, where all the other children were born, Nina raised her brood in an idyllic setting. They enjoyed all that upper-middle-class life could offer. They fondly remembered their villa's lovely lake and the spacious grounds of the "Ardesley" estate. They also benefited from their father's position in Switzerland's social world. Herbert had done more than found a business. He made himself part of Switzerland's and Europe's elite social set. He had also established ties with the diplomatic corps of many nations. He was especially close to America's ambassadors to Switzerland.

The Field children had the finest educations. But there were no schools with the religious commitments found in the old Brooklyn Friend's Meeting House. Herbert tried his best to pass down the basics of Quaker Quietism to his sons and daughters. Some of Herbert's teachings took hold. Noel quite early in life took up the Quaker's spirit of social reform, with an emphasis on the internationalism that attracted Herbert and Paul Otlet. For example, confronting his shyness and his own speech difficulties, Noel formed a peace organization at his high school and committed himself to the ideals of his father's good friend Henri La Fontaine.

More Concilium Problems: A Decline in Users and a Faulty Business Plan

Although a family man with responsibilities, Herbert's career remained unsettled. There was some momentary upbeat news, however. In 1908, the subscription list was rebounding from troubles caused by the economic panic of the previous year. Herbert had 160 full purchasers and near five hundred subscribers to parts of the file. He had sent millions of cards to customers. The Concilium had indexed more than 175,000 individual articles. Nevertheless, competition remained as a threat as more and more scientific societies were investing in bibliographic work.[10]

Much worse, there was a sudden decline in income. The year 1908 had been an anomaly. The number of subscribers began to drop in 1909 and 1910. An especially great 1909 disappointment was the letter from the public library in Springfield, Massachusetts. It stated there were so few users over the previous decade there was no justification to continue the subscription. Herbert's hopes to tap the public library market dwindled. His early decision to tailor his system to specialists had precluded it from serving the general public. A patron who wanted information on a popular topic such as elephants needed the help of an experienced librarian. For example, the Concilium's guide, *The Conspectus*, had no entry for "elephant" in its alphabetical list. A user had to know to go to the word "Proboscidea" (let alone the term "Ungulata vera") to find directions to card section 9, and then subsection 9.61. After that, a search had to be made for titles that might have words indicating types of elephants.

There were personal as well as business pressures in 1909. Herbert's rather dangerous customer-seeking trip into the center of the nationalist fury of Bosnia and Serbia proved a financial disappointment and a source of family discontent. Nina objected to such dangerous travel, but Herbert's journey was one that a pregnant and now none-too-happy Nina had to tolerate. In addition, Herbert was having another round of health problems. One called for a painful sinus operation; but even that did not end his reoccurring headaches.[11]

By 1910, Herbert had to admit, at least to himself, that the Concilium's business plan was not working. There was a seemingly insatiable need for subsidies. He had thoughts of putting another mortgage on the Concilium's building, and his lawyers were at work on a total refinancing plan. Conditions seemed so dire that he was on the verge of demeaning himself by asking old friends for personal help. High on the list was a childhood neighbor, Brooklyn Poly classmate, Harvard alumnus, and wealthy capitalist, Wendell T. Bush. Unfortunately, Herbert's old friend did not come to the rescue. Nor did others in New York's liberal circles step forward to save the day with grand contributions.[12]

The Concilium's financial status was precarious. It had survived its first five years only because Herbert took no significant salary, paid for all his business trips, and assumed the responsibility for the Concilium's debts. In 1900, he had finally begun receiving a paycheck, but it was less than one-third of a clerical worker's income. The only way the center had made it through the next five years was the Swiss government's agreement to increase its annual subsidy five-fold, to approximately 20 percent of the Concilium's operating costs. In 1910, in exchange for that, and for a salary that was increased to two-thirds of what a white-collar worker earned, Herbert made some difficult promises. He pledged to send all of the Concilium's journals and books to the Swiss national library, to present several Swiss centers with copies of his files, and to stand ready to make-up any Concilium financial defaults out of his own pocket.[13]

He needed more financial help, but it was not easy to obtain. In his search for support, Herbert had even found it necessary to turn his 1903 honeymoon into a donation-raising tour of the United States. It had not been very rewarding. As a fellow of the American Association for the Advancement of Science (AAAS), he and his Harvard friends had been able to convince the organization to provide that decade-long but small subsidy. It was a not very generous one, the equivalent of six weeks wages earned by a white-collar worker. The Elizabeth Thompson Fund contributed a tiny amount and the Microscopy Society gave a little.

Importantly, Herbert Field had been unable to integrate himself into an emerging new world of philanthropic giving. He failed to tap the bountiful resources of what had become the premier charitable organizations of the early twentieth century, those of the Carnegie interests. Even Henry Osborn's lobbying for a generous subsidy for Herbert from the Carnegie Institution had been unsuccessful.[14]

A Plea to Mr. Carnegie: The Concilium Is a Library for World Peace

However, at the beginning of 1907, there was reason to believe the Concilium could finally gain generous Carnegie support. The multimillionaire steel magnate Andrew Carnegie had been giving millions for information causes since the 1880s. That included support for the Royal Society's science information effort. He had also helped create more than two thousand public libraries in the English-speaking world. His interests then broadened. In the early 1900s, after prodding by the ambitious Columbia University president, Nicholas Murray Butler, and Butler's friend Elihu Root, Carnegie began an initiative for world peace. In 1903, Carnegie funded the grand internationalist Peace Palace in The Hague and his managers were finishing the plans for a multimillion dollar permanent Endowment for Peace. There was also important Carnegie involvement in international health and medical programs.[15]

To Herbert Field and his allies, the Concilium seemed to be a perfect fit with the Carnegie policies. And, Herbert was a well-connected potential supplicant. For example, on one visit to America, he stayed on the old family estate on the edge of what had become a millionaires' retreat, Long Island's Gold Coast. The Whitney, Barnes, Morgan and Dodge families had built magnificent, near baronial, estates there and hosted America's old and new wealthy. In addition, Herbert had colleagues and supporters at the nearby Brooklyn Museum's famed Cold Spring Harbor Laboratory that was on the verge of receiving a huge Carnegie grant for studies of evolution and genetics.[16]

Herbert had prepared a proposal for Carnegie just as the first of the Carnegie supported yearly volumes of the Royal Society's *International Catalogue* were being sent to its subscribers. Despite the publication of the *Catalogue*, Field's allies thought his modest proposal was a shoe-in. Carnegie's advisors approved it.

Herbert was surprised and depressed when the foundation rejected his well-reviewed 1903 request. That meant continued stress on the Concilium and on himself.[17] As a result, Herbert had to keep tapping his own funds. He was also at the edge of paranoia. His emotional reactions were triggered by the cancellation of some critical subscriptions and the loss of professional support. When the U.S. Department of Agriculture ended its subscription "by mistake," when John Shaw Billings of the great New York Public Library (and the acclaimed Surgeon General's Library) did not submit a request, and when the Smithsonian's men hesitated in announcing their approval of his methods,

Herbert contacted his influential "best friend" Charles Davenport. He asked "Charlie" to identify who was "working against him" in Washington.[18]

Despite all the worries and frustrations, Herbert could not end his search for American money. He immediately launched another attempt to gain Carnegie's attention. He soon came to believe, yet again, that people were plotting against his cause. He tried to arrange a face-to-face meeting with Andrew Carnegie in London, while at the same time submitting another proposal to the Carnegie foundation's board of directors. The one-on-one conference fell through, as did the second grant proposal.[19]

Is Herbert His Own Worst Enemy?

Herbert soon heard the rejections were due to what some regarded as his not quite high-enough professional ethics. The reason, he was told, was that he had yet again refused an offer from the Royal Society to be cooperative. He was infuriated, especially after he was informed that any Carnegie funding in the future would probably be contingent upon his finally agreeing to become part, and a very pliant part, of the Royal Society's information project. After the U.S. Congress refused to contribute to the society's work, only Carnegie's contributions kept the great project alive. The foundation did not want to support an uncooperative competitor.[20]

Herbert was deeply hurt by the Carnegie foundation's demand and by the rumors about his ethics—but he stood fast. He rejected the Carnegie requirement and then had another bout of suspicion. He wrote his friends and asked them to find out who was at the center of the "intrigues" in London, not Washington. In 1909, Herbert became so worried about the Royal Society's actions that he requested that Charlie Davenport have him made an official representative (arbitrarily, of *any* large American library) to the Royal Society's project so that he could exert some direct influence to block any British information policies that might harm the Concilium. Herbert had to swallow some pride when he asked Davenport to make the appointment a paying one.[21]

Soon placing less faith in charitable subsidies, Herbert explored yet other ways of increasing the Concilium's revenues. He began advertising bibliographic files for yet more specialized subjects, most of which were related to zoology but which, with a few changes in the items to be included, could be sold as innovative offerings to new types of subscribers. Herbert announced novel or enhanced specialized bibliographies for geology, embryology, ecology, faunistics, evolution, anthropology, and even human anatomy. He thought they might attract customers willing to pay twice as much per card as those who were subscribing to the huge zoological bibliography.

However, the development of such new offerings would, he knew, demand what he did not have at hand, more money and time. He postponed their development. Herbert did not sense that some of his new projects threatened others' bibliographic offerings. There were soon objections to his proposed new services.[22]

Charity, or More Personal Debt?

The few new products Herbert developed were not successful, and the Concilium continued to need subsidies. Herbert again toured the United States seeking funds to reduce the remaining debts and, importantly, to upgrade his aging equipment. He approached possible donors with refurbished pleas. He spiced his description of the value of the Concilium with a bit of a threat. He began to hint that he could no longer remain at the Concilium, so that securing permanent funding was an absolute necessity. In 1904, he had pleaded with America's librarians to create a $2 million endowment so that if he had to leave Switzerland the Concilium could continue operating. Even that ploy had failed to work.[23]

Much worse, the Concilium's best market, the card-loving United States, was becoming unresponsive. As well, Herbert was encountering signs of a resurgence of professional opposition to numbers and to a central bibliographic service. A once close institutional friend of Paul Otlet and Herbert was even signaling it would not continue to give blind adherence to any European database plans: the American Library Association was backing away from moral as well as financial support for any radical information initiatives. By the 1920s, the association made that clear by rejecting a plea to help rescue Otlet's Brussels operation.

Despite his threats of abandoning the Concilium, Field did not forsake his center. He always found something to reenergize his faith. In 1904, by asserting that the number of subscribers was increasing, he had been able to borrow enough money to finalize plans for his new building. Herbert had to pledge his own assets for the loan. The future had then looked bright enough that Herbert turned down yet another offer from London's Royal Society to work for its *International Catalogue*—even though the job would bring a steady income and residency in London, Nina's hometown.

Unfortunately, in 1905 there was a string of bad news. While awaiting his building's construction, Herbert's landlord refused to renew the lease on his offices. There also were strong hints that German interests might block Concilium subscriptions, and, there were indications the French government was working against his center.[24]

In 1906, the Concilium was forced out of its quarters, even before the plans for the Concilium's new building were completed. The classification work was disrupted as Herbert quickly moved his staff and equipment to temporary rooms. Under those conditions, Herbert could not keep up with the indexing. To add to the problems, his main assistant, Marie Ruhl, became sick and was unable to work for six weeks. With that, Herbert faced a truly major business and personal crisis. A new great hope, the anatomical series, was almost abandoned, and even the zoological file had fewer additions than in previous years. Customers were inquiring why their packets were smaller and why he was not sending them as frequently as before.[25]

Worse things were happening. Zurich's city government made a startling decision. Herbert almost lost all hope when he read the official documents from the zoning office. They stated that the government had denied a building permit for his ideal offices and information center. Herbert was near frantic. He had to spend much of his energy convincing the officials that his building conformed to the residential zoning guidelines. It took some doing, but he and the city agreed upon a design that made his information center appear to be a residence.[26]

That did not mean an end to his problems. The year 1907 was very, very bad. Herbert's always-marginal health failed once again. He experienced another "fever" during the first months of the year. At the same time, his physiological editor resigned due to that rather trying battle over wages. The Concilium's production of classified cards then dropped by one-half. In addition, his best customers in America faced the financial panic at the year's end.

The situation was so bad that Herbert temporarily closed his center. One consequence of the closure was a severe decline in income. Another was near-fatal damage to the center's reputation. The credibility of the Concilium faced its greatest test as it attempted but failed to meet its schedule of deliveries. Then, in summer, as there was some return to normal work, and as the new building was in its final stages of construction, Herbert traveled into yet another disappointment.[27]

The Great Plea and the Disappointing Reaction

With hopes that his fellow scientists in America would come to the Concilium's rescue, Herbert sailed across the Atlantic to attend the 1907 International Zoological Congress in Boston. He attended as the official representative of the Swiss zoologists. He used all his influence to try to persuade the congress members they should step in where the philanthropists had failed. He was extremely frustrated and embarrassed. He realized that people knew he was begging for money. Even while burying his pleas in the long technical report needed to satisfy the congress in its oversight role, he could not disguise his aggravation.

The opening arguments in his report went a bit too far and weakened his case. First, he stated that contributions to help scientific information would be as noteworthy as gifts to an expedition in search of evolution's "missing link." He then admitted that while his center was deep in debt, the responsibility was not his. Undiplomatically, he blamed some of his center's weaknesses on his scientific colleagues. His customers, he stated, tended to pay only after their bills were past due. He did not mention the expansive and expensive plans Voge laid out, his own illnesses, the disruptive shift of Concilium offices, or the recent resignation of his best classifier.

Herbert told the Boston gathering that the expedient of closing the Zurich operation for a time (in order to save wages to meet the most pressing debts) had not reduced its liabilities to a level he considered manageable. He avoided giving a precise statement of the total Concilium debts. He did not tell the group they had risen to over one-half million dollars.[28]

Field then added an important and, he thought, motivating and positive postscript to his report. He stated that despite the present financial problems, the Concilium was on the verge of being able to cover most of its costs and to regain its level of output. He also pointed to his impressive recently published seventy-page list of the journals read and classified at the Concilium. He claimed it demonstrated that scientists could trust his center to report on all the literature. He emphasized he was even surveying journals from Malaysia, Costa Rica, Peru, and Indo China.[29]

Despite that reassurance, little was accomplished during the visit—except gaining a promise by the Congress to advertise the Concilium's need for money and an indication that it was forming a supportive committee to aid in fundraising. Just as bad, another attempt to enter the world of large-scale philanthropy failed. A request for a pittance, just $7,000, from another new giant foundation, Rockefeller's Medical Institute, was refused even though Herbert had made brief but personal contact with its administrators, and with a rising philanthropic star, Abraham Flexner.[30]

After the presentation to the congress, Herbert had quietly let his closest friends know more about his financial woes. He revealed that he had taken a $50,000 bank loan for the Concilium and said he could not pay it. He warned that he was facing foreclosure; but that, and his assertion that his cards were now self-supporting, did not persuade any sponsors or donors to come forward. Herbert's pleas came at the wrong time. The nation was on the verge of a grave financial panic and economic downturn. America's banking system would collapse after the Knickerbocker Trust Company of New York failed in its bid to corner the world's copper market.

The very best Herbert could foresee was the Concilium hobbling along, sustained by hopes that increased subscription income and minute subsidies would meet the new mortgage, salaries, and the old debts. Then, he and his American banker friends discovered the debt burden was greater than Herbert had calculated. While half of it was in mortgages and other long-term obligations, the remainder was a sizeable and immediate threat.[31]

Herbert Loses His Temper

Herbert lost his perspective and almost abandoned his commitment to his Calling. He made some more not-too-veiled threats. The Concilium almost died. Now a married man with responsibilities, and frequently ill, he informed his American supporters that he should no longer have to tolerate a salary less than a low-paid clerk's, nor should he have to use his own money to subsidize the Concilium. He displayed a balance sheet showing his significantly negative personal income over the previous decade. He hinted that he required close to a 400 percent salary increase to stay on at the Concilium.

Herbert made slightly more guarded statements. One hinted that he was thinking of moving the Concilium. He tried to gain support by mentioning that he was considering changing the Concilium into an American-chartered institution run by an exclusively American board. But that was a bluff, as indicated by Herbert's reaction when Charles Davenport suggested he might be able to arrange a buyout of the Swiss interest in the Concilium. Herbert responded by declaring that the Concilium had to be in a neutral European country.[32]

Even after his outbursts and threats, Herbert's American friends did not abandon him. They tried to calm him through assurances of at least their moral support. Charles Davenport did more. For example, Davenport heard that a zoologist at the University of Pennsylvania had just received funding for some sort of information project. Davenport pledged to contact him and see if he could help finance the Concilium. Unfortunately, that Pennsylvania man, Milton Greenman, avoided the meetings Davenport had arranged—but he eventually wrote to Herbert in Zurich asking about the Concilium's work.[33]

Charlie Davenport then promised to form a permanent committee to follow up on the congress's hints of aid to the Concilium, asking only that Herbert again give sincere consideration to moving back to America. The committee that Davenport created would later morph into one that became central to the Concilium's life after World War I, the Committee of Bibliography for Science of the American Association for the Advancement of Science. That name did not reveal its main agenda item, the Concilium's welfare.[34]

Herbert was not satisfied with his friends' assurances of an eventual rescue. His emotional state worsened when he learned his American bank was refusing to renew a large loan for the center. Because of the legalities of the Concilium's charter, he was personally responsible for it and all of its other debts. When the bank's official notified Herbert that the loan was due, Herbert knew that his youthful commitments had put his family in serious danger.

Herbert took another near-humiliating step. Not willing to impose on his mother or his older half-brother, Herbert used his social connections to contact the bank's president. He argued that science and a family such as the Fields deserved special consideration. Herbert thought it might be the end, but he was finally told the loan had been renewed, and the bank's president gave indications that the bank would never refuse such a noteworthy gentleman as Dr. Field.

That, and the promises by Davenport, had not quieted Herbert's emotions. The crisis frightened him and taught him some basics about legal obligations. He recognized that he had put all of his assets at risk for his ideal and that he and Nina might lose everything. He was now a family man with dependents.

The Swiss Connection: Establishing the Priority of Family over Information

Herbert again made what appeared to be a threat to abandon the Concilium. On his return to Switzerland in November he immediately informed the Swiss government that he might desert the project unless some significant changes were made to protect his family. Lawyers and the Concilium's European overseers were consulted. The outcome was a compromise. It was not too favorable for Herbert.

The nonprofit Concilium was changed into an unusual type of freestanding for-profit corporation. With that, Herbert thought he was relieved of the threat of losing all his personal holdings. Moreover, he believed the Concilium might be able to receive some tax relief. Importantly, Herbert thought he could now solve the Concilium's funding problems by selling shares in it on the market.

Not one of the beliefs was justified. Herbert's only direct compensation was stock in the new entity. He accepted a large percentage of the issue in return for his previous investments and loans, although he predicted the stock could never pay dividends and that it was unlikely to appreciate. He soon learned there was little chance of selling any of the shares in America, but he persuaded a colleague in Vienna and one in England to take the remaining 10 percent, perhaps calling on their sense of obligation to science.[35]

Herbert had to accept a particularly burdensome demand. The Swiss required him to write a new will so that in case of his death all of his stock and other interests in the Concilium would go to the government's agency, the Swiss Society of Naturalists. Herbert thought he was to get something in return. The society promised to pay the Concilium's debts and return to his family the stock and bonds he had used to secure any new Concilium bank loans. There was a very, very important qualification to that, however. His family would receive the stocks and bonds only if the society had ample funds when Herbert died.

With the new Swiss arrangement completed, Herbert launched more travels. He soon went to England, again searching for backers and customers. That trip came at the end of another round of subscriber hunting in Eastern Europe, one that caused Herbert to suffer through weeks of dysentery. He was exhausted when he reached England, and very lonely. On top of his physical and emotional difficulties, this 1910 excursion was a demeaning, hat-in-hand, experience. It was embarrassing enough that Herbert had to stay in his father-in-law's house to save money. Worse, he needed to ask his relative to provide the funds necessary to hire a replacement to continue the classifications in Zurich. Herbert also found it necessary to borrow money from friends for his other travel expenses.

Herbert seemed so sick, so tired and, especially, so depressed, that his father-in-law insisted that he be allowed to pay for a long sea voyage for Herbert. Herbert refused.

Then, he quietly went to another London friend for a personal loan to keep the Concilium's work going. During all this, production at the Concilium slackened. On his return, Herbert found it necessary to pay for additional personnel to catch up on the unprocessed manuscripts at the center and other expenses. That led to yet another search for ways to raise money.[36]

Herbert and his oversight board had endured many disappointments in their attempt to raise the funds needed for additional staff, and for refurbishing the Concilium's near decade-old equipment. Now, their search became desperate. They announced another public sale of stock at one hundred Swiss francs each. That, like the Davenport campaigns in America, failed and led to a very great disappointment. Herbert could not replace the worn-out Concilium printing equipment.

A new press was critical. Without it, there seemed no way to continue the Concilium's work. In response, Herbert once more submitted an application for another $140,000 bank loan. It was rejected. The only alternative Herbert and his business advisors could find was not attractive. They had to cast off another part of the Concilium's original design. Herbert had to contract printing to a local firm that took outside work while occupying the lower floor of the Concilium's new building (figure 8.1).[37]

A Bit of Optimism, Some Bad Advice

In 1908, the completion of the building had briefly raised Herbert's spirits, and the financial situation had begun to show an unexpected upturn. Although he had not obtained an endowment or large subsidies, there were new subscribers who accepted his higher pricing policies. Finally, it appeared that he might be able to meet operating expenses, mortgage, and debt payments from the increased operating income. Much to Herbert's pride, his staff was reducing the backlog of articles. The production of entries was returning to where it had been in 1906, so Herbert could advertise that his service was again a "current awareness" operation. There was other good news: the rising young American commercial publisher of book-form indexes, Halsey William Wilson, let Herbert know that he was never going to compete with the Concilium's card system.[38]

In addition, Charles Davenport's efforts in America seemed to have the possibility of contributing to the Concilium's welfare. Davenport had kept his word about forming a committee to further the Concilium's interest in America. He persuaded the administrators of New York's great American Museum of Natural History to meet to consider ways to aid the Concilium. Led by Herman Bumpas and Henry Osborn, the temporary committee held its first meeting just before Herbert left for the continent in 1907. In 1908, the committee offered its first suggestion: ask the twenty-five or so major Concilium subscribers to donate over $2,000 a year, it recommended. It did not take long to learn that was unrealistic.

Figure 8.1
Concilium building, from *Annotationes*

The committee almost died, but Davenport took over as its head and called for more suggestions. The one put forward by the museum's administrators had a grating tone.

They intimated that the Concilium's yearly deficit showed that Herbert was not a good businessman because he was selling his subscriptions for one-fifth of their cost of production. Then, as a palliative, the museum's librarian Ralph Tower suggested the museum might provide Herbert a bit of help by purchasing a full set of the "author" file-cards and take a yearly subscription to it. When it was realized that would cost at least $20,000 the committee backed away from the suggestion. Worse, and surprising, a negative view of Herbert's cards was emerging at the museum. Osborn and Bumpas wrote Herbert that the cards were too hard to file and too expensive. They suggested he consider just publishing his pamphlets.[39]

Frustrated by the lack of progress by the museum's committee, Charlie Davenport again tried to come to the rescue. Under the mantle of the American Association for the Advancement of Science (AAAS), he pulled together another group of influential zoologists and biologists. The group deliberated on how to keep the Concilium alive. Unfortunately, it did not have productive new ideas.

One of its feeble recommendations concerned the deficits. Some committee members again urged that Herbert raise the price of an American subscription to $2,000 a year. The committee contacted American subscribers to test their reaction. But almost all the institutions the committee reached rejected the suggestion.[40]

Herbert was relieved when he heard the idea of raising American subscription rates had been abandoned. He had opposed the committee's ideas of increasing rates from the beginning for ethical reasons, and because he feared he would lose his European customers if they heard of the new American charges. He sent rather emotional letters stating his objections to increasing the fees. His short-lived alterative idea was to offer a deluxe edition of the cards, using the highest-grade linen stock, to his affluent American subscribers. Herbert thought the upgraded card stock (more elegant than Dewey's linen cards composed of materials from shirt factories) would justify a higher subscription fee while allaying any fears of price increases among other subscribers.

Herbert also strongly objected to the suggestion, probably put forward this time by Henry Osborn, that the American Museum of Natural History buy out the Concilium. That proposal included a salary near ten times what Herbert was then receiving, as well as permission to continue his work in Zurich. But Herbert was giving no ground on his idealistic commitment to the Concilium as being an embodiment of internationalism and neutrality.[41] He had additional reasons for rejecting the generous offer. It would diminish his control over operations, and he thought other nations' scientific societies disliked cooperating with an American company.

Mad Dogs: Rants, Outbursts, and Discouraged Friends

Soon Herbert's resistance discouraged the museum's men and even Davenport's AAAS committee. They reluctantly backed off from an active role. Fate, as well as Herbert Field's intransigence, had played a part in reducing the role of the committees. One important member Herbert had tried to court, Milton J. Greenman, resigned because a mad dog in Philadelphia bit him. He had to undergo the then-painful medical treatments and suffer through months of recovery.[42]

Unfortunately, the committee became less interested in the Concilium's problems just as the center's short period of relative prosperity ended. Subscriptions had increased after the 1907 financial panic and then stalled. They slowly declined as 1910 approached. At the same time, any remaining hopes of entering the electrochemistry market died of atrophy, as did the intention to expand the Concilium into another possibly lucrative applied field, forestry. For the next four years, the best Herbert could do was to keep the Concilium from declining more than it had, hoping that he would not be forced to again increase fees.

Herbert's temper and mercurial emotions again made his situation worse. Despite Charles Davenport's advice to be careful about political sensitivities, Herbert endangered the Concilium when he wrote a nearly ranting letter to Henry Osborn demanding that he immediately contact the Smithsonian Institution to have its representative recant a statement he made at a 1910 Royal Society's *Catalogue* meeting. The Smithsonian's man, Herbert wrote, had testified that cards were a burden and should not be used for the *Catalogue*. Herbert accused him of "stabbing him in the back." Herbert demanded that Osborn take action. Next, he informed Osborn that the Royal Society's project was surely on its deathbed because of financing problems. It deserved its demise, he wrote. It had always been the society, not he, that had killed attempts at cooperation.[43]

With that outburst, Herbert lost a friend. Such words did not sit well with Osborn. Osborn sent a reply stating that he knew the *Catalogue* would continue, that the Smithsonian had no obligation to Herbert, and that Herbert should abandon the Concilium's card system!

Failed Internationalism and Information Reformers Left Adrift: What Can a Quaker Pacifist Do in Wartime?

Osborn's comments hurt, but the gravest blow came with the seeming failure of Herbert's loftiest ideal, internationalism, as World War I broke out.

Herbert had not foreseen the debacle. A bit more sensitive to world events than Herbert, Paul Otlet had sensed an impending disaster in 1914 and rushed to the United States to secure new funding and perhaps a new home for his service. But the Americans turned him down just as Germany invaded his neutral Belgium. His center and his UDC system fell into neglect. He fled to other neutral countries. At the outbreak of the war, Henri La Fontaine was on a speaking tour in America and remained there until the conflict ended. Although nearly destitute, he found many sympathetic Americans such as the scholar-diplomat Stephen Duggan, who would become critical to the Concilium's fate. Duggan found La Fontaine a temporary position at City College of New York and helped to secure a Carnegie grant. As well, La Fontaine and Duggan may have aided those running the radio listening-station at the college that was monitoring European broadcasts, even forwarding encrypted ones to America's secret code-breaking organization run by Herbert O. Yardley.

Meanwhile, Herbert had made few plans for moving his center, or for what he and his family should do if hostilities began. He should have known better. Switzerland avoided immediate invasion, but the Concilium faced the cutoff from its supply of journals, card stock, and customers. The battles on land, blockades of the sea-lanes, and the threat of submarine warfare meant the sudden loss of three-fourths of its subscribers.

Herbert soon feared that even the Swiss might become involved in military action, but, still wishing to oversee the Concilium, he decided not to flee to England or America. He sent his wife Nina and the children to the resort town of Lugano in the Swiss-Italian Alps for safety. He then turned to the Concilium's plight. There was little income, not even enough to make the payments on the outstanding debts. More bad news came when the Swiss government informed him it was going to end its subsidies.[44]

Herbert struggled to keep producing cards, although it appeared senseless. Against great odds, he and the remaining staff did fairly well. In 1915 he was, for example, able to send the New York State Library over 17,000 cards and follow with 23,000 in 1916. But the Concilium had to bow to its fate. Field had to lay off almost all of his remaining workers and he again tapped his own funds to pay the Concilium's ongoing overhead costs.[45]

The financial situation was so dire that Herbert left his villa in Zurich and moved into a tiny flat. He rented out most of the Concilium's rooms. Yet, he did not fully retreat. He began to muster the support of Swiss scientists to reverse the government's financial decision, but the hard-pressed government bureaucrats resisted.

Herbert then took another hard-to-swallow step for a proud man. He called on his many friends in America for more help. He pleaded for a quarter-million dollars to meet the emergency. They responded, but with only small contributions. At best, their assistance helped to pay for some of the center's minimal costs. It kept the offices warm, protected the files, and paid the mortgage. There was nothing for any employee except the secretary.

As he tried to save the Concilium, Herbert began to weigh critical career options. Where should he be, and what role should he play in the war? Should he finally relent and take his family to safety in America, perhaps helping his brother Edward with the family's business? Should he stay in Zurich and continue the Concilium's classification work by himself despite a probable lack of materials? Could he afford to do that and to support his family in the Alps?

He soon made an unpleasant decision. Herbert put what was left of the Concilium into the hands of his secretary. He then looked to his Quaker social commitments as he searched for a way to continue his work for internationalism.

9 War and Its Aftermath: Working for Peace, Competing Ideologies, and the Beginnings of Intrigue with Allen Dulles

As a man in his mid-forties, Herbert Field did not have to serve in anyone's army—even the U.S. Army. He was not required to participate in any capacity. Although his salary from the Concilium was suspended, he was not utterly desperate for a job. He needed some income to help maintain the center and his two households—one in Zurich, one in Lugano—but he made his first step toward a wartime career without a financial motive.

U.S. representatives in Switzerland had asked him to help with the aid efforts of American charities. There was good reason for that. Herbert was a then-neutral American, a man with a fine reputation throughout Europe, and a known Quaker pacifist. He agreed, and in 1914 secured a special document from the U.S. Department of State allowing him to travel within the war zones in Germany and France. While some of his travel may have involved Concilium business, he took on ever-increasing responsibilities for relief and refugee agencies as they moved into devastated and dangerous areas, including those occupied by German forces.[1]

Both Field and the Quaker-backed organizations he served held near-ideal credentials for projects that took them over contested borders. Quaker groups had an impeccable long-term record of staying out of politics and being truly impartial. The new American Friends Service Committee, the organization formed to allow the Quakers' pacifists and conscientious objectors to play a role in World War I, upheld the old traditions.[2] Herbert had his own unblemished reputation, and he had other advantages as a relief administrator. Because he was multilingual, he was able to communicate with government representatives throughout Europe.

His now gray hair and beard gave him the look of a mature man who could be trusted. Along with all that, he knew important people throughout Western, Central, and Eastern Europe. That facilitated the delivery of supplies and provisions to refugees and exiles.

After trips to Paris and London to be briefed on relief procedures, and while barely escaping bombardments, Herbert returned to Switzerland. He ran the agency through its new quarters in Berne. He did not confine himself to his office. He took aid to the displaced and needy throughout the continent. He traveled across Europe and negotiated with many political leaders, including those who had plans for a postwar realignment of the old empires.[3]

Herbert, Meet Allen Dulles / Odd Intelligence Beginnings and Idealism at Its Highs and Lows

Because of his relief work and his status in the international scientific elite, Herbert came to the attention of the U.S. Department of State's intelligence section and to a young American diplomat assigned to neutral Switzerland to gather political and economic information. That Princeton graduate, Allen Dulles, was one of the first American espionage agents. Four decades later, he became America's chief spy when he headed the Central Intelligence Agency.

Before World War I, geographically isolated America had little need for and experience in foreign intelligence; so, in his mid-twenties, Dulles had to teach himself how to gain hard-to-get information for America's possible war effort and for the coming reworking of the national boundaries of Europe. He had little help, as the United States was an "innocent abroad" when it came to spy craft. It had no secret bureaus with strings of HUMINT informants around the world. Its human intelligence assets were a handful of military attachés, missionaries in Asia and Africa, some anthropologists who looked for German bases and influence in Mexico and the Middle East, businessmen engaged in foreign trade, and a few of the State Department's more adventuresome employees.

Other means of unlocking secrets were also undeveloped. America had only a tiny and still less-than-effective radio-snooping agency. Its code-making and code-breaking capabilities were less than acceptable. Also, the United States had not imagined it would ever need to create a group to train undercover operatives how to play what came to be politely called "dirty tricks." Yet, in 1914, it needed information, and quickly, as it seemed likely that America might enter the war, certainly if the nation's Eastern intellectual-internationalist elite had its way.[4]

Soon, America's men were going into battle, and that other Princetonian, President Woodrow Wilson, was determined the United States would play a major, if not dominant role in Europe's peace settlements. Knowing that America's allies were shaping information to fit their nationalistic and colonial interests, Wilson demanded his own sources of intelligence. Like Henri La Fontaine and Herbert Field, Wilson was an idealist. He demanded reliable information for his grandest hope, the creation of a world organization to secure permanent peace and to bring democracy to the world. In response, the government began to throw together an intelligence network.

Switzerland: The Very Best Site for Intelligence and Sources

Allen Dulles was assigned to the best of all places for European intelligence work. Switzerland's location and its neutral status meant that it housed embassies, relief organizations, and refugees from all nations. It had even welcomed the soon-to-be world revolutionary, Russia's Vladimir Lenin. As soon as Dulles arrived in Berne, he began to use exiles and relief workers to build a core of informants. Herbert Field had an ideal profile for being such a "source" for the fledgling spymaster.

When the twenty-five-year-old Dulles contacted Herbert, he found an idealistic and pliable man who desired to help those interested in peace and future world harmony. One of Herbert's associates described him as "a burly man, with bushy grey hair and beard, heavy grey eyebrows behind which lay the gentlest, bluest, most candid pair of eyes I ever saw on an adult man. They were the eyes of an unsophisticated and loveable child." A young American diplomat who met Herbert in Zurich thought he was a retired minister.[5]

When Herbert began helping Dulles, he did not consider himself a spy. He began by playing the role of an information resource; only later did he become more like an agent. As a first step, Herbert provided Dulles with much hard-to-get information on conditions throughout Europe, including Russia and the broiling areas of religious-ethnic conflict that became the postwar "Slavias." Such actions, Herbert felt, were not a breach of his status as a neutral relief worker, nor were they against his Quaker beliefs.

He soon gave the appreciative Dulles more. He established contacts with current and potential leaders of Germany and Central/Eastern Europe, including many who had fled to neutral Switzerland for the duration of the war. That had not taken much effort. Herbert was well known and he was a good host. His apartment was a gathering place where Dulles and young State Department officials, such as Christian Herter and Robert Murphy, were introduced to many potential sources of invaluable information, and to people with ideas as to how to reshape Europe at the war's end.[6]

Herbert kept expanding his intelligence role—as well as showing that he had the makings of a diplomat. European nationals who worried about America's possible entry into the war saw Herbert as someone who could act as an intermediary. As early as 1917 the Munich-based Socialist leader, Dr. de Fiori, asked Herbert to broker an acceptable peace between America, its allies, and Germany.[7]

When the United States formally entered the war Herbert felt less restrained. He helped with specifically American problems and policies. When he noticed that an influential American consul was having too many confidential meetings with Germans, Herbert informed Dulles about the get-togethers. Dulles had so much confidence in Herbert's judgment that he helped to have the man suspended, even before the State Department conducted an investigation. Herbert also used his food-aid powers to help the Allied cause by threatening to cut off supplies to the Swiss government unless it suppressed Germany's anti-American propaganda.[8]

Herbert did more for the war effort. He served on the United States' purchasing board in Switzerland. It aided more than relief agencies seeking supplies. It acquired strategic military items and helped prevent the enemy from gaining them. Herbert also prepared long papers on economic and political conditions in Germany, ones that went beyond what was necessary for his relief work. They were usually sent to a mysterious "Mr. X." Some of the reports were quite sophisticated, such as one that applied the methods (for the first time, Herbert claimed) of the brand-new science of nutrition to the impact of the embargo on Germany and her allies and, importantly, on neutral nations including Switzerland. The report had the flavor of analyses that were produced by scientists and academics employed by U.S. intelligence agencies during the Cold War.[9]

Another assignment was of extreme importance. Herbert developed a detailed report on Germany's poison gas capabilities. Perhaps he did not realize it, but that put him in danger of the Germans believing he was a spy. Herbert made contact with a noted German chemist who had moved to Switzerland because of his opposition to the war. Herbert convinced him to reveal the details of gas warfare in general and Germany's capabilities in particular. The professor knew much, including what resources Germany had for the manufacture of a wide range of new gases. The Germans had not yet deployed some of them.

In addition to reporting the professor's information to the Americans, Herbert arranged to have the head of the International Red Cross attend a special secret meeting in the first days of 1918. At the gathering, the professor provided facts and figures; he also demonstrated the nature of different gases by exploding bomblets on the tile floor of his laboratory. The demonstration had an impact. The Red Cross quickly intensified its campaign to outlaw all use of poison gases. At the same time, Dulles forwarded Herbert's report to America's scientific teams working on gas-warfare.

A Nod from the British and an Early Warning of the "Red Menace"

Herbert Field became an intelligence asset for others than Allen Dulles. In mid- and late 1918, when he was on brief trips to London, he used his poison-gas reports to bring himself to the notice of those in the upper levels of the British government. On receipt of his gas-related papers, the Ministry of War notified him his documents would be sent to the Office of Special Intelligence (and probably to England's own gas center at Porton Down) to be used for both technical and propaganda purposes. As a special nod, British intelligence placed Herbert's name on its list of significant persons.[10]

Herbert played a critical role in October 1918. When he returned to Switzerland, the American Embassy entrusted him with an extremely delicate matter. A mistake in handling it might have caused the loss of thousands of lives. He was assigned to translate and encode the secret offer by Alfred Hohenlohe-Schillingfürst, of a separate

Austrian peace and to forward the confidential telegram to Washington. There was more to that telegram than a peace offer. It contained an emotional warning about the dangers of the Bolshevism that Lenin was bringing to Russia. Herbert met and worked with another American expatriate intelligence "asset," the wealthy defrocked Congregational minister George D. Herron. Herron, a radical socialist who distrusted the Reds, aided Dulles and British intelligence and became a significant advisor on the Soviet problem at the Paris peace conference.[11] (Ironically, the color *red* had positive associations in Russian culture long before left-wing Europeans in the mid-nineteenth century waved red flags during their revolts, and prior to the Russian Revolution and the rise of Communism. In addition, post-Revolution Communists, even in the United States., referred to themselves as *Reds*.)

The Bolshevik threat deeply worried Herbert. Soon, he wrote his own long report for Dulles on the nature and likelihood of a Red takeover in Central and Eastern Europe. It contained an exploration of the possibilities of a Bolshevik-sponsored Rumanian revolution and of Soviet Russia's expansionist policies. His report was balanced and proved useful to American intelligence. It clearly indicated that Herbert already had misgivings about the revolutionary tide—something he was unable pass on to his children.

Juggling Family Crises and Working for Peace: Herbert's Stress and His Great Honor

Meanwhile, Herbert coped with the stresses of long-distance parenthood, including a frightening episode when the authorities confined little Elsie in an isolation hospital. Her long, tender, and innocent letters to her father, which included heart-tugging illustrated stories she composed to entertain and please Herbert, almost led him to turn away from his charity and intelligence work. He felt guilty that he had left Nina alone to deal with the children's crises. The emotional burden increased as he and Nina had to handle other family-related frustrations. When they were informed that Herbert's mother Lydia passed away in late 1917, they could do little but send a belated letter of sorrow to friends and relatives in America. Travel on the continent, or over the Atlantic, had become a near impossibility for civilians.[12]

Herbert continued with his work for peace as well as the relief and intelligence-related work. Significantly, he treated his efforts as "most secret." In all his letters to his friends during the next decade, he never mentioned his labors for Dulles, nor did he say much about the relief work.

Herbert's role in world affairs deepened when a grateful Allen Dulles arranged what Herbert considered to be one of the greatest honors of his life. He received an invitation from President Woodrow Wilson to participate in the team effort to prepare America's plans for the historic peace settlement at Versailles. The group was to redraw the map of Europe and help create Wilson's internationalist vision, the League of Nations. The appointment brought Herbert in contact with the greats and future greats of American diplomacy, with the feared revolutionary force that threatened to overturn the world, and with some of the most important men in America's developing science infrastructure.

Herbert's first task for the American Peace Commission was a dangerous one, although well paid. He was to survey and report on conditions in Germany and on the options for its postwar restructuring. His initial step was an orientation session in Paris where he met many of the influential liberals and diplomats of the era such as Herbert Hoover, Joseph Grew, Walter Lippmann, Vernon Kellogg, and even T. E. Lawrence (known in popular culture as Lawrence of Arabia). Significant for Herbert's sons' futures, the young Whitney Shepardson was at Versailles. He soon became a director of Carnegie biological research programs and, later, the head of the Secret Intelligence branch in the Office of Strategic Services and, then, director of the post–World War II anti-Communist Radio Free Europe.[13]

After the orientation session, Herbert was sent into Germany by way of an automobile trip through some of the most war-ravaged areas in Europe. The carnage he saw made him emotional; but he did not retreat to Paris. Arriving in Munich, he soon became America's expert on Bavaria's new politics, including the threat of a Soviet-style revolution. He was also responsible for reporting on the condition of prisoners of war held by the United States.[14]

Herbert the Agent: The American Peace Commission

Herbert had a "cover" for his intelligence assignment and he denied any reports that he was an important American official. His papers only specified the affiliation "American Peace Commission." "Peace" was a not a fitting term. Munich was in its own version of a war. Herbert was in danger from the moment he arrived. Like most of Germany, Bavaria was in chaos, Munich even more so. Food shortage was a polite statement for impending starvation. Public safety had broken down. Gangs of refugees and demobilized soldiers and sailors were on a looting spree. One political faction after another had its militia, always ready to seize a government headquarters or a public utility. There were threats of insurrection from radical Bolsheviks, and from the old landed German aristocracy and its military allies.

Herbert correctly feared for his life—and for the future of German-American relations. Despite having a young and armed army officer to act as his protector and courier, Herbert was frightened. There were rumors that his hotel was to be attacked. It was more than a rumor. A mob soon tried to free the hostages some leftists held in retaliation for a political assassination. The captors threatened to kill three hostages for every man their enemies would harm.[15]

When a right-wing cadre began preparations to rescue the hostages, Herbert took the advice of a friend and fled to her house to hide for the night. In his emotion-driven haste, he managed to stumble and fall, nearly breaking his arm. Despite the injury,

Herbert again refused to leave Munich. Oswald Garrison Villard, a visiting American editor in Munich, feared that because Herbert was the only official American representative then in Germany, his death or severe injury would endanger desperately needed relief work, as well as undermine a sensible postwar settlement. Villard advised the distraught Herbert to move to the suburbs and away from the mobs and rioting. Herbert refused.[16]

In spite of what some considered his foolishness, Herbert survived the crises. During the next month, he submitted more coded reports, ones recommending what policy America should have concerning the reconstruction of Germany. He correctly predicted that any harsh peace settlement would lead to a later disaster. He saw grave consequences from the unemployment and suffering any severe reparations policy would bring. Poverty would lead to an eventual takeover by either the far left or the far right. A liberal or even a moderate left-wing government, Herbert advised, had little chance of survival unless there was immediate and long-lasting help.

The "Red Menace," Again

Germany desperately needed help. A coalition of socialists, who had only a tenuous grip on power in Munich, was the only force holding Bavaria together. For example, the last shreds of stability were nearly destroyed in February 1919. The government's eccentric agnostic-socialist leader Kurt Eisner was assassinated in the street—a short distance from Herbert's hotel. A pool of blood was just outside the hotel door. In reaction to Eisner's assassination, the most aggressive of the Bolsheviks seized the Munich government and initiated a program of confiscation, nationalization, and "cleansing" of the middle and upper classes. It was a brutal month. There was more looting. Herbert was again warned to leave the city for his own safety.

The Munich Red takeover frightened Europe. In his reports, Herbert reinforced the fears. He found the Bolshevik-Sparticists to be brutal, uneducated, and introducing what he termed "stupid" policies. And, he emphasized it was more than Germany's Reds who were misguided: Bolshevik policies had ruined Russia's own economy. Herbert repeatedly advised his State Department contacts that something had to be done to stop the tide of Red—perhaps even the military occupation of Germany. Herbert's disgust with radical politics was magnified when he was informed that he had been spied on since his arrival, and that stories had been planted in local newspapers improperly picturing him as a man of intrigue. Like the Reds' actions, the articles convinced Herbert that any populist movement was untrustworthy.

The situation in Munich grew worse. The State Department ordered Herbert to leave the city. He refused. He delayed his departure so he could witness the conclusion of the looming political crisis. An alliance of socialists and right-wing groups did muster their forces and overthrew the Communists, installing a relatively moderate but shaky republican government. The new coalition was unsure of itself, partially because word was received that a Soviet-aided revolutionary regime under the merciless Bela Kun was ravaging Hungary. There were fears of an invasion of Germany from the east.

Herbert then learned of a surge of other Red threats in Central and Eastern Europe. His contacts were also informing him of conditions within Russia itself, some of which made him write emotional letters to friends around the world. What he had learned he did not like. Herbert discovered, for example, that the famed psychologist Ivan Pavlov was destitute. Herbert asked for donations for Pavlov and inquired about the possibility of resettling him in the West. At the time, there was no indication that Pavlov would soon become a Soviet hero, honored and consulted by Lenin. During the early years of the revolution, he was just one of the uncounted thousands of "bourgeoisie" intellectuals the Reds took their vengeance on, depriving them of jobs and even food. Certainly, Herbert could not have predicted that Pavlov's methods of psychological conditioning would become the foundation for the tortures the Soviets applied to his sons in the late 1940s.[17]

Herbert sent off many additional reports from Germany and, later, Switzerland. Some included his negative accounts of secret Soviet agents in Munich who, he said, were posing as political negotiators seeking American approval while they were encouraging local radicals. Field reported they were involved in deep espionage and many assassinations. Of special concern to Herbert was a man he identified as Dr. Herplant Parvous who had a history as a double agent and, Herbert reported, was using blackmail and bribery to swing both Bavaria and Switzerland to Communist rule.[18]

Herbert also advised the State Department of the growing distrust of Jews in Germany. He traced the rise of the hatred to Germany's military and diplomatic defeat and to the role of Jews in the leftist radical governments that were seizing power in many areas of the nation. There were also warnings from Herbert that Germany might secretly resume the production of poison gases. He insisted the making of such gas be prevented and that the country's army should be permanently reduced to a skeleton force. Herbert went on to recommend an early version of a Marshall plan for Germany. Massive aid, he implied, could benefit the American as well as the German people. He also urged the United States to create an early version of the Voice of America to counter the "menace."

In his last reports, Herbert added some additional insights into what the Russian Revolution might mean for Europe's future. For him, the spreading revolution meant disaster. What his many contacts in countries such as Hungary revealed to him reinforced his earlier fears about Marxism in action. He also returned to the subject of anti-Semitism in Germany, but he did not report on the growing presence of small political groups that were exploiting it, such as the one soon led by a young Munich oratorical marvel, Adolf Hitler.[19]

Internationalism's Inner Circles

Herbert Field's efforts for the American Peace Commission, and for Allen Dulles, were well known in Washington. His work was so admired that he was asked to donate his papers and diaries to the World War I collection being assembled at what became Stanford University's Hoover Institution. More important, the State Department sent him another request to become active in the American diplomatic inner circles in London and Paris. Herbert put aside for a time the question of possibly reopening the Concilium and accepted that offer of work at the Versailles Peace Conference.

When he returned to London, Herbert began to work with the cadre of America's new university-trained progressive academics and statesmen who were advising Colonel Edmund House and President Wilson on the remaking of Europe, as well as on the disposition of Germany's colonies. Herbert's abilities and his Harvard background immediately led to friendships with the cream of America's young diplomats. Adding to his relationship with Dulles, Herbert became closer to Christian Herter, whose diplomatic and political career led him to become U.S. Secretary of State during the Eisenhower administration, and to Joseph Grew—a future ambassador to Denmark, Switzerland, Turkey, and (at the time of Pearl Harbor), Japan.

Herbert was excited about the future League of Nations. With its formation, he wrote his young sons, nationalistic barriers to progress would come down, and there would be a greater role for international science-information systems. Herbert and the new academics recruited into Wilson's Inquiry for peace planning also considered themselves experts in the vital subjects of ethnicity and nation making, issues critical to a viable postwar Europe. They felt able to accomplish what the Ottoman and Austrian empires had not been able to do: build stable nations out of many ethnic, class, and religious groups that had hated and fought each other for centuries.

As did other rather naïve Americans participating in the group, Herbert foresaw the redesigned Albania, Bulgaria, Rumania, Hungary, and Poland becoming peaceable kingdoms. The envisioned Czechoslovakia and Yugoslavia also seemed to be embodiments of the hopes of those liberals who had bet on social science and the new learning to be able to conquer social and economic evils. Herbert could not have realized that his 1919 participation in unstable nation building would have much to do with his sons' terrible fates thirty years later.[20]

Assessing Commitments at Versailles: What about the Concilium?

Herbert had personal affairs to deal with while the future-shaping decisions were being made at Versailles. When he was between the Paris negotiating sessions he traveled to England to try to solve his in-laws' health and financial problems. The chore included lending them some of his dwindling savings. Personal matters also pushed Herbert close to abandoning the Concilium. Britain's diplomatic office almost persuaded him to take a position in London. An even more tempting offer came from a British scientific board. It wanted him as its chief information classifier. He met and was entertained by some of England's highest science officials. The British also tested Herbert's worth as a science diplomat. He made a quick trip to Trieste on England's behalf.[21]

Herbert decided to continue with the American Peace Commission. However, he used part of his time to lobby, as did Paul Otlet, to have information systems placed high on the agendas of those establishing the League of Nations. Herbert also contacted the French Havilands. If he had any hopes they could provide funding for the Concilium, they were dashed when he learned what the war had done to them. He encountered Adolph Voge in Paris but wanted little to do with him.

Herbert returned to Zurich when his duties were fulfilled. Although he feared that his fortune and future were doomed, he quickly moved his family back from the mountains and into a new four-story handsome villa with a bucolic estate. Then, he turned his attention back to his information work.

Herbert reenergized his great expectations for the Concilium. But he was overly optimistic. He quickly learned that little help would come from the Swiss government. It rejected his pleas to restore the subsidies. Switzerland, although neutral, faced hard times because of the disruption of the world's economy. There was little for niceties like libraries. In fact, the once-rich Swiss were in dire need of financial help. They even turned to Herbert for aid. Although the Swiss government had disappointed him, Herbert sympathized and agreed to help prepare a delegation that was on its way to America seeking relief for the beleaguered Swiss economy.

Herbert was more than a bit alienated when he learned that while Paul Otlet just received a very generous grant from the Belgian government, the Concilium was not to have a share in those funds. The Belgians gave Otlet's operation significant financial maintenance allowances and the use of 150 rooms in a great palace. Herbert may have also been a bit jealous when he heard that Melvil Dewey and the American Library Association had explored bringing a copy of Otlet's files to America and providing support for the Belgian center.[22]

There were some rumblings about other international information initiatives in Europe that might help the Concilium, but Herbert could not await word about them or possible League of Nations support. He decided that he had to revive his bibliographic center immediately or it could never be brought back to life. He was convinced that unless he could begin shipping cards his subscribers would never return. He began another fundraising sojourn. After more emotionally trying weeks dealing with the financial affairs of Nina's ailing parents in England, Herbert booked passage to New York City. He felt the only hope for his Zurich service was again American largesse.

He also thought that his own financial situation could be saved only by a trip across the Atlantic and by an investigation into what happened to his investments during the war years.

Although the late 1919 voyage began another "hat-in-hand" tension-filled sojourn, Herbert looked forward to at least parts of his visit. However, the trip ended in bitter frustrations. Instead of a warm reunion with his relatives and what he expected to be a month-long successful fundraising campaign, Herbert spent six difficult months of exhausting travel and worry over the Concilium while having to deal with family disappointments. In return, he got little to relieve his stressed personal finances, or those of the Concilium.

10 The Postwar Liberal Establishment and the Family in Postwar Decline

Herbert Field disliked leaving his wife alone, but he expected he would be back in Zurich with Nina and the children in a few weeks. He departed from England on a luxury steamer, booking a beautiful cabin despite his depleted savings. That put him in the company of America's brightest and most influential people. Colonel Edward House, President Wilson's trusted advisor, was on board. House introduced Herbert to some of America's most powerful men. For example, Herbert spent much time with the Supreme Court Justice and American Zionist leader Louis Brandeis. Brandeis had gone to Germany and Paris to rescue what was left of the Zionist movement in Europe, transfer it to America, and to lobby for recognition of a Jewish state in Palestine.[1]

Herbert's ideas on world peace, and on the reconstruction problems in Germany, impressed Brandeis. Perhaps that is why soon after landing, Herbert found himself the guest of William Howard Taft, ex-president and chief justice of the United States, who hosted an afternoon's conversation at his home. Taft, like Herbert, was a great advocate of arbitration in world conflicts and had done much to forward the movement while he was president. He also had been influential in shaping Andrew Carnegie's Endowment for World Peace. Herbert never fully appreciated it, but the meeting indicated that he was being ushered into what a later generation called "the American liberal establishment," a set of people, institutions, and beliefs his sons, surprisingly, soon rejected.

Herbert was on the verge of reaching the pinnacle of that "establishment" through a personal conversation with President Woodrow Wilson. Then, when Wilson suffered his "mysterious illness" (a stroke that kept him paralyzed and near death for some seventeen months, but which was kept secret from the public), the scheduled meeting was canceled. Secretary of State Robert Lansing's rather personal letter to Herbert (perhaps because he was Allen Dulles's uncle) informed him of the cancellation, but did not detail Wilson's condition. The letter contained only an invitation to visit the State Department to have a discussion of Herbert's plans for peace and for science information.

There was another contact that Herbert could have, but did not make in late 1919. Herbert Hoover, the fellow Quaker who was America's czar of postwar relief and reconstruction efforts, knew of Herbert's work and dreams through his assistants who had encountered Herbert during their years in Europe. Herbert did not pursue a meeting with Hoover because he thought he did not have the time to linger in America. He felt he had to conduct his personal and Concilium business quickly before his funds ran out. The war had destroyed much of the value of his investments in German and English securities. Worse, Germany had confiscated all his bank accounts as soon as America declared war. In addition, the disruption of European trade led to a threatening reduction of the value of his American assets.

When he reached New York, Herbert put himself on his version of a poverty budget. That hurt his pride but it saved money for Nina and the children who remained in Zurich hoping for Herbert's quick return so the family could celebrate Christmas together. When planning his latest American trip, Herbert had calculated that he did not have enough to stay in hotels or to lease a transient's apartment, even for one or two months. With much distaste, he accepted relatives' offers of room and, at times, board. He also had to depend upon them for much of his transportation. Truly embarrassing, he needed to borrow an overcoat from one of his cousins so he could buy clothes for Noel, Hermann, his "dear" Letty, and "his little girl" Elsie.

He thought his first American stop in September, in northern Long Island, would be quite pleasant. He was to visit his half-sister and her husband at the old family retreat. Herbert always had a warm affection for his older stepsister, Fannie, and had relied upon her advice and love since childhood. He felt so close to her that in 1904 he wrote a sizeable sum for her into his will. He expected that Fannie would bring him up-to-date on the rest of his extended family.

When he arrived at the estate, Herbert became disheartened. Fannie looked old; so did her husband. Although they were only in their mid-sixties, they appeared and acted much like aged and defeated people. Fannie and Henry had been able to hold onto their home in Brooklyn, as well as the summer retreat, but life had not gone well for them. They were part of a family in decline.

Trouble had marked parts of Fannie's earlier life, but the problems of 1919 were different. She had conquered difficulties in her early years when she lost three of her seven children, but at that time her husband Henry Griffen (another Quaker related to the Havilands) was doing very well as a stock and commodities broker. They had lived with grace in Great Neck, where three servants, including a coachman, cared for her and the children. The purchase of a home in Brooklyn in the early 1900s, so the children could have access to Quaker schooling, reduced the need for servants to only one, but there was no threat to the family's upper-middle-class lifestyle. Fannie was even happy to bear her last child when she was fifty years old. Money was still there for nannies and other servants, if needed. Her family prospered during the next decade but the war and the economic depression that followed hit the family hard. Henry was no longer a broker and had found only a series of low-paying jobs.[2]

Fannie gave Herbert some hints that he would soon discover to be on the mark. Other members of the Field clan were also enduring what they considered financial hardships. However, "hardship" was a relative condition. Although she believed, as would Herbert, that the Fields had fallen into near poverty, Fannie's decline was limited. The estate still maintained a private tennis court and a swimming pool.[3]

Calling on Charlie Davenport, Grantsman par Excellence

Herbert could not stay long in Great Neck. He had to turn to his main objective, finding a way to revive the Concilium. He arranged for someone to drive him the short distance to Cold Spring Harbor where his ex-neighbor, colleague, and close friend Charles (Charlie) Davenport was conducting his newest version of eugenics research. He was becoming famous (and soon infamous) because of it. Davenport had shifted much of his attention from plant and animal studies to questions of human evolution. Direct research on human genetics was then infeasible, so Davenport was building one of the world's great science databases, perhaps the world's largest genealogical file. His center was using a huge store of family histories to explore the influence of genetics in human development.[4]

Davenport expected Herbert and had been devising plans to rescue his friend's service. The key, he told Herbert, was finding a way to finally tap the funds of the huge new American philanthropies. Davenport was a skilled grantsman and willingly shared his knowledge of the "who and how much" for science funding in postwar America. Davenport also had a network in action and was reviving the old pre-war "information committee."

Davenport told Herbert that he arrived at the right moment, thinking he had the key to the pockets of the great foundations. The trick, he said, was Herbert's better use of the science institutions to pressure the philanthropies. He did not use the term, but he explained that it was now a "corporate age," no longer one of lone gentleman. In addition, he stated, because of America's new involvements in Europe, the foundations were more receptive to ideas of how America might help continental reconstruction.

For years, Davenport had been the driving force in the American Association for the Advancement of Science's (AAAS's) Committee on Bibliography for Science. He had selected its members from the zoological-biological community. And, they all were friends of the Concilium. At the conclusion of World War I, Davenport reenergized the committee. He foresaw a great expansion of its role, and America's, in world science information. He correctly sensed the war had changed the nature of international science. A rich United States now had the opportunity to take a major part in shaping world science and its institutions. He was not going to tolerate a return to British and

German domination. Davenport put his old friend's Zurich service at the center of his information visions.[5]

As soon as Herbert appeared at Cold Spring Harbor, Davenport brought him up to date on the changes in the infrastructure of American science. He advised Herbert that some of the Concilium's old contacts were no longer major players in philanthropy and recommended that Herbert search out friends among the managers within the new and huge Rockefeller foundation. He also advised Herbert to court the members of a group founded at the beginning of World War I to wed science and the American government's needs, the National Research Council.

Courting the National Research Council and the "Carnegies"

Herbert had learned of the National Research Council while in Europe and had previous fleeting contacts with both Carnegie and Rockefeller managers, but he pressed his friend for details. Davenport spoke with authority. He was one of few academic scientists favored by the administrators of the Carnegie or other large foundations before World War I. In addition, he was a leader among those making the recent scientific breakthroughs that motivated the major foundations into making academic science research a priority.[6]

Davenport was well acquainted with the National Research Council and its ambitions to become the center of America's early version of Big Science. The council was already involved in well-funded projects that served national purposes and looked forward to an international role. In fact, Davenport was just completing a large and very expensive study of the physical characteristics of World War I soldiers as part of a broad council and Surgeon General's project. Because of his contacts in the council, Davenport knew it was planning to take an independent role in science information. He already had made the council aware of Herbert's information contributions and encouraged its leaders to contact Herbert.

Davenport's description of the National Research Council's work and its power convinced Herbert to follow up on a probing letter he had received from a council representative asking if Herbert was interested in cooperating in a just-forming initiative for world science information. The letter stated that both the council and Clement W. Andrews of the innovative Crerar Library in Chicago had a great deal of money available for information projects. Most importantly, the letter implied Herbert's European science-information work and professional connections were relevant to the task of defining the council's part in rebuilding all of world science.[7]

While Herbert thought about how best to respond to the council he took Charlie Davenport's advice on grant hunting. Davenport's list of donor prospects had not ended with the Rockefeller foundations. Among his recommendations, despite Herbert's previous sad experience with the Carnegie Institution's medical research grant makers, he told Herbert he should draw up a new proposal for them and the other Carnegie-funded organizations.

Davenport then added something that Herbert did not want to hear. He again urged Herbert to bring his family and the Concilium to America, and soon went further toward his own objective of Americanizing the Concilium. He mentioned such hopes to the foundations and National Research Council leaders when he sent them letters introducing Herbert. Despite Herbert's unease about Davenport's wish to have the Concilium transferred to America, the meeting with his friend Charlie lifted Herbert's spirits, as did his next stop.

Fleeting Hopes Dashed: You Really Can't Go Home Again

Herbert traveled to Connecticut to visit another part of his extended Quaker family, the Stablers. They had maintained the old family holdings at Greenwich. Herbert was pleased with the visit, and with how it connected him to his family's long history in America. The solid hold the Stablers had on their lives counterbalanced what Herbert heard about his immediate family while he was in Great Neck. He was so impressed that he penned a letter to Nina about the Stablers. It included a special section for Noel describing in detail the fascinating family burial ground that had the gravestone of Noel's great-great-great grandfather.[8]

Herbert left Greenwich for what he foresaw as a pleasant but very short stopover in Brooklyn. He was a in a hurry to get to a meeting in Boston. He envisioned only a few enjoyable hellos to his family and an overnight stay at his childhood home in Columbia Heights. But on his arrival in Brooklyn he encountered what Fannie hinted at—and more. Herbert quickly sensed that he could not count on his immediate family for help.

On the short visit he did not learn how much his family had declined. The bad news would come in spurts. However, it was already clear that the Fields were no longer affluent. Herbert had learned that his older half-brother Edward and his wife had moved a few blocks from their old home to the townhouse next-door to where his mother Lydia had lived with Hamilton at 106 Columbia Heights, but he did not know the reason. As soon as Herbert opened the door to number 104 he knew why. The family bedrock, Edward, had fallen on very hard times.

Edward had been his father Aaron's hope for the continuance of the family businesses and had not let Aaron down. Although he never attained the great riches of his father, he became a respected businessman because of his dry-goods import business. Edward did very well and opened an upscale New York City office. He managed the family's funds, even helping Herbert with his investments and with the North American accounts of the Concilium.

Unfortunately, the war had cut off his supplies. His business floundered. The warehouse was empty and he and his wife Lilla were reduced to living in one bedroom of the number 104 townhouse that Hamilton Easter Field once planned on using as a gallery and housing for his hoped-for live-in students. There never were many students and the townhouse had soon become a regular boardinghouse.

With Edward nearing sixty and overwrought about his finances, the burden of maintaining the family fell on Lilla. She had to care for seventeen boarders; she was the one who had to stoke the coal furnaces at number 104—as well as the furnace in Hamilton's house, number 106, where he and three "companions" lived.[9]

Lilla's boardinghouse was so crowded, and she was so busy, that Herbert had little time to talk with her or Edward. Herbert had to sleep next-door where Hamilton had left a servant to care for the house, but not the dirty furnace. Herbert knew there would be no chance for serious talk the next day because he had to leave early in the morning for an important meeting. Before going to sleep, Herbert wrote another letter to Nina. It had a noticeable tinge of bitterness. He said he had few details about the situation in Brooklyn, but it was it was clear that dear Lilla had been forced into a demeaning life, even though Hamilton could afford a summer place and someone to look after an empty house.

Hurtful Advice from George Parker, a Successful Old Friend

Herbert's next stop was very important, and emotionally trying. He knew he would have to listen to details about the family that he did not want to hear. As well, as he rode the train to Boston, then took a short car ride to Cambridge, he must have ruminated about his decision in 1893 to heed his information Calling rather than seek an academic position. He was on his way to stay with and ask for help from a former Harvard zoology program classmate who had become a highly successful academic and a relative.

That man, George Howard Parker, had continued working with Edward Mark after graduation, becoming himself a leading Harvard professor, and a world-renowned expert in zoology, marine biology, and the social consequences of evolution. He was also on his way to high office in the nation's leading science organizations. Herbert and Parker had been close since their youth, and Herbert had stayed in Parker's home many times. Herbert knew that Parker and his wife Lulu, Aunt Louisa Stabler's daughter, had done well financially, and that both were happy living in their spacious Cambridge home. Lulu, who was in the first graduating class at Columbia University's new women's college, Barnard, had intellectual interests and activities that compensated for the lack of children. She had gained a reputation for social and political involvements that went beyond the usual "ladies" activities of the era. By the mid-1920s, she even had a reputation among some Bostonians as being very radical.[10]

In contrast to Parker's shining career, Herbert's information service was near death. He was in debt and his investments were at risk. But Herbert voiced no bitterness while he stayed at the Parker house. They discussed family affairs (especially Nina and the children), the latest research in zoology, and the condition of science information. Then, the two men turned to the Concilium's needs. Soon what had begun as a friendly discussion became a disappointment, an almost hurtful one.

Parker very gently tried to convince Herbert to abandon his card system and to center the Concilium's revival on book-based indexing. Parker told Herbert that he was not alone in the belief that the grand Concilium experiment had not worked. Another of Herbert's friends who was visiting Cambridge soon joined Parker in giving that advice. T. H. Morgan, the famous Columbia University geneticist who pioneered the use of flies for research, also tried to convince Herbert to alter his approach.[11]

Both Parker and Morgan were sensitive to Herbert's feelings and softened their statements about the cards being too hard to file and too cumbersome to use. Fortunately, Herbert never saw Morgan's comments that he and some of his fellow scientists at the American Museum of Natural History wrote in the early 1920s. The museum had been the Concilium's most important client and was using the Dewey/UDC to classifying its own articles, yet many of its staff came to feel that cards and long classifications were a nuisance. Although more and more business and progressive government agencies were turning to card files, the museum staff concluded that Herbert's system was "in-Efficient."[12]

In 1919, after Parker and Morgan mentioned the book-from alternative, they were very careful about another difficulty with the Concilium's products. They delicately hinted at something that tore at a fundamental assumption of Herbert Field's Calling. They told Herbert that working scientists were not using the files and had little need for them.

Herbert Keeps the Faith

Although he was dismayed, Herbert was not swayed by that (or other friendly but negative) advice. Herbert remained committed to the card system and to the Dewey/UDC classifications, and he let Parker know it. Although Herbert was Parker's houseguest he argued with Parker, citing how, among other things, European scientists, who lived in a multilingual setting, were devoted to the Decimals. George Parker bowed to Herbert's convictions and finally agreed that the Zurich effort deserved to be revived.[13]

Parker decided to arrange a grand meeting of Boston area zoologists and biologists to discuss how to rescue Herbert's service. Parker and his wife also agreed that Herbert should return to stay with them as soon as the gathering's date was set. Herbert then graciously let the Parkers know how happy he was to be reunited with caring relatives, and to be received by others of Harvard's faculty. As he left on a short train ride to Maine's coast, Herbert reenergized his hopes for the Concilium's future, but he soon became deeply worried about his family's condition.

What Has Life Done with Hamilton Field—or—What Has Hamilton Done with His Life?

Herbert was on his way to visit his brother Hamilton. Herbert stepped off the train at Cape Neddick, Maine, and rode to one of the new farms Hamilton had recently purchased. Herbert was unaware of what had happened to his younger brother. Although it was October, Hamilton had remained in Maine because, so far, the weather had been mild. Despite Herbert's upbeat expectations about this reunion, it was more of a disappointment than the one with his stepbrother Edward.[14]

Several years before, Hamilton founded an arts colony in nearby Ogunquit, but he grew tired of that settlement. Ogunquit was too primitive, he told Herbert, so now he was seriously thinking of giving it to the townspeople. Besides, he said, a very special friend had settled in nearby Neddick. Hamilton was busy fixing up the new farm. He wanted it to be a more hospitable site than Ogunquit for entertaining his artistic friends; perhaps he had in mind even the soon-to-be revered modern artists Edward Hooper and Samuel Rothbart.

Among the recent visitors at Neddick was Hamilton's long-time protégé Robert Laurent. Laurent had just returned from France. In 1917, Laurent became the only one of Fields' circle to join the military. He had just become a U.S. citizen, and because he was of draft age and did not want to be in the army, he enlisted in the navy's new air corps. He served as an interpreter and carpenter at a base in France. To Hamilton's very great surprise, Laurent had returned to America with a wife, Mimi Caraes.[15]

Hamilton exhibited much pride while showing Herbert how he had just created a dam so his friends would have a lake where they could pose their nude models. Herbert was not much interested in that, so he tried to coax Hamilton into discussing such important matters as the status of the family home in Brooklyn and the pressing issue of the settlement of mother Lydia's estate. Lydia died almost two years before, but Hamilton and the other Fields had yet to tell Herbert anything about her will. Hamilton managed to turn attention away from those matters and to focus instead on discussions of how he was going to fix his battered Ford automobile and his progress, or lack of it, refurbishing the old run-down, but historic, farmhouse.

Herbert soon wanted to leave. The encounter was generating a feeling of bewilderment. What had happened to his brother? Urbane and sophisticated Hamilton now looked like a country bumpkin! Hamilton dressed in old overalls and was unkempt and dirty; he seemed unable to deal with the practicalities of life. So Herbert did not think it was the correct setting for pressing him about finances, and—because they now had little in common besides family matters—as soon as he found an excuse, Herbert left for Cambridge.

Dark Hints of Trouble Despite Bright News

The Parkers were ready for him on his return. Lulu had prepared a room. Parker had arranged a meeting with all the "best men" attending the zoological society meeting.

He also arranged great surprises. George scheduled a dinner with their mentor Edward Mark, a meeting with representatives of the World Peace Endowment, and a conference with the local chapter of the Friends of the League of Nations. Also, while Herbert awaited the special gathering about the Concilium, George managed to set up something very, very special: a meeting with Harvard's president, Abbott Lawrence Lowell.[16]

Lowell's ancestors were the ones who industrialized Massachusetts in the early nineteenth century, and whose great fortunes allowed Harvard to become a university. At the meeting, Lowell and Herbert discussed the politics of the U.S. Senate's debate over ratification of the League of Nations treaty. That motivated Herbert to send off another round of letters to Washington urging senators to approve the measure.

The conference with the zoological leaders came next. It became, in Herbert's view, a triumph. They approved his plans for reviving the Concilium and promised to support his search for financing. Herbert's follow-on letter to Nina was buoyant about the information service and, unexpectedly, the family finances. He told Nina he now felt better about a few of his old investments. But there were some dark hints in his letters.

Herbert was noticing some recent threatening trends in America and Europe. His writings contained sharp criticisms of the anarchists who, among other actions, had sent bombs to leading Americans a few months before, and, of Communists in Germany who continued to use violence and intimidation. He wrote that he approved the recent American roundup of suspicious aliens. Nina had a habit of reading Herbert's letters to the children, but these admonitions about radicalism, violence, and those who felt ideologically anointed may not have registered in their young minds.

Just as he was leaving Cambridge, Herbert received some very encouraging news from Charles Davenport. He said meetings with representatives of the major philanthropic organizations were being arranged, but told Herbert that it would take some time to confirm a schedule. Although he was disappointed, when he realized that he would likely have to remain in America longer than he had planned (and thought he could afford) Herbert decided to return to Brooklyn. He could not put off dealing with his and the family's financial affairs and with the apparently strained personal relations among its once harmonious members.

A Gloom Grows In Brooklyn

Herbert's joy over the possibility of gaining large sums for the Concilium soon turned into deep frustration and bitterness. One problem after another arose. They began to surface the day he returned to Brooklyn. The trip to Columbia Heights from the train station did not signal good fortune. A train hit Herbert. His clothes were ripped, his body bruised, and his skin torn. He downplayed the incident, writing to Nina and the children that it was a minor episode. But, he had to admit that he was hobbling around. However, after he was finally able to gain some details from the family, he did not try to hide his disgust about what his relatives had done to the family assets.

As soon as he reached number 104, Herbert began to draft proposals for the Concilium's grants, but his focus soon shifted to his personal finances. That required knowing what happened to the rest of the family's investments. While he awaited Hamilton's return from Maine, he confronted Edward. What happened to your business? What is the state of my and the Concilium's money that you were managing? The answers were difficult to accept because Herbert had always had the deepest faith in Edward's honesty.

Herbert was startled when Edward confessed that to try to save his business he had put up much of Herbert's investment fund and other property, and some of the Concilium's income, as collateral for loans. Much, if not all, he predicted, was going to be lost. He also admitted that the pressures on him had been so great that he had neglected much of the Concilium's bookkeeping. Edward even failed to submit the required federal taxes on Herbert's U.S. income. Herbert was frightened when he realized that little might be left to support Nina and the children.

It took two months to straighten out Edward's books. Then, Herbert had to warn Nina that, at best, he might be able to bring only twenty cents of each invested dollar back to Europe. Only time would tell if the other investments had been lost forever, he wrote. Herbert called upon his Quaker background to muster forgiveness for Edward, but he could not hide his frustrations.

While awaiting Hamilton's return, Herbert's Quietude was severely tested when he carefully looked over his parent's old home at number 106. He found the beloved and once-elegant homestead uncared for, a shambles. Although he realized he again had to sleep in number 106 because of his finances, Herbert almost lost his temper, and his free room, when Hamilton returned from the farm with a retinue of artistic companions. They were all slovenly dressed. Hamilton Easter Field arrived with a young Negro boy and Robert Laurent and his wife. They tramped into the house and immediately plopped down to listen to music, paying little attention to their guest. Herbert was outraged when he discovered that Hamilton had given Lydia's bedroom over to Robert and Mimi Laurent.[17]

Soon, there was a constant stream of more unusual people. There was much conversation among them about art and alternative cultures. Despite their proclaimed radicalism, there was little concerning the intense practical social issues of the times. The wrenching postwar unemployment, violent strikes, and race riots in the United States were not the focus of discussions at number 106, although one of the bloodiest episodes in American labor history had occurred just below the Heights a decade before.[18]

More than the strange conversations puzzled Herbert. Why had Hamilton surrounded himself with so many young men, and, while poor Edward and Lilla were living in near poverty, why was he spending so much money on the youths? Was Hamilton really putting one of those boys through college? Herbert could neither ask nor get answers. With Hamilton's habit of working in his pajamas from four in the morning to noon, then sleeping the rest of the day, Herbert had little chance for a mature discussion with his brother—even about important family matters such as Lydia's will. After Herbert made some inquiries around town, and when Hamilton could no longer find an excuse to keep silent about family matters, what Herbert learned brought him to a state of un-Quaker-like anger.

Before confronting Hamilton, Herbert visited his Haviland relatives. They told him that Lydia, for unstated reasons, decided not to follow common practice and leave shares to each of the children. She gave everything, everything, to Hamilton! Poor Edward and Fannie got nothing; Herbert got nothing. A bachelor without family responsibilities got it all. There was a bit more news that Herbert picked up on the visits. The Havilands informed him of just how much the chinaware businesses in America had suffered during the war. Any lingering hopes Herbert might have had of again tapping into the American Haviland fortunes were dead.

Herbert could not think of a valid reason for Lydia's decision, but he must have wondered if someone had wielded undue influence over an eighty-year-old widow. When Lilla told Herbert that Hamilton had convinced Lydia to buy up the several townhouses surrounding number 106, so that Hamilton could have adequate gallery space and room for a grandiose arts school campus that failed, Herbert became even more concerned about his younger brother and his friends.

After some time, Hamilton began to talk to Herbert about his relations with Lydia. He began by trying to soften the impact of Lydia's willing everything to him. He told Herbert that he did feel an obligation to give Herbert some share of Lydia's money, but perhaps only 20 percent. He then apologized. Unfortunately, he told the cash-strapped Herbert, he had no money to spare. A very disenchanted Herbert wrote Nina that given the way Hamilton was living there was no hope of ever receiving any funds from him. That belief was amplified after Herbert asked to look into Lydia's accounts.

Hamilton had not kept the files up to date, but what was there indicated that he had not stayed within precise legal boundaries. He had dipped into Lydia's not yet fully settled estate—and he could not explain how the money had been spent. Herbert confirmed that a part had been used up in acquiring the farm in Maine and by putting at least one young man through college. But he could not fathom what had happened to the yet undetermined remainder. That did not please Herbert. At the time, he was trying to put away enough money to buy a package of Christmas clothing for his children.

Herbert soon grew frustrated, then furious. He wrote Nina that Hamilton Easter was clearly irresponsible and self-centered. The situation was tense as Hamilton continued to be evasive about other details concerning the family's wealth and properties. To Herbert, Hamilton had become a dark, selfish, strange man. Just a few years later, Hamilton's s actions conformed to that judgment. When Hamilton died, he left everything to his protégé.

The Old Columbia Heights Disappears, and the Special Librarian Tests Herbert's Limits

Herbert was dismayed. He wanted to get away from Hamilton and Columbia Heights. Not only his brother, but also his old neighborhood, had changed. He wrote Nina that it made him shudder to realize that everything he had loved about the Heights was disappearing. Herbert was correct. The old gentle elite Quaker neighborhood was declining into a cluster of rooming houses and artists' colonies. One of the reasons was the result of the progress Herbert so admired. A subway line from New York City had opened in 1908, speeding a rush of commuters into the once exclusive neighborhood.

The changes to the neighborhood and numbers 104 and 106, and to Hamilton's behavior, were not all that was aggravating Herbert. In late 1919, another series of contacts again tested his patience and understanding.

Adolph Voge unexpectedly appeared at 106 Columbia Heights. Voge did not come to offer help to the Concilium; rather, he came to ask for assistance from Herbert. Voge had been employed at the Library of Congress, but after his World War I military duty, he did not return to Washington. He stayed in Paris well after the peace conference had ended, and much longer than any library or information business demanded. He was hospitalized there for a time, and he told one friend that that he stayed on until he was sure he could find a job in America. But, Herbert believed most of the delay was due to Voge's womanizing in France.[19]

On his return to America, Voge learned that Herbert was in Brooklyn. He quickly sought him out. During their meetings, Voge played on Herbert's sympathy by saying he had been to San Francisco to see his wife and children in hopes of reconciliation—although it had been at least three years since their separation. Adolph then tested Herbert's generosity. He asked for a recommendation for a job with a salary that was five times what Herbert had ever received from the Concilium.

If the salary was not enough to test Herbert's Quaker vow not to be envious, the nature of the job certainly was. The position was exciting and professionally cutting-edge, and it held out the promise of making Voge a major player in the library and information professions. The giant DuPont chemical company planned to establish an Intelligence Department with a large staff and a budget that allowed experimentation in information methods. It was a dream job for a Special Librarian especially if, as Voge believed, he could both shape and manage the entire bureau.

Voge placed Herbert in an awkward position. Herbert did not think his old friend was the type of man for such a job. He composed a very carefully phrased letter that would not alienate Voge, but that would lead a careful reader to be cautious. Whatever Herbert's letter contained, it did not prevent Voge from the opportunity to establish one of the most advanced library-information centers in corporate America, or in the world.

Herbert Revisits the "Better Sort" and Meets an "Angel in Armor"

After that, Herbert wanted to rush back to Europe. He certainly was more than ready to get out of the now-uncomfortable townhouse at number 106 and leave its bothersome guests behind. But he could not leave. He could not afford to. Only his continued faith that he would soon be able to board ship for Zurich with grant money saved his composure. He again pressed his academic friends to put him in contact with more people likely to give support. Charles Davenport and others responded and provided another list of influential contacts. Among those Herbert was told to contact was Bashford Dean, a respected member of New York City's social, artistic, and intellectual elite. He was also a noted zoologist and a professor at the college Herbert had turned down as a youth, Columbia.

Bashford Dean immediately impressed Herbert. Dean's lifestyle matched the culture of the old Columbia Heights, not Hamilton's number 106. Dean could afford to be a "gentleman" scientist. He drove Herbert out to the gorgeous Wave Hill House he was leasing on a Hudson River estate—so they could talk, and so that Herbert could see Bashford's magnificent collections of zoological specimens and medieval armor—and Herbert felt he was back with "his kind of people," or better. After Herbert's tour of the estate, and a long conversation about the future of science information, Herbert believed he found an "angel," someone with the interest and riches to save the Concilium.

Herbert hoped Dean's generosity would rescue his center without the necessity of humbling himself by submitting additional grant proposals. So, when Bashford offered to drive Herbert back to Columbia Heights it seemed the Concilium's future was assured. Herbert felt he found a true friend. Dean even made a courteous detour on the way. He stopped to give Herbert a tour of the farm in the city that he and his wife had donated so that New York's children would always have contact with farm life.

Herbert was impressed with the farm and the Dean family's civic generosity, but he should have been more sensitive to Dean's parting words, the early 1920s version of "I'll get back to you."[20]

11 To the Centers of Science and Philanthropy: Political Power in a New Information World

While awaiting word from Bashford Dean, Herbert Field learned that initial meetings had been arranged with the top leaders of the Carnegie interests, as well as with the science information group at the National Research Council. Unfortunately, the Carnegie's men wanted detailed descriptions of the Concilium project beforehand. Herbert stayed in his room at number 106, polishing his proposal while tuning out Hamilton and his artistic friends' strange conversations. Herbert sent it off to the Carnegie Peace Endowment, with hopes of immediate action because of its relevance to education and world peace.[1]

Then, he composed a letter of inquiry to the Rockefeller Foundation using Charles Davenport as a reference. Herbert may have played on his previous personal contacts with Mrs. Edith McCormick, John D. Rockefeller's daughter. Herbert had worked with her husband to convince the United States and England to allow food shipments to Switzerland during the war. Herbert also had previous social contacts with George Vincent, the president of the Rockefeller Foundation.

Conundrum: Can Zurich's Concilium Be an American Operation and a Worldwide Center?

Herbert was puzzled when he failed to receive an immediate reply from the Rockefeller representatives.[2] He was unaware the Rockefeller men had taken a step that almost led to the demise of the Concilium, and of Herbert's information career.

Instead of replying to Herbert, they sent a long confidential letter to Charles Davenport asking for his advice. Did scientists respect Herbert Field? Was the Concilium valuable? Was its work overlapping with others, such as that of the scientists (the Royal Society) in England? Was Herbert the right type of man to develop a new international science information system? Was he the man to manage a center perhaps even larger than the old Concilium?[3]

Without telling Herbert, or the science information advocates at the National Research Council, Davenport and his bibliographic committee at the American

Association for the Advancement of Science (AAAS) began a series of meetings. They forged an ambitious proposal to make the Concilium "the" center of a worldwide and comprehensive science and perhaps medical information network. Davenport also envisioned his committee becoming the heart of an international science information supervisory group.[4]

The Davenport plan was much, much more ambitious than Herbert's or the National Research Council's. Herbert's proposals to Carnegie and the Rockefellers were relatively modest ones aimed at continuing the Zurich center much as it had been before the war. He was asking for a subsidy, one-third of a million dollars, spread over several years. Part of his proposals, unlike Davenport's, was a promise, despite the earlier frictions, to continue to be cooperative with other information efforts, especially with those of Paul Otlet.[5]

In contrast, Davenport's team was contemplating a request for more than three million dollars for an institution that might eventually encompass all science information. And, there was no mention of Paul Otlet, the Royal Society, or any other international effort in the plans they soon dispatched to the Rockefeller men.

Additionally, Davenport's group thought in terms of an American operation, even if it was based in Zurich. For them, the United States had to be the heart of a huge postwar world science information system. Davenport believed the Rockefeller Foundation would approve that, and that it would put Herbert in charge of the entire project.

Many of the members of the AAAS committee were long-time admirers of the Concilium. For years, they had informed their colleagues of their faith in Herbert's work. Even Henry Osborn, who had become frustrated because of Herbert's behavior a decade before, was brought back to the Davenport team.

The AAAS group was powerful and well connected. Clark Wissler and R. W. Tower represented Osborn's American Museum of Natural History in New York. The man who was central to bringing the AAAS back to life in the late nineteenth century, James McKean Cattell, was also a group member. To make sure the committee did not seem to be a creature of just a New York clique, Davenport had added to the group Henry Ward of the University of Illinois Zoology Department and C. W. Andrews of the University of Chicago's marvelous John Crerar Library.[6]

Unfortunately, none of them, not even Davenport or Henry Osborn, kept Herbert or the men at the National Research Council aware of the nature of the proposal they were to submitting to the Rockefeller managers. Meanwhile, Herbert followed some of the previous advice from his closest friends, Davenport and Parker. They had urged him to contact the council directly. Fortunately, Herbert had some acquaintances and ex-colleagues there. His first step was to write to a fellow student of Edward Mark who had become an important member of the council, Robert Yerkes.[7]

Yerkes had become a force in America's new applied science. Although he was eight years younger than Herbert, he had already built a career as a Harvard professor, a founder of American experimental psychology, and as a leader in top-level professional organizations.

During World War I, he led a large and path-breaking council project to develop and administer intelligence and aptitude tests for American soldiers. It was a prime example of large-scale science and Efficiency. More than a million soldiers were tested. The results soon became the basis of many controversial evaluations of the relationship between intelligence and national and racial origins—and a handmaiden of eugenics.

Yerkes was another example of the "scientific" liberals of the era who were redefining American culture. While he supported the seemingly conservative sorting of humans based on intelligence testing, he led a radical long-term council study on human sexuality, one that challenged the established moral values of the time. After the testing project, Yerkes worked as an officer for the council and became a successful grantsman, gaining generous funding for his famed primate research center at Yale University.[8]

In 1919, Field's hopes were raised when he learned that one of the major assignments Yerkes had taken on for the council was to be the administrator of its Research Information Service. Herbert believed at the time that this council service was dedicated to aiding international science bibliography in only a general and hands-off way.

Yerkes responded to Herbert's letter. He told Herbert that an appointment at the council's temporary (and not very impressive) offices in the Smithsonian's "castle" in Washington was being arranged. Then, Herbert learned the council was moving to larger but still mundane quarters at 16th and M Streets while it awaited the construction of the brand-new multimillion dollar monumental quarters in the Foggy Bottom section of Washington, D.C.

Herbert did not know it, but the new building reflected the revised orientation and greatly magnified ambitions of the National Academy of Sciences and the National Research Council. Unlike the AAAS, which was oriented to individual scientists and their small projects, the council wanted to organize and lead an American science tailored to an age of large and coordinated projects. The council embodied the hope for Big Science in America. Its new visions were different from Herbert's.[9]

New Goals and Ambitions for the Academy, the Council, and the Service: Cosmos Club Connections

Herbert was excited when he received the council's invitation. He thought he had found new Concilium allies and, perhaps, sponsors. Then, he was surprised and somewhat disappointed to learn that he was not to visit with his old acquaintance, Yerkes. He had been scheduled to meet with a new group whose administrative head was a less famous but important man in science and science information, Charles Erwin McClung. McClung was making his scientific name through his evolution-related discovery of the role of chromosomes in biological inheritance, and through his efforts to turn Philadelphia's relatively small Wistar Institute into a major medical-biological research center.[10]

McClung had become associated with the council's Research Information Service during the war when its main assignment was to provide the government with information about scientific resources and researchers. It had focused on mustering scientific manpower for the government, as well as on gathering intelligence about scientific-technical advances in foreign countries. It had even arranged to have the position of "scientific attaché" created by the State Department. The department posted its scientists in foreign nations to gather science information and to establish contact with laboratories and manufacturers. The council's representative worked with America's other intelligence gatherers. Its attachés were forerunners of the CIA's Cold War science agents.

By 1919, the Research Information Service project was changing. Near the end of the war, Robert Yerkes had been asked to review the council's information policies. One of Yerke's first contacts was with Charles McClung. His focus was on something very different from the service's old operations.

Along with a botanist from Cornell University, Jacob Richard Schramm, McClung sought the creation of an American-based and council-driven international information system for biology—one that was to be housed at his Wistar center. He already had begun efforts to build a new cooperative international biological society that would create a huge network of volunteers to support the proposed information system. McClung and Schramm now hoped for a more inclusive system.

Herbert did not know of McClung's ambitions before the November 1919 meeting. After he returned to Europe in 1920, Herbert still did not comprehend the scope and implications of McClung's information objectives.[11]

No Longer a "Gentleman's" World: The New Culture of Science and Philanthropy

Beyond that, Herbert also never understood how much American science and philanthropy had changed since he had settled in Zurich some two decades before. The days were gone when a gentleman's word was sufficient for obtaining a grant; when a scholar needed only a few influential friends to vet a project; when articles were accepted without anonymous peer reviewing; and, when competition in science appeared to be confined to the research bench.

The changes and conflicts he would encounter after meeting with McClung's group bewildered Herbert. The now fifty-year-old man who thought he had established himself as an unassailable figure in world science faced a new reality. Since he was a boy, Herbert knew first-hand of charitable reform and the drive to expand science. But what was happening in postwar America was strange to him.

Herbert was a part of the old "small" system of science. Before World War I, the most powerful component in the United States was the National Academy of Sciences. It was born at the beginning of the Civil War as a special organization to bring the expertise of America's best scientific men to the service of the nation. The academy continued after the conflict ended, but not as a democratic or populist body. It was a group with member-appointed, rather than popularly elected, inductees. Members chose only those who were well established in their fields and who represented the cream of the new research universities—as well as the best of their faculty members. The "best men" selected the "best men" for the academy. It was not an organization for scholars "on the make," or for those who labored in the small old-fashioned colleges.[12]

Despite its talented and influential members, the academy accomplished little after the Civil War. The government rarely called upon it for advice and it did not have funds needed to sponsor or direct American science. That changed when Woodrow Wilson called on the academy for assistance in preparing for World War I. It responded by creating an operational branch, the National Research Council. Many new academics were brought to Washington to staff the "NRC."

The council's job during the war was to coordinate scientific research and efficiently use it for national purposes. Its goal was to integrate the scientific work of the largest academic, private, corporate, and government research institutions. While Thomas Edison's Inventors Council solicited help from those who some critics called "tinkerers," and while the AAAS looked to the work of individual scientists, the NRC favored established professors who worked with others and with corporate research centers. It doled out the government research funds it administered only to those willing to work on council-approved cooperative projects.[13]

The National Research Council gradually moved beyond being a coordinator. Significantly, it went from working with the existing older scientific societies to attempting to guide them. Such a new type of "coordinating" work was a thrilling experience for the many young men who served with the council. They felt powerful and thought they were bringing Efficiency to America's scientific infrastructure. They were able to convince the nation's experts to work together on important large-scale projects, such as submarine detection and chemical warfare. That type of collaborative, yet top-down directed work, seemed much more effective than the old traditional academic approach of isolated researchers laboring on individual tasks.

The council's young scientists desired to continue their campaign after the war. They were determined to bring order to American science and, as did a new breed of educational reformers, turn American universities into competitors to Europe's best research centers. They were resourceful, and they were pushing America toward achieving both goals. By the late 1930s, the United States became a powerful competitor in several academic fields and had enough strong graduate programs to begin to avoid having to send its brightest students abroad for their advanced degrees.

That was more than a decade away. To the energetic council's leaders of 1919, there was much to do and to overcome. American science badly needed reordering, they argued. They emphasized that science was not and could not be a domain of individuals pursuing abstract ideas. They argued that science had many of the aspects of an industry, in which each scientist made a small contribution to a final result, and that it now required expert-defined organization.[14]

The Struggle to Develop a New World of Science

Not all researchers agreed with the National Research Council's views and methods. Despite the council's members being appointed as representatives of the various old scientific societies, there were growing complaints that the council was doing more than facilitating older organizations' work. As the war was ending, some scientists feared that the council was about to undermine the independent discipline-based societies. However, the council's' advocates won President Wilson's approval and the government granted it permanent status.[15]

In 1919, the council began an attempt to bring the same or more degrees of coordination over American peacetime science as it had achieved during World War I. Its leaders were not shy. As soon as the war ended, they published a very readable work extolling their organization's accomplishments and its desired future, *The New World of Science: Its Development during the War*. Their public relations campaign continued.[16]

In a move of later importance for the history of science information, the NRC joined the older science umbrella group, the AAAS, in sponsoring the creation of a publicity-lobbying organization for science, the Science Service. The council felt no qualms about using money from the Scripps newspaper fortune for the promotional enterprise. However, the council's emphasis was on reorganizing America's science establishment, not just making it popular.[17]

Restructuring American science was not easy. To the disappointment of many council leaders, it was difficult to continue the cooperation with industry and government scientists after the war. In response, the council's attention shifted to reforming and expanding university-based academic science and its institutions. But even the narrowed focus called for hard-to-get funding. Only a few institutions, such as the Massachusetts Institute of Technology (MIT), were able to forge new links with industry, and the flow of funds from the military stopped at the war's end.

One consequence of the scramble for funding was very important to Herbert Field's career. The council's members had to shift from being relatively passive recipients to being aggressive fundraisers. With the cutoff of government money, and little likelihood of business contributions, there were just a few places to turn. The targets became the only two large foundations that had shown indications of possibly supporting science. The council's men determined they had to convince the Rockefeller and Carnegie foundations' managers to support the council and its plans for American and international academia.[18]

The council wanted funds for itself and, importantly, to play a dominant role in determining what men and projects received any science-related grants from the foundations. They claimed that only central control could eliminate the growing wasteful competition for research funding. The council's leaders sought to establish themselves as "the" science advisory body to the new giant philanthropies. In addition, a principal goal of the postwar council was to convince donors that large-scale projects could and should be handled by university-based research teams—not by independent institutes or corporate research centers. Following from that, a central aim was to provide America's researchers with tools their universities could not finance with their own funds. The council lobbied for the huge Mathematical Tables projects, the construction of costly measuring and computing machines, and various types of science-information products.

The university focus was not all of the council's strategy. It wanted to conduct its own research. It leaders had grand ambitions for it becoming a high-level think-tank and soon secured grants for projects on problems ranging from sex and race issues to ecology.[19]

Important for Herbert Field, the council's members were first in line when the new giant philanthropies hinted they would support scientific research on a regular basis. Their ambitious crusade got off to a good start. Among the first council successes were a multimillion-dollar grant from the Carnegie Foundation for the construction of its impressive building, and funding for council-administered grants to establish and direct fellowships for advanced study by the brightest-of-the-brightest young American scholars.[20]

Designs on the Council: The McClung Plan

Herbert Field was not a member of the new National Academy of Sciences / National Research Council group, but he had joined earlier leading professional organizations in the sciences. He had always been a member of the American intellectual elite and had benefited from the small older philanthropies that supported individual scholars. The visit to the council's offices in November 1919 was his first introduction into the new types of scientific organizations, academic reform initiatives, and philanthropic policymaking—all of a nature and scale that he and few others could have imagined in the nineteenth century.

For Herbert, the trip to Washington was a burdensome financial investment. On all of his previous stops, he stayed at the home of one of his relatives, but he had none in the capital city. Fortunately, his colleagues arranged accommodations for him at the center of Washington's intellectual life, the Cosmos Club. When Herbert settled in for his visit, the club hosted the most important liberal American leaders and writers. Of course, Herbert's friend Charles Davenport was a member.[21]

Herbert was impressed with the caliber of the Cosmo's members and, at first, by those he encountered at the council. Its staff were of the same breed as the Cosmos and academy's men—but younger and with a new level of vibrancy. The term "academic entrepreneur" was not yet in fashion, but it fit their approach to making America's universities and research equal those of Europe. That meant the need for unprecedented amounts of money and for breaking with old academic customs and, perhaps, civilities.

Some older academics found it questionable that the council's members were not afraid of diverting attention away from "pure" and theoretical research or to taking a political stance, including support for America's entry into World War I. The council's men also welcomed becoming involved with applied research, if that was what was needed to attract funding. Importantly, they were not blue-sky purists. For them, science was more than thought and classification. It was an empirically based, laborious, experimental, and expensive activity.

The original leaders of the council, such as the outstanding physicists Robert A. Milliken and George Ellery Hale, were institution builders as well as researchers. They looked to the future and brought a new generation of academic policy makers into their fold. The council's venture into anti-submarine technologies, for example, employed MIT's young engineering professor Vannevar Bush. He went on to become perhaps the most important science policymaker in American history, as well as a leader in mechanizing scientific information systems and in demands for radical reform of information methods.[22]

The council's involvement with war research, and thus its break with academic neutrality, had not been to the liking of some scientists. For example, the German immigrant and Columbia University anthropologist, Franz Boas, resigned from his council post because he thought its work was blatantly anti-German, and because war-related research ran against his ultraliberal principles.[23]

Despite complaints, the council's leaders had gone far toward achieving their goals. By late 1919, the council was close to realizing its hope to become the one filter between science and the foundations. At minimum, a negative recommendation by the council meant an unkindly evaluation from foundations' program managers. But with men like Yerkes on his side, Herbert was sure he could pass the test with the National Academy of Sciences and the National Research Council. He also expected his visit would lead to invaluable advice as to how to convince the foundations to rescue the Concilium.

Herbert had prepared a presentation that fit what he knew of the council's information ambitions. He came to Washington armed with an outline that was tailored to what he expected to find, a welcoming and receptive group. Unfortunately, Herbert did not know how much Charles McClung and his allies had changed and expanded the visions of Yerkes's initial postwar information group. In coordination with the various branches of the council, such as its medical and agricultural divisions, the council's new Research Information Service leaders had constructed an ambitious plan to deepen its role in American and, importantly, world science. Its information goals went far beyond what they had been during the war.[24]

The McClung design had several components. There was some amplification of the older service's tasks. Yearly lists of all American research laboratories, researchers in every field, research in progress, and newly awarded PhD degrees were to be published. The council's staff was also going to: build a library; publish a comprehensive list of all bibliographies and guides to scientific literature; keep an updated listing of scientific societies and their meeting times and places; build a large file of all foreign research reports solicited by the council (already some twelve thousand items); and ensure the publication of important papers that would not normally be available in established journals.

Then came the great additions to the council's information agenda. They were of extreme importance to the Concilium's future. First, the council's Research Information Service was determined to take the lead in developing and imposing international standards for scientific abstracting and indexing. That was a first step toward the ambition of the council to be the leader in establishing and coordinating abstract journals and indexes in every field of science—for the entire world.

High on its agenda, and critical for Herbert, was the council's intention to establish itself as "the" leader in rebuilding the entire international information infrastructure, and to do it in coordination with the American internationalists who had committed to developing a cooperative world system for all-important aspects of society.

Significantly, and despite what Herbert had been led to believe, the new Research Service was not thinking of giving a high priority to bibliographic services such as his. The initial McClung plans made no mention of the Concilium or Paul Otlet.[25]

Frustrated by Philanthropy: Funding Science Information

There was a problem facing all of the National Research Council's information goals. It frustrated men like McClung. He and his colleagues had found little money for their projects, and they had become convinced that philanthropies such as the Carnegie and Rockefeller institutions had little or no interest in science information. In reaction, the council's men were not ready to share any of the funds that might become available. The McClung group was also frustrated because it had few places to turn for financing.

McClung knew that government funding was not an alternative. Although the United States had long before established and supported information services for agriculture and medicine, the council's leaders believed the time was not ready for any acceptable form of subsidization of information for academic scientists. Because of that, and the disappointing news about the foundations, many in the council were extremely apprehensive about anyone with a competitive vision for the future of science information.[26]

Herbert did not know it, but the council's men had feared him as a possible competitor. Fortunately, he assuaged the fears during his presentation in Washington. In late 1919, after the conferences with Herbert, the council's leaders believed he had agreed that the Concilium would play a part, but only a subordinate one, in the broader systems the council was hoping to establish and direct. They also believed he would not become a funding competitor. He had told them about the very modest request to the Carnegie Peace Endowment. It did not seem a threat.[27]

There was another threat, but the council's men, and Herbert, had no information about it. Extremely important for the Concilium's future, neither Herbert nor the council had heard of Charles Davenport's all-encompassing Concilium-centered Rockefeller proposal, and the council continued to believe the Rockefeller planners had no interest in any type of science information project.

At the time, Herbert Field appeared to be a valuable ally for the National Research Council.[28] A coordinated relationship with the Concilium seemed a way to gain a foothold in Europe, and a way to allow the Research Service to maintain its role in international science once postwar reconstruction was over. There was a consciously shared goal. Both the service and Herbert agreed that they had to make sure Germany could never again dominate science information. Some council members took that resolve to the extreme. A few even supported France's demand for the banishment of German scientists from international associations.[29]

But there were others in the council who were true internationalists, as was Herbert. Some of their visions conflicted with the Research Service's goals. Those others were attempting to create a democratic worldwide body for the coordination of science funding, research, and bibliography. For them, the council should only be a cooperating part of an even larger and truly international organization. At the time, they were helping to change the older International Research Council from a hastily constructed wartime alliance among Britain, France, and the United States into a fully international association. They hoped that it would replace the previous world scientific coordinating groups that had, they claimed, been hapless captives of German dictates.[30]

Some of the council's top leaders and foundation policymakers, such as the Rockefeller and Carnegie foundation-managing Flexner brothers, had played central roles in reforming that International Research Council. They hoped trust and equality among nations would be the foundation of its success. They pressured the National Research Council's activists to avoid anything that might indicate America's intention to control the international council or the new subject-field unions it was to establish. Significantly, that list of new organizations included a proposed International Union of Bibliography and Documentation.[31]

Cooperation in the Abstract(s), Harmony for a Time

Conflicts over the international organization had not surfaced when Herbert Field met with the McClung group. All their meetings had been harmonious. Herbert thought he had found information soul mates. And, for a while, so did the council members think they'd found theirs in Herbert. After Herbert sent McClung a copy of his modest Carnegie proposal, the council signaled that Herbert posed no danger to their plan to soon establish what had the potential to be a serious threat to the Concilium, a new international abstracting service for biology. Herbert had indicated that he would cooperate with the "abstracts."

Herbert also felt secure. After the long November meetings, some lasting from morning through dinner, Herbert wrote Nina that he was not only pleased but also excited because of the warm reception he had received.[32]

Unfortunately for the Concilium's future, Herbert was unaware of the ways of the new science enterprise in America. He had been lulled into a false sense of optimism and security after his meeting with Bashford Dean, and now, after the November meetings, Herbert was not sensitive enough to interpret the real meaning of hints that he should stay in touch. Herbert had also failed to notice how important one type of information service, abstracting, was to the council. He thought their proposed abstract service would be just another of the type of publication that he had dealt with for a quarter-century. He also did not foresee the implications of his agreeing to have the Concilium share in the creation of the new abstracting organization. Nor did he appreciate that his work was regarded as only a minor part of a larger strategy.

Herbert's excitement about the next items on his Washington agenda may have caused his lapse in awareness. First, he was about to meet with one of the most influential men in America, and to do so in one of the nation's most remarkable structures. The Carnegie Institution of Washington was a shrine for science, and its leader was one of the greatest men in the nation.

Courting Elihu Root and Carnegie's Endowment for Peace

Elihu Root was a Wall Street lawyer (representing J. P. Morgan), a diplomat, internationalist, administrative genius, foundation director, and a confidante of presidents and foreign leaders. He modernized the War Department. He was also a path-breaking secretary of state, a special envoy to the new Soviet Russia, a U.S. senator, and a prime mover among those who guided Woodrow Wilson in the struggle to restructure Europe and create the League of Nations. Root's labors had been critical to the establishment of many institutions for international peace and arbitration, efforts that won him a Nobel Peace Prize.

Root was not a lone wolf, however. He had been a central member of Wilson's Inquiry group, which was how Herbert first came in contact with him, and a founding member of the group that shaped American foreign policy into the twenty-first century, the liberal-internationalist Council on Foreign Relations. It was the permanent extension of the Inquiry and an East Coast organization that paid little deference to America's long tradition of staying aloof from European entanglements—a tradition some called "isolationism."[33]

In 1910, Andrew Carnegie chose Elihu Root to be the first leader of his massively funded Endowment for International Peace. Root helped establish the endowment's influence by purchasing an imposing building at Two Jackson Place in Washington, DC, for the peace foundation's headquarters. It was just around the corner from the White House and near the Cosmos Club.

The managers of the some two-dozen other Carnegie-sponsored programs called on Root for advice on more than the peace initiatives. In 1919, although in his mid-seventies, Root accepted the call to preside over all the Carnegie philanthropic interests, including the Washington center for the support of high science, the Carnegie Institution, and the programs dealing with libraries, international education, and higher education in America.[34]

When Herbert met with Root, he knew his grant proposals touched on several Carnegie interests, and he thought that Root would guide him through the Carnegie organizational maze. Herbert was warmly received and carefully listened to. Root provided counsel and encouragement about the Concilium proposal Herbert had just submitted, indicating that it did fit with the Carnegie initiatives. Root indicated that he approved other of Field's ideas. He liked Herbert's thoughts about creating American institutes to familiarize Europe with American culture. Root also agreed with Field that the United States should help European libraries to catch up on the books they had been unable to acquire during the war. Root soon arranged to have hundreds of volumes sent to Zurich's main library.[35]

Diplomatic Connections, Career Change, Opportunity Overlooked—and a Confrontation

Herbert's next visits in Washington also were pleasant experiences. He contacted Edward House to say hello and to learn about American political trends. He told Herbert more about Wilson's mysterious illness and described what the chances were for America's entry into the League of Nations. After that, in response to the letter from Secretary Robert Lansing, Herbert was on his way to the State Department. Field took advantage of Secretary Lansing's invitation to bring his ideas to him because he wanted to put his expertise concerning Central and Eastern Europe to the service of the department. Lansing's staff gave him a grand tour of the offices of the department's leaders. Such attention was flattering, but it led to another narrow escape for the Concilium. Herbert was asked to work for the State Department.[36]

It did not take much prompting to entice Herbert to consider applying for the surething job. Herbert's eloquent detailing of his plans for the American Institute, a precursor of the post–World War II United States Information Agency, was one of the reasons for the opportunity. As well, the responses to his many letters to U.S. senators, and his contacts with the leaders of organizations such as the World Peace Movement, made Herbert think he might have a future as a professional diplomat rather than a bibliographer.[37]

Herbert again came near abandoning his information career. Only at the last minute did he turn down the offer. Thinking of the long-term needs of Nina and his children, and the chance that he still might rescue the Concilium, he informed his diplomatic friends that he was putting the offer on hold. He told a Swiss banking friend, who had broached the possibility of Herbert becoming a bank officer, to keep the position open while he explored the future of the family's—and the Concilium's—finances. Herbert did not cut his ties to diplomacy or important diplomats, however. After the face-to-face-meeting with Secretary Lansing, Herbert visited Christian Herter, a young department officer he had met earlier in Europe. The young man had not yet fully made his impressive mark on American life, but Herbert had sensed he would become a major force in government.

Christian Herter and Field were much alike, although Herbert was much older. Both were from well-to-do families, were highly educated, and were, in many, ways, European. Herter's father was an American expatriate, an artist who resided in Paris until it was time for Herter's higher schooling. He was enrolled in a private school and then attended Harvard. He, like Herbert, graduated before he was twenty. He soon had a State Department posting in wartime Europe where he met young Allen Dulles, as well as Herbert. Herter then served with the Versailles delegation and returned to the United States as a man deeply committed to internationalism and peace.[38]

Herter had great talents and he was a congenial host who could reminisce with Herbert. He said Herbert could call on him for Concilium recommendations even though he was about to serve as an assistant to Herbert Hoover. Herbert then received more recognition in the diplomatic sphere. He was invited to the British Embassy where he was entertained and introduced to such greats as Lord Grey, the ambassador, and the leaders of the Swiss delegation to the United States.

Thrilled but exhausted by his rounds of meetings, Herbert packed to leave the Cosmos Club and Washington to attend a new set of conferences. Although he remained short of money, being able to endure the November weather only because of his borrowed overcoat, he was more optimistic about the Concilium and his family's future. But, he was not so sure about his relationship with his brother Hamilton.

Georgetown University: Future Impacts for Noel Field

There was an important thing Herbert did not do during his November stay in Washington. He failed to take a short trip to a then-small Jesuit college in the capital, Georgetown. His failure to visit Georgetown University limited his perspective on world events. A young priest named Edmund Walsh, whom Herbert had probably encountered at the peace conference in Versailles, had established a foreign-service training school and, later, a policy center that, among its other contributions, was a leader in the post–World War II development of methods for the automatic translation of Soviet scientific documents. Under Walsh's guidance, Georgetown evolved into one of America's most influential international policy and intelligence institutions. It became important to shaping American Cold War policies, and Senator Joseph McCarthy's efforts, which contributed to Noel and Hermann Field's bizarre encounters with the Soviet Union in the 1940s. Georgetown even housed centers to train Cold War agents.[39]

By 1919, Father Edmund A. Walsh had made two very long trips into the new Red Russia. From his direct experiences, and his Vatican and other sources, he knew of the repressive measures Lenin took against all religions. If Herbert had talked with Father Walsh, his own rather negative view of the results of European radicalism would have been greatly amplified. And Herbert might have given even more profound cautions to his children about extremism.

Back to the Heights: Bad Times for Herbert's Temper, Emotions, and Health

Before his next round of meetings, Herbert needed to stop at the Heights for clothes, mail, and rest. On his return to Brooklyn in late November 1919, he was again forced to live at 106 Columbia Heights. He had no choice. His travel budget was nearly exhausted. Fortunately for Herbert's finances, and his emotions, members of his extended family frequently took him on short auto trips and invited him to their homes for dinner. But even with those diversions, it was even less pleasant than before at number 106. Hamilton was making it clear, although slyly, that it was his house and that Herbert was just a visitor. With that kind of treatment, and suffering from two months of frustration because of Hamilton's dodging of questions about Lydia's estate, Herbert became exasperated.[40]

There was a confrontation. It was one that forced Hamilton to finally reveal what detailed financial facts he had. Herbert realized his hunches were correct, but found they were underestimates of Hamilton's squandering of the family's money. Hamilton Easter Field estimated that he had withdrawn more than $500,000 from Lydia's accounts in the last year alone! The best estimate Hamilton could give was that what remained in Lydia's estate was worth less than $60,000. Hamilton made no indication he would share that with the rest of the family; he did not give Herbert any assurances that he would ever provide him with the once-promised 20 percent share of the Fields' wealth. He did not inform Herbert about one very important fact.

He did not let Herbert know the old townhomes contained many, many thousands of dollars' worth of saleable art that Hamilton purchased during his career as a dealer. He could have put those paintings and artifacts on the market to rescue the many family members who had fallen on hard times rather than purchase farms and support his friends. Even without knowing of the valuable artworks, the stress caused by the realization of Hamilton's irresponsibility led to another resurgence of Herbert's migraine-like maladies.[41]

Receiving news that Nina had to have a serious and costly operation intensified Herbert's painful headaches. Herbert was severely depressed because he could not return to Europe to be at her bedside. He went through a personal hell, pondering whether to return immediately to care for her and the children while she recovered. He decided to stay in America, but there was a price for that. Herbert sank into a period of unusually deep melancholy in December.[42]

As at many other times in his life, Herbert's circumstances had left him susceptible to a series of emotional lows—and extreme highs. Despite the many friendly receptions he had received in the last months, he saw little hope for the Concilium. He wrote Nina that he thought his life's work was doomed and so was his career as a bibliographer. He bemoaned the state of the family finances and again considered beginning a new career. He was uncertain as to what it might be.

He began to imagine the very worst. The shock caused by Nina's health also led Herbert to realize the seriousness of his situation. If the Concilium could not meet its debts, all his stock would become valueless, and his other assets would be at grave risk. Only his Quietism saved him.

Herbert dusted off his emotions and found the strength to ignore the declining neighborhood and Hamilton's dark, nonconformist friends. He wrote the recovering Nina that he now had forgiveness in his heart for both his brothers, Hamilton and Edward, despite their negligent actions. And, he emphasized, he would not give up his crusade for the Zurich service.

On Behalf of the Concilium, Charlie Plans a Rescue and Herbert Appeals to Abraham Flexner

One reason for Herbert's rebound was a brief message from Charles Davenport. Davenport, while still not revealing the nature of the great information proposal he had drawn up, wrote Herbert that the AAAS committee was going to submit a very positive report on the Concilium to the Rockefeller administrators. Davenport gave no details, only stating that Herbert's salary was to be very generous. Davenport did not reveal that his group was recommending a subsidy for the Concilium ten times what Herbert had asked from the Carnegie Foundation. Nor did he say that the committee was describing the Concilium as "the" center for world scientific information. Importantly, considering what soon became a struggle that nearly ended the Concilium, Davenport failed to tell Herbert that the AAAS was submitting its proposal without any communication with the men planning the National Research Council's role in the new science information world.

Unaware of the possible repercussions of Davenport's actions, Herbert began another round of contacts with some of the nation's most influential people and institutions in America's liberal establishment. His first stop in a series of New York City contacts was with a true mover-shaker in American higher education, Abraham Flexner of the Rockefeller and Carnegie foundations. Herbert had finally received a response from the Rockefeller men. It included an invitation to meet with Flexner for a general discussion. Significantly, Davenport's grand proposal had not yet reached the foundation. Although they had met before only briefly, when the foundation refused Herbert a small grant in 1907, and although they came from different family backgrounds, Herbert and Flexner shared many things—including the advocacy of a rational, liberal social order.[43]

12 More Conflicts between Old and New Science

The Flexner brothers, Abraham and Simon, were critically important to the development of modern American medicine and higher education. Sons of middle-class Jewish-immigrant parents who settled in Kentucky, they became influential progressive reformers. Simon, born in 1863, chose medicine for a career. Abraham, born in 1866, selected education. It took some time for Simon to find his life's path, but Abraham was a quick starter. He completed his undergraduate work at Baltimore's new Johns Hopkins University at, amazingly, age nineteen. Then, he arranged to have his older brother enter its medical school. With that, Simon found his way and he quickly became a professor, a world-renowned university bacteriologist, and head of the Rockefeller Institute for medical research.[1]

Simon gained international standing in his profession, and he married into the East Coast's economic-social elite. He wed the daughter of the aristocratic Baltimore Quaker who founded both Johns Hopkins Medical School and Bryn Mawr College. Bryn Mawr, begun as an Orthodox Quaker yet true college-level institution for women, became one of the elite, ultra-liberal Seven Sister eastern U.S. women's colleges. The "best" young ladies went to the school and left it as committed advocates for modern individualism. Simon's sister-in-law became the second president of the influential school. Simon's marriage thus brought him into the worlds of extreme feminism, liberal social reform, the nation's Quaker elite, and even high art and music.[2]

Meanwhile, Simon's brother Abraham took a different path. That was partly because he did something unusual for a Hopkins graduate. Although Hopkins was the first American college founded on the model of the German research university, and although its graduates had a leg up on others in gaining high-status employment, Abraham returned to Louisville, Kentucky, to become a high-school instructor. The new learning he encountered at Hopkins quickly led him to create a soon-to-be-famous college preparatory school. Mr. Flexner's School, as it was called, relied upon the teachings of the latest scientific educational-psychology theories. The school became renowned for reshaping troublesome students who came from wealthy families. Their parents were willing and able to pay the high tuition that ensured admission for their sons to Ivy League colleges.

Abraham did not remain a local schoolmaster for long. One reason was that he also married into high society, at least the South's. He had tutored a young lady from a family with a highly distinguished lineage. She later gained admission to another of the elite Seven Sisters, Vassar College in New York. After graduation, and a stint as a theatrical reporter in New York City, in 1898 Anne Crawford consented to marry Abraham. She brought social status and, eventually, her family's money.

Although living happily in the South, something prompted the couple to change their lives. After twenty years of success in Louisville, Abraham closed his school in 1904. He then enrolled in Harvard University's new graduate education program. The experience was, to his surprise, not rewarding. He left Harvard after only one year without an advanced degree because, he said, he was disappointed in his education professors. Then, as had Herbert Field, Abraham began using his accumulated savings for a crusade. His Calling was for the reform of graduate education in America.

Abraham Flexner: Setting New Education Standards in the Names of Reason, Science, and Internationalism

With the financial help of his brother and major philanthropists, Abraham toured Europe to study higher education. His writings, which condemned much of American education, came to the notice of the leaders of the Carnegie foundations who were already determined to modernize America's college system. Their goals were like those of the National Research Council: to turn higher education into an efficient and well-ordered system. Carnegie's men asked Abraham to create a ranking system (later to become the Carnegie Classifications) that helped channel funds and recognition only to, perhaps unintentionally, already rich and modern research universities. All of the favored institutions would be what the large-scale universities on the East Coast claimed they already were: nonsectarian. Carnegie's men believed the old liberal arts colleges should be relegated to secondary status, if not ignored, especially so if they were tinged with religious sectarianism. The term "sectarian" seemed not to apply to those institutions that had, for example, liberal Quaker or Unitarian beliefs in their backgrounds. The Carnegie higher-education efforts, including its funding policies, played a key role in shaping faculty members' career expectations, in speeding professionalization, and in deepening the requirements for accomplishments in research.

Carnegie soon called on Abraham Flexner for another task, to reshape medical education. By the time Herbert met with him, Abraham was a major executive in the Carnegie and Rockefeller foundations, serving as a critical decision maker over millions of dollars devoted to educational programs. Vital for the Concilium, as World War I was ending, was the decision of the Rockefeller as well as the Carnegie managers to play a major role in rebuilding Europe's scientific and educational systems. Their policies for Europe followed the same philosophy as their American commitments: expertise, efficiency, and order marked by a deep obligation to internationalism.

Meanwhile, Abraham and Anne Flexner became part of the liberal social-intellectual set of New York City. They enjoyed the arts and even took a prominent place in a great 1915 social event, the march for women's suffrage. Of course, they were internationalists, they were evolutionists, and they were self-proclaimed social progressives. Anne was very busy as a Progressive and Feminist. She also carefully prepared her children for entry into exclusive colleges such as the now-liberalized Quaker's Swarthmore.

She also helped Abraham with plans for what was created in 1930, an educational institution where the greatest scholars of the world were allowed to do what they wished, without any duties or imposed expectations about productivity. That Institute for Advanced Study, at Princeton University, was unique in America. Ironically, as Abraham began heading that bulwark of liberalism, his daughter was beginning a long career as a communist author. But, in the 1919, the Flexner family's link to communism and its relationship to the lives of Herbert's children were in the future.[3]

The December 1919 New York City meeting between Herbert Field and Abraham Flexner seemed foreordained to be pleasant and positive. How could it not be? Both were trying to fulfill the promises of reason, science, and internationalism. Herbert's plans for the revival of the Concilium even fit with the needs of Abraham's brother Simon, who was working for improved medical and public health–related bibliographic services because he viewed them as an integral part of the rebuilding of Europe.[4]

Abraham assured Herbert that the foundation was on his side. However, he asked that Herbert be patient while the foundation's experts had time to look over the report Davenport's group was about to submit. Then, Flexner dismissed a problem that worried Herbert. When Herbert admitted that he had just been told that the Carnegie Foundation was likely to provide a grant, Abraham told him that the Rockefeller boards would cooperate with the Carnegie interests and that dual submissions were acceptable. That eased Herbert's fear that he had committed a breach of new philanthropic etiquette.

However, something of importance went unsaid during the meeting. Herbert did not tell Flexner of a rather emotional letter he just sent to Charles McClung. *Please support my grant proposals because I cannot go on with palliative measures*, he wrote, in essence. Flexner could have interpreted that as an insulting prediction his foundation was not going to appreciate the Concilium's worth.[5]

One Omission Compounds Others

Herbert had waited rather patiently for word from the National Research Council, but there was no reassuring reply. There was not even a message that spelled out the details of possible council-Concilium cooperation. And, McClung delayed informing Herbert that he and his fellow University of Pennsylvania faculty member, Milton Greenman, were becoming the driving force behind a movement that undermined the Concilium's position. McClung also failed to inform Herbert that circulating within the council were highly critical reviews of his modest Carnegie grant proposal.[6]

Abraham Flexner, in contrast, was impressed with Herbert's information goals and was interested in his initiatives for international peace and education. He had already passed on word of Herbert's ideas to the Rockefeller president, George E. Vincent. Meanwhile, the foundation's bureaucrats had begun to process Davenport's first proposal outline. Vincent immediately decided to re-contact Charles Davenport and asked him to respond once again to a long list of questions about Herbert's character, his abilities, and the value of his bibliographic work. He also requested a copy of Davenport's plan.

Then, both men acted less than transparently: Vincent did not tell the council what he had done, and Davenport, again, did not tell the council or Herbert what he was doing. Davenport composed another laudatory reply to Vincent's inquiry about Herbert. He next called in the members of the AAAS Committee on Bibliography for Science and began to buff up its Concilium-related proposal for international science information. Of course, it allotted Herbert and his Bibliographicum the critical role. Davenport worked hard and fast. The refined proposal was in George Vincent's hands within three weeks. Again, there was no direct communication about the proposal with the council. Soon, all parties distrusted each other.[7]

American high science remained a small social world. Before Vincent could even consider Davenport's proposal requesting millions of dollars for the Concilium, word of its contents reached the National Research Council. On the same day that Davenport's packet arrived at Vincent's office, a letter from the council arrived, one that later caused Herbert to sink into yet another period of almost sullen despair. The council's letter reflected the belief that it had been subjected to ungentlemanly behavior by Herbert Field—a dire accusation that set off an immediate flurry of emotion-charged activity.[8]

It took some time before Herbert learned of the Davenport / Vincent / National Research Council complications. He had continued his own search for funding, still unaware of what his old friend "Charlie" was doing. Herbert was focusing on his own plans for linking information and peace, and he was about to meet with two influential internationalists—men who were also major players in the world of liberal higher education.

The Great Nicholas

Nicholas Murray Butler had been a student, then professor of philosophy, at New York's Columbia College. While he was abroad in the 1880s gaining additional degrees, he met and became a lifelong friend of the young Elihu Root. That helped Butler in his phenomenal rise to power as the college leader who brought Columbia into the upper ranks of the modern research universities. His first step toward the presidency of Columbia was his creation of its Teachers College. It spawned the schools attended by the Flexner and other children of the liberal elite, and it remained the backbone of the liberal's Progressive education movement.[9] Teachers College brought Butler fame and influence. In 1901, he assumed the presidency of Columbia and became adept at courting the rich and powerful. He raised millions of dollars in gifts, helping even its biology department, and Herbert's friend, the zoologist Henry Fairfield Osborn, to achieve international recognition. He was so good at his work that he remained president for forty years.

Butler had a cause that went beyond the university: international peace. He was active in establishing arbitration organizations and drew plans for turning his passion for internationalism into powerful institutions. He courted Andrew Carnegie. In 1910, with Elihu Root's aid, Butler convinced Carnegie to give more than the twenty-first-century equivalent of one hundred million dollars to establish the Carnegie Endowment for International Peace. When his friend Root assumed the presidency of the endowment, he made Butler the head of its important education and communications division. Later, in 1925, Butler was given the responsibility for the entire endowment, a post he held until he was replaced by a young Johns Hopkins and Harvard graduate, Alger Hiss, in 1946.

In 1919, when Herbert Field met with him, Butler was at the top of his form. He had successfully weathered a trying battle with his faculty and student radicals over the university's support of America's entry into World War I. Butler's stature was not diminished by the bitter conflict, although such renowned Columbia faculty as Charles Beard resigned their professorships and formed the independent left-progressive, New School for Social Research.[10]

Butler was an impressive man who commanded attention and who could intimidate even the now interview-skilled Herbert Field. During their meeting Butler seemed receptive to Herbert's ideas about science information and to his peace initiatives, and he indicated that he intended to say positive things about the Concilium. He also suggested that Herbert speak with the supervisor of one of the endowment's spin-offs, the International Institute for Education. Butler had selected a New Yorker, Stephen Duggan, to head it.

Herbert said his "thank you" and left the meeting with somewhat renewed optimism about Carnegie funding. But the experience left him with another feeling, one shared by many who commented on Nicholas Murray Butler. Herbert wrote Nina that he never met a man he disliked so much, and he expressed some suspicion that Butler might undermine the Concilium's grant.[11]

Stephen Duggan, in contrast, was a man hard to dislike. He had an ingratiating Irish "way" that compensated for his not being born into the right social set. He also had a special talent: he could make a denial of help seem an acceptance.

International Education and Stephen Duggan: A Link to the Future of Herbert's Children

In the 1880s, Stephen Duggan attended the academically rigorous institution for New York City's working and lower-middle-class students, City College, where tax dollars subsidized the rise of hundreds of its talented students into prominence. While it would be a generation before City became a hotbed of socialist (and Communist) intellectual fervor that generated espionage agents such as Julius Rosenberg, Duggan emerged from his undergraduate studies as a committed peacenik. After graduation, he taught in the New York schools, obtained a PhD from Columbia University, and became a noted and well-respected professor of political science at City College.[12]

Duggan was not an ivory-tower scholar just writing about world justice and peace initiatives. He was an active organizer of a wide range of institutions. He was also an expert on the crises in the Balkans as well as an early advocate of international organizations for peace. In 1917, he sketched a plan that caught the attention of Nicholas Murray Butler, Elihu Root, and the Carnegie Endowment's staff.[13]

It took several meetings with Duggan before Butler agreed to create the Institute of International Education, and to appoint Duggan as it leader. Butler gave Duggan a mandate to create a system to further the circulation of ideas and people through college student and faculty international exchange programs. With the Carnegie and, later, Rockefeller funds, Duggan stewarded the exchange programs while continuing his research, writing, and advisory duties.

Stephen soon became an expert on the League of Nations, a leader among those working for an understanding of the Soviet social experiment, and an advocate for international cooperation. Duggan was even a driving force behind demands for the diplomatic recognition of the Soviet Union. He founded the American-Russian Institute and became an educational advisor to the Soviet government as soon as the United States gave it formal recognition in the mid-1930s. That organization and Duggan's other activities gained the attention of anti-Communist groups that suspected Duggan of more than mere academic friendship with the Soviet Union.

Meanwhile, his student exchange program blossomed and his institute was fast becoming the center for all American cultural exchanges. The program, and Duggan, became so close to the State Department that it was difficult to tell if there was a distinction between them. Even the young diplomat Noel Field worked with Duggan on European cultural matters. Meanwhile, Duggan had become an influential member of the East Coast liberal establishment and his son, Laurence, became a Philips Exeter and Harvard man.[14]

Duggan was no stranger to controversy. As World War I was ending, he criticized the proposals for Europe's postwar partition and what he thought were overly negative reactions to Russia's new social experiment. However, he was never a Red or even a fellow traveler. After one of his many trips to the Soviet Union, he penned a chapter titled "The Passing of Capitalist Civilization," but his comments became increasingly critical of the Soviet Union's political policies and some of its "fanatical" minions. He tellingly protested the Soviet version of an "affirmative action" program in which middle-class families were near starvation because they were not allowed employment—all in the name of equality. He commented negatively on the growing thought conformity in Russia, but he also became critical of the other great European social experiment of the post–World War I period, Fascism.

Duggan's criticisms were usually subtle and his polished and objective-appearing demeanor made him an often called-upon resource for American newspaper and radio reporters. However, Duggan's major contribution to the history of broadcasting and popular liberalism came through his protégé. In the early 1930s, he hired a bright young Quaker to help him with the expansion of the student exchange and summer programs in Soviet Russia. Edward (originally Egbert) R. Murrow went on to achieve radio and television fame and international influence. After serving with the International Institute for Education, Murrow moved into the radio field when Duggan secured him a post with one of the popular channels for liberalism, the Columbia Broadcasting System (CBS). Murrow repaid his debts to Duggan. He became a trustee of Duggan's institute in the 1940s (along with Allen Dulles's brother, John Foster Dulles) and he stood by his close friend Duggan's son Laurence in the loyalty crises of the late 1940s.[15]

During the mid-1930s, Murrow was Duggan's point man for Soviet initiatives. Just before he left for the post of education director at CBS, Murrow set up an expanded student program and reenergized the American-Russian Institute. It hosted speakers like the Quaker- and Unitarian-aligned Bryn Mawr student and ultraradical, Anna Louise Strong. None of the activities seemed politically out of place at the time. There were thousands of American experts in the Soviet Union, hundreds of Russian students were in American schools, and many American liberals were substituting a Russian excursion for the old grand tour of Europe. Even the authoritative *New York Times'* correspondent in Moscow made Stalin appear to be a kind-hearted liberal.[16]

Meanwhile, Stephen Duggan was playing additional roles in internationalism and in a new round of reconfiguring American higher education. Among his many works was support for the League of Nations' efforts to facilitate the international sharing of statistical and mathematical data. And, during the 1930s, his organizations were at the center of a program that, with millions of dollars from sources such as the Rockefeller Foundation, brought persecuted academics out of Germany and other fascist nations. Duggan's organization placed them in American colleges, even in Flexner's Institute for Advanced Study at Princeton. The New School for Social Research in New York City was especially welcoming to the refugee scholars.[17]

Duggan's 1930s rescue program was limited. His organization focused on bringing out intellectuals, rather than aiding average refugees. Importantly, many of those receiving aid from Duggan's rescue committee were on the ideological left. Also, because Jews were being targeted in the fascist countries, they formed a dominant percentage of those allotted fast track status on immigration lists and were among the first to be given permanent positions in U.S. universities. That brought complaints of favoritism that later haunted the Duggan family and its friends. Among his refugee intellectuals were some devoted Communists who would link Noel, even his mother, to espionage in the 1930s.

And yet Duggan did not intend to bring in "agents." In his lifetime he compiled a long résumé of great liberal-humanitarian achievements. He was also open, affable, humble, and congenial, even when he encountered Herbert Field.

Back to the Meetings of Late 1919: Other Misread Intentions

Duggan was especially amiable when he met with Herbert Field at the end of 1919, perhaps because Herbert had an "in." In addition to Butler's introduction, Charles Davenport had worked with Duggan on commissions attempting to form policies for the eugenics threat of the "feebleminded."[18] The December meeting was pleasant; and, because Herbert's two causes matched Duggan's, Herbert was sure he had made another good information and internationalist friend. Herbert's optimism heightened when Duggan reminded him that Andrew Carnegie was one of the major sponsors of John Shaw Billings's path-breaking index of medical literature, the Index Medicus.

Herbert and Duggan did seem to be an intellectual match. And Herbert made the trip over the Brooklyn Bridge more than once to visit with Duggan, each time to reinforce his request for Carnegie funding. He was sure he had Duggan's support. The visits with Duggan did pay off, at least in one respect. Duggan gave the final approval needed to ship another huge donation of books for Zurich's libraries. Unfortunately, Herbert again misread some hints: Duggan was not the devoted Concilium supporter Herbert thought he was.[19]

Despite what he believed about Duggan's support, when Herbert returned to 106 Columbia Heights and Hamilton's odd companions, his rollercoaster emotions again took hold. A fit of Christmastime melancholy overwhelmed him. It was partly because he had been away so much longer than he expected. He desperately missed his family. The malaise was also the result of news from Zurich. Although Nina was better, the Concilium was rapidly going downhill.

Herbert had left the Concilium's affairs in the hands of his caretaker-secretary, Marie Ruhl, the only staff member to stay at work throughout the war. She continued to do classifications, but after 1917 she was unable to send out any of her work. The war also halted her effort as the editor of Europe's entomological journal, *Socitas Entemologia*.

With Herbert devoted to his relief and intelligence activities, and with Nina uninterested in the Concilium's details, Ruhl also had to oversee the financial matters of the center—despite having no business experience. She tried to carry the same burdens after the war.

Because Herbert was not paying her an adequate wage, she was justifiably discontented. Herbert sympathized, but he could not afford to raise her salary. He feared she might find another job. Worse for his emotions, Herbert learned that a delegation of bibliographers from Britain's Royal Society had just arrived in the United States. They were asking for grants from Carnegie and Rockefeller. With that, he wrote Nina telling her the Concilium might well be doomed.

Fortunately, Herbert received news that the Rockefeller Foundation was now following through on Davenport's proposal. The foundation's message had not contained some vital details, however. In early January 1920 there had been an important luncheon at New York's Century Club, Herbert learned, to discuss the Concilium's future. Rockefeller, Carnegie, and council representatives (who had not yet digested the details of Davenport's proposal) were able to overcome the objections of one member (surprisingly, Stephen Duggan) and pass on a strong recommendation to the Rockefeller's program managers. Herbert was disappointed to later hear about Duggan's resistance, although Herbert had recently received some indications that Duggan's smiles during their previous conferences had not been true signs of approval of his ideas. Herbert learned that, probably at Duggan's request, he had deliberately not been invited to that special luncheon at the Century Club.[20]

The Grand Visions Unveiled: Herbert Learns of Charlie's Plan and Wistar Compiles an Anatomical Index

It was well after receiving the news of that luncheon-conference that, for the first time, Herbert learned the details and enormity of the ambitious Davenport-AAAS proposal. Herbert was told by an "insider" that the foundation was planning to pay him a salary more than twice what he had ever asked for, even in his moment of pique in 1910 when he demanded a salary equal to high-status professor. Moreover, he finally realized the proposal contained grand visions for an expanded Concilium.

It was to be the center of a fully international system, with scientists from around the world serving as advisors. Under Davenport's prodding, the Rockefeller men were thinking in terms of Big Information. They proposed to give Herbert over $1 million to get the center restarted and $240,000 a year as a continuing subsidy. The Concilium's average expenditures before the war were $70,000.[21]

Davenport contacted Herbert and told him the final decision awaited the recommendations of two experts, and then a vote by the Rockefeller trustees at one of their regular meetings. With that, Herbert again became dejected. He had expected an immediate decision and then to be able to return to Nina and the children. He wrote Nina that no matter what happened with the Concilium grants, he would be glad to finish with the center. He was again so close to abandoning his bibliography that he had re-contacted old friends in the business world to see what his options might be.

Something else was seriously disturbing Herbert. He was hearing rumors that some people at the National Research Council were about to bring their own plans for biological information systems into the competition for the coveted foundation money. There had been the mentions of abstracts at the November meeting in Washington and Robert Yerkes had recently told him of the council's interest in supporting the development of abstract journals for fields ranging from electrochemistry through all the natural sciences. However, Herbert did not know that steps were being taken to establish challengers to the Concilium.

Biological Society Competitors

Herbert became deeply worried after he heard about the early 1920 meeting of the American biological societies. It led to an organization devoted to finding the support needed for what became *Biological Abstracts*. Several very ambitious men in the council were the driving force behind that North American Union of Biological Societies. The new organization, they hoped, would have a membership large enough to provide the subscription base and the cooperation needed to support a first-rate American and, perhaps, international publication. They also played leading roles in the creation of post–World War I's International Union of Biologists and hoped that with help from the International Union the American service could merge all the existing small and struggling biology-related journals and, importantly, replace German and British biological offerings.[22]

The American Union of Biologists was also a part of the broader National Research Council's worldwide initiative to build sets of cooperating national subject-field unions. Once those were in place, they could be coordinated into sciences' version of the League of Nations. The council hoped that would eventually provide an international market for their proposed abstracts services for all subjects. Significantly, they seemed to assume they did not face considerable problems by treating English as a new universal scientific language.[23]

Central to the efforts to "organize" biological information were Charles McClung, Jacob Schramm, and Milton Greenman. Although they had not yet built a record as innovators in solving international problems, they were effective high-energy academic entrepreneurs in the worlds of emerging American research universities and science publication. They were driving forces behind the University of Pennsylvania's program to become a nationally acclaimed institution. The university's medical research arm, the Wistar Institute, was vital to achieving that goal.

Advocates for "Current Awareness" Cards and Abstracts

The institute had its own press and had taken over the publication of a half-dozen American academic journals. It managers were innovative and efficient businessmen, and they had their staff constantly seeking cost-cutting, efficient methods for production, printing, and distribution. In 1924, Wistar would proudly announce that it worked an academic publishing miracle when all its journals became self-supporting—none required subsidies. The center's staff members were also aggressive marketers. Just as World War I began, Wistar started a card service of its own, sending out "current awareness" cards with abstracts of forthcoming articles.

The cost effectiveness and innovative character of the Wistar's programs impressed the academic world and emboldened it managers. It was one of the institutions the Rockefellers soon called on to help European universities and libraries. Europe desperately needed aid to catch up on the scientific literature it had been unable to acquire during the war.[24]

Wistar was entering into competition with the Concilium by compiling an anatomical index. It was sure that its business acumen would lead the index to be another pay-for-itself operation. It also sought to play a larger role in all facets of science information, including the forging of new methods and technologies. McClung, Greenman, and their ally Jacob Schramm placed an emphasis on carving out a place in the publication for a new and increasingly popular science information format, the long abstract, which was a contentious aspect of direct importance to Herbert's Concilium work. They were also advocates for, and leaders of, a movement to develop international standards for abstracts.

In addition, they were the most visible American supporters of two ideas that were threats to the fundamentals of Herbert Field's service. The Wistar's men were firm believers that volunteer indexers could do their work using only a one- to two-hundred-word abstract. To further that, and with Council support, they were building their own expertise in the technicalities of science indexing. The books by Gordon S. Fulcher on new indexing methods would soon appear, giving a theoretical basis for the Wistar claims.[25] Using abstracts rather than full texts, the research showed, reduced the cost of production for science information, and it did so without lowering the quality or usefulness of indexes.

As unacceptable to Herbert as the focus on abstracts was the Wistar administrators' position as leaders in challenging the need for any complex classification or indexing. Of course, they had little use for number systems. They declared that "old fashioned" classification and indexing were superfluous for working scientists. They asserted that traditional classification products were, at best, useful in only a tiny handful of the world's teaching universities. They did not keep such views to themselves. As influential members of the most important biological societies, and even the National Research

Council, they were spreading their radical information message within America's, and when possible, the world's science institutions.

McClung and Greenman were also techno-advocates and, unlike Herbert, kept up with the latest advances. They were open to any innovation that could lower production cost and speed information delivery. They accepted offset printing, although it might lower quality. They also envisioned the use of electric tabulator/sorter technologies for both journal production and data retrieval. At times, they even condemned those, like Herbert, who were not exploring their use, although such equipment was far too expensive for any academic data center or library of the 1920s or 1930s.[26]

In early 1920, McClung, Greenman, and Schramm became deeply involved in heated National Research Council responses to Charles Davenport's proposal to have the Rockefeller Foundation fund a huge international science information service centered on the Concilium. The Philadelphians' intense reactions to the proposal reflected their ties to Wistar's goals and, with great significance, indicated their views on how to conduct scientific research. Their responses also were the result of their coming from different social and educational backgrounds than Herbert's, or even that of their council colleague, Robert Yerkes. None were graduates of elite universities and none were oriented to theoretical science. All were focused on the problems with, and financial and information needs of, applied science. Their devotion to the cause of abstracts reflected that.

Herbert's Grand Friendship Renewal Tour

In early 1920, Herbert Field did not yet know of the National Research Council's protests against Davenport's plan or the depth of the commitment to abstracts. However, the lack of decisions about his Carnegie and Rockefeller proposals saddened him. Another visit with Stephen Duggan, who was acting as the Carnegie's lead man on Herbert's request, was pleasant but failed to result in an expected firm commitment. Then, Herbert began to hear disturbing rumors about the council's new information directions and protests. Herbert sought consolation. He rushed to Charles Davenport at Cold Spring Harbor.[27] Davenport's encouragements revitalized Herbert and he once more called up his optimism and energy.

Herbert ached to return to Zurich and had reserved a cabin on a London-bound ship, but the Concilium's cause kept him from heading for home. At the end of January 1920, after learning that a final Rockefeller decision would not come for some weeks, but not fully aware of the seriousness of the opposition arising within the council, Herbert rescheduled and booked a later passage. He then risked his dwindling savings. He began a final American tour to mend hurt feelings, to recruit supporters, and to deflect other threatening developments in American bibliography.

His first stop was in Washington, D.C., where he talked to the Carnegie Foundation's representatives, met with his State Department friends, and dined with some of the European royalty he had negotiated with during the war, such as Prince Hohenlohe. He learned much about the progress of European reconstruction and some additional inside information about what was going on in the White House. He then refocused on the Concilium's needs and attempted to enlist additional influential people.

He dined with the leader of the Smithsonian Institution, visited the staff at the federal government's biological survey, and had a discussion with the experts at the weather bureau. The librarian there, O. L. Fassig, had been a fan of the Concilium and the UDC since the 1890s. After that pleasant visit, Herbert boarded a train bound for the Midwest intending to make stops in many cities on the way. In Baltimore, he courted Dr. William Welch of Johns Hopkins University. Welch was one of America's most influential public health reformers and medical bibliographers. Thinking he had another friend, Herbert moved on.[28]

In Philadelphia, Herbert took the chance of walking into another "lion's den" when he took the short cab ride from the train station to the University of Pennsylvania. He wanted to renew his ties to the Wistar Institute. A decade before, the group at Wistar had seemed ready to act as agents for the Concilium. At Herbert's request, its leaders had served on Davenport's earliest bibliography committee and Herbert had met most of them at professional gatherings. Although it had the potential to cause Herbert to plunge into one of his increasingly frequent depressions, the Philadelphia stopover was a pleasant experience. The Wistar staff warmly received him. He was not told of the deepening objections to the Concilium, or of what was developing within McClung's group.

Herbert moved west in February 1920. Pittsburgh housed an important destination, the Carnegie Museum. While there, Herbert pushed for another endorsement for the Concilium, for the museum's renewed purchase of his cards, and for the continued use of the Dewey/UDC by its classifiers. After that, it was to the true Midwest where the state universities were turning into major applied research centers. At the University of Michigan, Edward Mark's alma mater, there had long been a vigorous zoology department and it had become one of the Concilium's staunchest supporters. The department had taken a full subscription to his database.

During World War I, while the Concilium was shut down, the faculty launched its own zoological bibliographic project. Professor Lee Dick's new "Zoological Abstracts" lists were available on one-sided paper so the entries could be clipped and pasted on cards.[29] Herbert talked with the members of the Michigan faculty, assuring them the Concilium would soon reappear and asked that they not compete with it—for the good of science. He was guaranteed support for Zurich's effort. Ohio's university also received a visit and a plea to trust in the Concilium's rebirth. The most important stop on Herbert's February junket was, however, Chicago, Illinois.

Decades before Herbert began his bibliographic crusade in the mid-1890s, the midwestern merchandising and manufacturing center's leaders decided Chicago should become an educational and cultural center, one to equal Boston and New York. With a generous gift from John D. Rockefeller, a local Chicago college turned itself into a research institution, the University of Chicago. It lured the nation's best scholars with high salaries and amenities. Within a generation, the school gained a reputation as a national leader in the sciences and humanities. It also quickly became as "nondenominational" as had Harvard, and, like the major private East Coast universities, committed to liberalism.[30]

As the university was emerging, the city of Chicago sponsored the magnificent World's Columbian Exposition of 1893. It brought donations of thousands of natural science specimens from around the world. The fair's leaders decided the collection should form the core of a new permanent museum. The result, the Field Museum (named in honor of a major benefactor, Marshall Field), had been a valuable Concilium customer. The museum's neighbor, the innovative John Crerar Library, was also on Herbert's list of critical Concilium ex-subscribers.

Herbert had already spent some time attempting to reestablish the relationships, knowing that if the Field Museum and the Crerar did not believe the Concilium would soon resume shipping cards he had little chance of regaining other subscribers. Another reason for Herbert's visit to Chicago was the need to court an old acquaintance, the president of the Rockefeller Foundation, George E. Vincent.

A Critical Meeting with George Vincent: The Crisis Begins When a Gentleman's Honor Is Questioned

Vincent was the son of a famous Methodist Episcopal Bishop who founded the Chautauqua movement that brought the best of modern-liberal thought to middle-class America through lectures, summer camps, and attractive publications. George assumed Chautauqua's leadership when his father, John Heyl Vincent, retired. The elder Vincent had a residence in Zurich, from the time when he was in charge of his denomination's European operations. His son and Herbert had encountered each other several times when George visited his parents.[31]

George Vincent was more than a popular educator-organizer and adept administrator. He was also one of the new modern, scientific, and Efficient men. He was a Yale graduate, and, then, a member of the first postgraduate class at the University of Chicago. Although Vincent believed in science, his years as director of the Chautauqua adult-learning movement, his academic specialty (sociology), and his career as a professor and university administrator did not turn him into an unquestioning sponsor of the expensive natural and physical sciences.[32]

Charles Davenport had sensed that Vincent would need some convincing about financing his information plan. He advised Herbert to meet with Vincent in New York, and to do so before the final Rockefeller meetings to evaluate any Concilium proposal. But, Vincent was on a tour of the foundation's recipients in the West and the Chicago get-together seemed the only opportunity for Herbert to persuade Vincent of the worthiness of his zoological bibliographic service—and the need for its expansion. In mid-February Vincent received him cordially and they had a long chat. Herbert could not tell how well he had done, but as he took the train back to New York he was on one of his upward emotional spirals. He was unaware of the consequences of his accidentally being seen in Vincent's office by the leader of the National Research Council.

When Herbert Field arrived in Brooklyn and settled in as best he could at number 106, he was exhausted. The trip west had been grueling. He did not get a chance to wind down, however. As he tried to regain his strength he received a message that caused him to rush back to Washington. The National Research Council was going to block any Rockefeller or Carnegie grant!

The reason, Herbert was later told, was that he had not waited to submit the applications before he received the full and formal blessing of the council. If a project was not "its" project, council administrators would not support it. These communications did not mention something as or more important: the council's men were questioning Herbert's personal integrity.[33]

13 Information Science Infighting, Round 1—and Herbert's Final Struggles

Herbert Field's latest troubles had begun just before he launched his Midwest odyssey in January 1920, and as Charles McClung began hearing rumors about Charles Davenport's grand Rockefeller proposal. McClung assumed that Herbert had written it. The news probably came through Columbia University's anthropologist, Clark Wissler, who was in both the Davenport and the National Research Council information groups. Or, it might have come from Herbert's friend Robert Yerkes. McClung, then the operating officer for the council's information initiatives, had quickly requested a copy of the Davenport proposal from the Rockefeller Foundation.

At the same time, McClung sent Herbert a polite letter. *We heard of your Rockefeller proposal, how do your plans fit with the Council's science information program,* he asked?[1] It was not difficult to tell that McClung thought the council's science information project was more important than all others. While McClung waited for details about the grant proposal, and for a response from Herbert, another man entered into what became a dangerous situation for the Concilium. James Rowland Angell encountered Herbert Field, by chance, in George Vincent's Chicago office.

James Angell: Another Take on Science Information and the Council

Angell had just assumed the chairmanship of the National Research Council and was supervising the construction of its new monumental headquarters. He had heard in October something about Herbert Field participating in the council's information project and, later, about what he understood to be an agreement between the Yerkes-McClung group and Concilium. When he saw Herbert in Chicago, Angell did not know of the recent Rockefeller negotiations concerning the Davenport proposal. So, he wondered why Herbert was in Vincent's antechamber. When Vincent informed Angell of the general nature of the Davenport-AAAS submission, Angell was surprised. He decided to contact McClung and Yerkes in Washington.[2]

What Angell learned from them made him furious, but for reasons somewhat different from those that had so upset the Wistar men. They feared the emergence of the Concilium as a central and near all-encompassing science information center, one that would make the council and American science publishers follow directions from abroad. There were different causes for Angell's outburst. Some had to do with his educational and social background, some stemmed from his character and personality, and others were related to his sense of gentlemanly duty to the long-term welfare of the council.

Angell came from a milieu more like Herbert's than like McClung, Schramm, or Greenman's. But there were some important differences. He and Herbert were both from respectable upper-middle-class families that had an intellectual bent. His father was a professor at Rhode Island's nurturing Brown College, and then president of the University of Michigan. He also served a few years as a diplomat. However, in the mid-nineteenth century, a college presidency did not provide the level of affluence that later educational administrators became accustomed to. In addition, while the Angell family was urbane, it did not have the deep international connections or perspectives of the Fields.

James Angell learned much about academia and its administration from his father, but during his teenage years he showed little inclination to become an academic. Then, near the end of his undergraduate days at the University of Michigan, he decided to become one of the new "psychologists" who were gaining university positions as parts of the humanities became "scientific." Angell made his decision after he studied with the soon-to-be famous logician-educator-philosopher, and leading liberal, John Dewey.

Angell followed his years at Michigan with a stay at Harvard where he studied with the giant of pragmatism and secular philosophy, William James. Angell, while retaining his humanistic bent, then desired additional training in the new learning. So, he pulled together some money and made the pilgrimage to Europe to study with the best of its naturalistic "psychologists." He did well. Then, he faced a near catastrophe. It was an experience that scarred him. Because of his difficulty with the German language, the faculty rejected his dissertation. He was offered another year to rewrite it. Unlike Herbert's family, his could not afford the expense. Angell returned to America without a PhD. He brought back a mild distaste for European academics, and some deep anxieties about his future.

John Dewey and William James's admiration of his talents rescued him. They found him good jobs despite his lack of "the" new academic credential. His writing helped their efforts. He became noticed as a functionalist psychologist. That was an approach tied to Darwinist-evolutionary-pragmatist premises. Angell gained a junior position at the University of Minnesota, but was immediately called to the grandly ambitious University of Chicago. He began as a junior colleague of John Dewey. He did not remain a "junior" for very long. He soon headed Chicago's psychology department, became the university's dean, and rose to be a leader of the American Psychological Association. He wrote a famous and often reissued work on psychology, and he aided Robert Yerkes on the World War I intelligence testing project. James Angell's ascent to national prominence came when he was asked to serve as the council's chairman.[3]

After his anger over his dissertation problem subsided, Angell developed skills as an administrator and as an academic diplomat. His managerial abilities and his reputation as an impeccable "gentleman" caught the attention of America's top-level scientists and smoothed his way into the council position. His climb did not stop at there. His triumphs as an academic ambassador soon led him to the presidencies of the University of Chicago, the Carnegie Foundation, and Yale University. It was Angell who turned Yale into a modern university with special graduate schools. He was also the man who firmly linked the institution to America's new wealthy classes.

Can Gentlemen Be Gentlemen in a New Age of Science?

In early February 1920, after he talked to McClung and others at the National Research Council about the Concilium-Rockefeller proposal, Angell's sense of duty overrode his diplomatic good judgment. His actions almost destroyed Herbert Field's reputation, and his own. As well, Angell unknowingly endangered the relations between the council and the world's richest foundation.

In midmonth, Angell sent a heated letter to George Vincent, one that barely avoided calling Herbert Field a liar. Angell informed Vincent that from what he knew about the Rockefeller grant proposal, Herbert had been telling the council one thing and doing another. In all the meetings with the council, Angell wrote, Herbert mentioned only his rather humble Carnegie proposal and, moreover, had promised to cooperate with the council's larger plans for international science information. Now it appeared the Field-Davenport group intended to turn the arrangement on its head. It seemed that Field, not the council, planned to be in charge of international science information.[4]

Angell did a bit of backpedaling near the conclusion of the letter. Perhaps, he wrote, Davenport (who Angell now suspected of writing the proposal) had not been informed of the council's ambitions. That qualification, Angell hoped, would avoid a clash with the Davenport-AAAS committee if his letter's contents were made known outside the foundation. However, Angell had not made a full retreat. He returned to the offensive. He soon wrote Vincent that before taking any further action on Herbert's grant application, the council must be able to present its information plans to the foundation.

Angell then pushed his argument and went a bit too far for Vincent. Angell stated that the council had always assumed that the foundation was uninterested in information problems and, significantly, that he felt that the Rockefeller managers might have misled the council. Vincent did not take long to react to Angell's accusation. His first step was to try to contact some of Davenport's group.

Then, before having a full discussion with Davenport, Vincent made his own allegation against Herbert Field and a damaging decision as well. He sent a letter to Charles Davenport stating he had been informed that Herbert was not cooperating, as promised, with the council. As a result, Vincent declared, he was refusing to send the proposal to the Rockefeller Foundation's board for consideration! He then issued a warning that had the potential to destroy the Concilium. He let Davenport know that unless he found some way to reconcile any differences between Herbert Field and the council, perhaps through the immediate formation of a joint committee, there could be no chance of Rockefeller money for Zurich.

Davenport did not appreciate the importance of the pronouncement. He interpreted it only as a "recommendation" rather than a threat. His response created more tension between the AAAS and the council.

Davenport did not form the committee. He took the offensive. He immediately let the scientific community know of his displeasure over the council's tactics and its drive to organize American science from the "top down." Word also got out about Davenport's heated reaction to Vincent's letter. Angell was dismayed. He had not expected, nor was he fully prepared, to deal with such an explosive situation and its many possible ramifications.

Matters became extremely complicated. Vincent began to think that gentlemen (Angell and Davenport) who accuse other gentlemen of ungentlemanly behavior might not be gentlemen. However, he began a new series of confidential inquiries attempting to restore professionalism. They unexpectedly led to a rush of communications among the Concilium's friends. The attacks on Herbert's honor had shocked them. Quite emotional versions of Vincent's accusatory statements were soon made known to a man who decided to take action.

Clark Wissler was another of the young stars-in-the-making at Columbia University and the American Museum of Natural History. He also was a respected member of both the council and Davenport's committees. Because of the dual membership, he felt he was in a situation that could damage his own reputation as well as Herbert's. Wissler was very upset. He wasted no time in contacting Angell at the council: "I have been receiving numerous disturbing inquires about the Council's accusations against Herbert Field and the Council's uncompromising opposition to his work," he stated. Wissler warned that the heated Angell/Vincent letters were undermining the council's reputation at a time when it needed respect. He advised Angell to take immediate steps to undo his mistake and make it known that the council did approve of the Concilium. Wissler was so upset that he made a follow-up phone call to Angell's office, one that made Angell apprehensive.[5]

Charlie Bends, Angell Takes a Different Tack

Meanwhile, Davenport had reconsidered his original reactions to the Angell and Vincent communications. He decided to bend. He sent a peacemaker to the council who assured Angell there had been no double-dealing by Herbert or Davenport. In addition, Charlie urged his old friend Herbert to rush to the council's Washington headquarters.

Angell, in turn, was realizing that he had been very undiplomatic and unprofessional when he wrote his fiery letter to Vincent. Angell tried to reestablish his own reputation as a "gentlemen" and, most importantly, ensure that the Rockefeller's managers did not blame the council for the dispute.[6]

On the same day that Angell received Wissler's message, he composed a face-saving letter to Vincent at the Rockefeller Foundation. He hoped it would also mend the National Research Council–AAAS rupture. Still not fully recognizing that Herbert had little to do with drafting Davenport's proposal, Angell stated that Herbert had not fully informed the council about his contacts with the AAAS. That lapse of full disclosure, not any action by the council, was what had caused the current embarrassing situation, he said. Angell then stated that he hoped the current letter would prevent Vincent from considering the council as rude and uncivil. He would be glad to see the foundation give some aid to Herbert, he wrote—but, then he emphasized that the council's information men must have a chance to meet with Vincent to give him a full description of their plans.

Angell's next challenge was to try to undo the damage within the scientific community at large. He immediately sent off a high-priority telegram to Robert Yerkes at the Chemists Club in New York City; his message played down any idea that Herbert had acted improperly. He asked Yerkes to contact Wissler at the Museum of Natural History and assure him the only concern the council had with Herbert was the possible "duplication" of the council's programs.[7]

Angell then injected a new point in the council's favor. He asked Yerkes to stress to his colleagues the need for efficiency in information services. Angell more than implied that Efficiency could be achieved only if the council acted as the world's central coordinator of abstracting and indexing work. Then, he repeated that the council did not have any objections to a Concilium grant. The hidden text was the role of the council in supervising the Concilium. Angell followed up by sending additional apologetic yet rather self-serving letters to Vincent.[8]

Next, Angell had to deal with Herbert Field and the repercussions that might come from Herbert's reputation being unfairly sullied. Angell correctly suspected that it would not take long for Herbert's friends to let him know what had been taking place while he had been on his trip to the West.

Herbert Faces Another Professional and Emotional Crisis

What Herbert learned was not pleasant. He was shocked when he heard of the council's actions and its accusations of his personal misconduct. He hurried to Washington in late February for a face-to-face meeting at the council's offices. He explained that the details of the Davenport proposal had not been revealed to him. After that, he dropped by the State Department to double-check that his reputation there remained untarnished. Herbert left Washington believing he had resolved all the issues, and that Angell would inform Vincent he had done nothing uncivil.

However, something went wrong. Herbert once more dropped into near despair.[9] A few days after Herbert left Washington he received a letter from Vincent; it intimated that Herbert had indeed not acted in a gentlemanly way. When Angell heard of this, he again tried to right his original wrong. He hurried another letter to Vincent stating that Herbert had not been disingenuous. But he reiterated that before the foundation gave any money to Herbert both the Davenport and the council programs should be the subjects of a joint review.[10]

Herbert was not going to leave matters to Charlie Davenport this time. He immediately reacted to Vincent's latest allegation. He sought out Vincent, but the results of the meeting were not reassuring. Herbert remained bewildered and fearful. Unfortunately, he still did not realize how formalized and contentious American science had become. It required some explanations of the new ground rules by his old friends like Yerkes to ready him for another set of meetings with the council's representatives in Washington.

Herbert knew how crucial the meetings would be, and he mustered all of his diplomatic abilities. Herbert did not take the offensive when they met. He was deferential and apologetic, again explaining that he had not been informed of all the steps in the new grant-approval processes or of the scope of the Davenport proposal. He pledged to wait before making any more direct contacts with the foundations. His apologies accepted, he returned to Brooklyn to await the Rockefeller and Carnegie foundations' decisions as he made the last-minute preparations for his long-delayed departure for Europe.[11]

Going Home: To What and with What?

After six months in America, not the expected thirty to ninety days, Herbert was more than ready to leave number 106 and return to a real home. Being with Hamilton had become extremely distasteful. As well, Herbert's calculations led to the conclusion that within a few months the Concilium's debt payments could no longer be deferred. The end of his dream seemed near. His center faced bankruptcy and the loss of its building and equipment. He and Nina faced the loss of the family's investments.

The only good news was that George Vincent at last decided to forward the Davenport proposal to the Rockefeller board of directors. However, as Herbert packed he had yet to raise any money for the Concilium. Its fate seemed dire as he awaited word from Vincent and from the Carnegie foundation.[12]

As Herbert scanned the papers for news about the delayed arrival in New York of the *Imperator*, the great ship on which he had booked passage back to England, he pleaded with Hamilton to tidy up the family finances and to stop squandering money. Herbert was not sure if he succeeded. On his round of farewell visits to other family members he asked them to watch and report what was taking place at Columbia Heights. On one of the calls, returning the overcoat he borrowed from his cousin reminded him of the state of his personal finances.[13]

Herbert made other last-minute courtesy visits. One eventually led to a temporary "save" for the Concilium. Herbert called on one of his old Brooklyn Heights neighbors. He had learned the Roebling family moved to New Jersey to be close to their huge wire and steel manufacturing plants. Their estate was not too far from Brooklyn. They warmly greeted Herbert. The Roeblings remembered him as a young boy and were pleased that he had turned into a respected scientist. Herbert's stories about his information efforts enthralled them. In turn, Herbert was fascinated with the Roeblings' rise to great fortune and power. After the completion of the famous Brooklyn Bridge they became millionaires many times over because of the huge profits their company made during the war by manufacturing metal products ranging from barbed wire to the huge steel nets that snared German submarines in the North Sea.

The visits with family and friends were pleasant, but when Herbert returned to number 106 he was bombarded with bad news. First, he was notified there would be no Carnegie grants, for either information or internationalism. And, Herbert had to conclude that Bashford Dean had been misleading him. There would be no contribution to the Concilium from the Dean fortune. He wrote Nina in a harsh tone about that. Then, he received the worst news of all. The Rockefeller board of directors and its managers declared they had decided to put off a decision. They wanted more time to conduct thorough studies of the Concilium, the world's science information situation, the National Research Council's plans, and other international science campaigns. After that, there were to be more negotiations about the Concilium even if the foundation decided in its favor.[14]

The impersonality of the process bewildered Herbert. He did not understand how deeply the Concilium was caught in bureaucratic version of a catfight between the council and the AAAS faction. When he rushed to Vincent's office in New York for a last-minute interview, he learned only that he might receive a "bridge grant" to help keep the Concilium afloat for a short time. When he hurried to Washington and appeared at the council's office, in what was described as a highly nervous state, he did not receive calming news.[15]

There were additional political complications. Herbert was not made aware that Angell had done more to compensate for his earlier emotion-driven gaffe. Angell had urged Vincent to send a letter to the council stating that the earlier accusations against Herbert had nothing to do with the Rockefeller board's decision. With that, Angell sent a letter to Charles Davenport assuring him that the council had no part in the Rockefeller rejection. Angell did not stop there. Important to Herbert, he succeeded in convincing Vincent the foundation's next "review" must take the form of the joint committee that Davenport had not bothered to create when Vincent previously asked for it.[16]

Angell Makes a Shrewd Move, Herbert Lashes Out

James Angell, taking advice from the Wistar men, went on to make a recommendation that served the National Research Council's interests. He said he had the best possible point man for any of the foundation's future information programs. That man was the council's chief operating officer, the geneticist-turned-international-science-diplomat-and-administrator, Vernon Kellogg. Vincent took the bait. He quickly demanded the creation of the joint National Research Council–AAAS committee and that Kellogg guide it. Knowing of Kellogg's negotiating prowess, Vincent expected that he could reconcile differences among the scientific groups and present a businesslike plan to the foundation. Angell, aware of Kellogg's allegiance to the council, was sure the committee would make the "right" recommendations.[17]

Immediately after Vincent's decision, Angell sent a memorandum to the McClung group asking them to prepare a detailed strategic plan. Angell wanted it to be ready to be presented at the initial joint committee meeting. Kellogg also got to work. He instantly contacted Davenport and began setting up the committee.[18]

As Kellogg and Davenport were organizing the first meeting, Herbert had to accept the fact that he was leaving with nothing in return for his six months of, at times, demeaning efforts. All he had was a hint that his advocates might be able to secure that short-term small bridge grant to tide him over until a final decision was made by the foundation. But even that was not a sure thing—and his financial situation seemed desperate.[19]

With just days before his sailing, Herbert had to do something very distasteful. He went back to the Roebling family and explained the Concilium's financial situation. The Roeblings were sympathetic. They made a last-minute donation. That gave Herbert enough money to pay the Concilium's most pressing bills through the end of the year. But, there was no money to spare to restart or reshape the service.

Just as frustrating for Herbert was the Rockefeller administrators' continued avoidance of committing to a date for a funding decision, even for the possible bridge grant. Herbert was upset about that, and about what he now believed about his profession. He had begun to comprehend that American science was no longer a simple world of close-knit and courteous old-fashioned gentlemen.[20]

That led Herbert to break with his own sense of propriety. He again did not wait for his friend Charlie Davenport. One of Herbert's final acts before leaving was to pen an angry note directly to James Angell. Written in a near scrawl on poor Edward's "104 Columbia Heights" stationery, the desperate letter declared that the Concilium and Herbert could no longer tolerate struggling on small handouts. Herbert included suggestions that the demise of the Concilium was imminent because of the council's actions. Reflecting his emotional state, it also contained a plea that, despite the past problems, Angell and his council immediately come to the aid of the Concilium.[21]

During his last days ashore, Herbert also wrote a letter to Vincent, one that rings of extreme anxiety. Herbert played on the hatred of Germany that led to one of the goals of the council's information plans, to make certain that nation never again dominated world science. Herbert told Vincent that America must lead the way in building a new worldwide science information system. It was needed to prevent a German resurgence, he said. He stressed that the Concilium was vital to that cause.[22]

Herbert warned George Vincent that unless the Rockefeller Foundation supported true information internationalism, a man in Germany named Springer might well succeed in his plans to monopolize scientific publishing. Herbert added an additional warning for Vincent, one that could not have pleased the council or even some of those in the AAAS. Herbert Field argued against any national information system, including one controlled or even heavily influenced by the United States. The cure for any possible nationalistic illness, he underscored, was to have the information center in a neutral country. Switzerland was clearly the only choice. Such ideas fit Davenport's proposal, but not the council's, nor those of some men in the AAAS who had remained in the background of the science information struggle of the previous months.

An Information Plan for the Forgotten Men of "Small Science"

Herbert was not aware of it, but there was a third American information alternative that would have an impact on the Concilium negotiations. Its representatives had only a policy direction, not a detailed plan, and they did not have a well-organized group. But they had a vibrant spokesman, Burton Livingston. Livingston was an AAAS official, an ex-editor of *Botanical Abstracts,* and one-time chair of the council's biological section. As such he was "of" the council. And yet he neither approved many of its policies, including those for information, nor was he engaged in the high-pressure applied science projects of the council's men. Rather, he was oriented to the "forgotten scholar," the typical professor in a small institution who was not engaged in abstract research as were Herbert's favored customers.

In contrast to the McClung group, Livingston advocated a system of individually managed and funded journals, abstracts, and indexes in each special subject field. He vigorously argued against any type of centralization—although he wanted the United States to lead the postwar information recovery. At times, Livingston also voiced objections to any type of card-based bibliography or anything that threatened his book-form abstract projects. Fortunately for Herbert, under prodding from Charles Davenport, Livingston remained silent as the Concilium's fate was debated in 1919 and early 1920.[23]

More Bad Luck for Herbert, Beginning at Sea

Herbert was not a happy man as he left the United States, but he looked forward to a restful trip across the Atlantic. He had decided to splurge a bit and booked passage on one of the great luxury ocean liners of the twentieth century, the *Imperator,* in early

March 1920. It had its maiden voyage just a few years before, in mid-1913. It was the second largest liner in the world and had lavish accommodations for its first-class passengers. The ship, once the pride of Germany's merchant fleet, remained docked in Hamburg during the war, was seized by the United States and used as a troop ship. Then, it was turned over to England as compensation for the torpedoed *Lusitania*. Despite his shaky finances, Herbert treated himself to one of the ship's better cabins.

He looked on that as an investment as he might be able to socialize with many important people. Among them were the opera and Broadway producer, director, and songwriter Arthur Hammerstein (son of Oscar Hammerstein I), Adolph Zukor (the Hungarian film mogul who founded Paramount Pictures), and several high-ranking British diplomats. But, Herbert's luck continued to be bad—incredibly so. Another ship hit the *Imperator* as it lay at its New York mooring. After some hasty repairs, the liner was able to leave port on time, although a struggle between an escaping felon and policemen on the gangplank upset the usual departing ceremonies.[24]

Herbert looked forward to some rest, but within a day came serious problems as the ship sailed into unexpected turbulence. The huge vessel took water. It soon listed so badly that passengers were thrown from their chairs. The ship's condition became threatening. It required the cutting of holes in the ship's side, and the use of all its pumps, to allow it to continue at greatly reduced speed. The ship was a full day late reaching England. Passengers' travel plans were disrupted. Moreover, Herbert could not immediately return to Nina and the children. There were family and professional matters to deal with in England, as well as on the continent. But, by mid-April 1920, Herbert was back in Zurich. Some letters from America were waiting for him. They made his family and business situations seem a bit less severe.

Concilium Woes Redux: Staffing, Funding, and Restarting

There were a few sprinkles of good news. George Vincent let Herbert know that the small bridge grant would probably be approved. However, Herbert would have to wait for any word about adequate funding while the foundation mulled the future of international science information. Angell then sent Herbert a personal apology. Never, he wrote, had Herbert done anything ungentlemanly in respect to the proposals and Angell saluted the past work of the Concilium. However, those two letters did not compensate for what had happened in Zurich during the war and during Herbert's absence in 1919 and 1920.[25]

Herbert found Nina overwhelmed and stressed. Although she had help around the house from a young German girl, Herta Vieser, who was Noel's schoolmate and had become much like an adopted daughter, Nina was anxious and tired. She had recovered from her 1919 operation and she was ready to aid the Concilium, but there was little she or Herbert could so without enough funds to hire new staff. The Concilium's condition had worsened since Herbert left for America. There was a frightening amount of work to be done. As well, Marie Ruhl was showing signs of deep weariness and discontent.

Ruhl was deeply frustrated by the classification work. She was not fully qualified in many subjects, despite her and her father's scientific background. The classifying was wearing and unsatisfying, particularly because she had access to only a few journals and because the work she completed was piling up unused. She could not deliver it to the few remaining faithful subscribers. Adding to the problems at the Concilium in 1920, the resident janitor was less than happy, and the printer had found the last six years tough going. Both had seen their incomes drop because of the suspension of the Concilium, and both were voicing their displeasure. There was a more ominous threat. Creditors were eyeing the building, envisioning its continued use as an apartment house.[26]

Herbert could not imagine a means of resolving the Concilium's serious financial crisis except for a large grant. He did not, could not, risk much more of his own money, and there was miniscule income from subscriptions. He had nothing more that could serve as security for bank loans. Meanwhile, debts were multiplying. In early 1920, Herbert had only the small Roebling gift and the promised Rockefeller bridge grant to keep his center alive.

When that Rockefeller "bridge" money finally arrived, the amount was a disappointment. That last-minute, rather condescending emergency grant of some $125,000 could only stave off a looming December 1920 bankruptcy. Although a second grant for the same amount was promised for 1921, neither seemed adequate to allow Herbert to return the Concilium to the operational level needed to attract subscribers.[27]

While anxiously awaiting the final word from the foundation, Herbert did the best he could to restart the Concilium. He wrote and visited many scholars and institutions, urging them to resume their subscriptions. He also made desperate pleas for back copies of journals. He kept up hope and, importantly, appearances. When visitors to Zurich, including George Vincent and Halsey W. Wilson, dropped by the Fields' villa in 1920 they were told the Concilium was on the verge of full recovery. Herbert also courted visitors from the League of Nations.

Among them was a council member who was serving on the league's medical information committee. Herbert wanted to enter the market for medical-related indexes and was very pleased when the representative sent a glowing account to America—even though the Zurich center had yet to resume operation. However, such testimonials did not stop Halsey Wilson from urging his friend Herbert to turn away from the card system and concentrate on a first-rate, book-form ongoing bibliography.[28]

Herbert Honors His Commitment to Cards and Cumulative Files

Wilson had the credentials to justify being an advice giver. He had founded a highly successful for-profit commercial bibliographic-indexing service in the United States, the

H. W. Wilson Co. Its imposing building on the Harlem River should have made Herbert sensitive to Wilson's counsel. Wilson was a master at understanding information markets and knew how to make library services turn a profit. In 1900, he had saved the foundering library cooperative–based *Poole's Index,* turning it into a thriving commercial operation.[29] By 1920, his company had successful book-form indexes covering engineering/industrial arts, agriculture/botany, and international periodicals. His famed *Readers Guide* to general literature was a perennial bestseller. In contrast, only two large nonprofit science bibliographic efforts, *Chemical Abstracts* and *Engineering Index,* could boast of a history of coming close to being break-even operations. The Wistar journals were self-supporting, but in 1920 the institution's offerings remained limited.

Herbert refused Wilson's advice. He stood by his commitment to cards and to the huge cumulative file system. He resumed his own classifying, including redoing much of Ruhl's work, and he made some additional attempts at raising contributions on the continent.

Herbert went on something of a "take on the windmills" tour because devastated Europe was unable to feed itself, let alone provide subsidies or take many generous subscriptions. The Swiss were not contributing, although they continued to have some control over the Concilium. But Herbert kept working as 1920 dragged on, although he became increasingly apprehensive about the doings in America.

He was very busy and frazzled. In 1920 and early 1921, he made trips into France, Germany, and England, sometimes with Nina and the children, to attempt to reestablish business and professional connections. In doing so, he traveled into some of the most war-ravaged areas of Europe.

A handful of times there were pledges of future support. He was beginning to succeed in finding back issues of journals, and he was gaining promises of a continuous flow of new ones, including items from Europe's colonies. That helped him believe he would eventually be able to claim the Concilium covered the entire world's literature.

One trip in October 1920 was bittersweet. He was invited to a special international meeting in London. Its purpose was to explore the resuscitation of his competitor, Britain's very ambitious *International Catalogue of Scientific Literature.* The various national societies had continued to send bibliographic entries to London during the war and were waiting for the Royal Society to resume its task of coordinating and editing the work. Some 2.5 million items lay in piles to be processed.[30]

The rebirth of the *Catalogue* seemed an impossibility given the continuing unsettled conditions in the British Empire, the economic woes in Europe, and the hesitancy of America's foundations to commit additional funds to a project that never was financially successful. Significantly, although England was launching a government-directed plan to rebuild its science information infrastructure, the *Catalogue* was not high on its list. More immediately useful to the effort to make England a world technical leader were publications such as the *World List of Periodicals.*

Herbert thought the demise of the *Catalogue* would be to the Concilium's advantage, but the failure of the *Catalogue* might also be a sign that scientific information was not a high priority to anyone. The latter seemed to be the case when the American representative from the council, whom Herbert shared meals with at the luxurious Carlton Hotel, said he could not assure Herbert the Concilium was ever going to be funded. That representative, Herbert's friend Robert Yerkes, had cast positive votes for the Concilium during mid-1920 when the Vincent-Kellogg joint NRC-AAAS committee was deliberating the Concilium's fate.

When the Council Frightens Herbert, Charlie Outfoxes the Council

Robert Yerkes, perhaps out of compassion, seems to have refrained from telling Herbert Field that the National Research Council, suddenly and unexpectedly, had a change of attitude. It now thought the Royal Society's effort was worth saving. The council was considering an attempt to raise a substantial American contribution. However, Herbert was soon receiving some hints that good news might come his way.[31]

The joint NRC-AAAS committee that George Vincent demanded on behalf of the Rockefeller Foundation had begun meeting shortly after Herbert left the United States in March 1920. Despite the council's goals, and Vernon Kellogg's stake in the council's own information ambitions, it seemed destined to save the Concilium.

Charles Davenport had worked an academic's version of a political coup. He had reenergized his old AAAS committee members and arranged to have the joint committee loaded with them. In doing so, he masterfully outflanked some of American science's best strategists. He also performed magnificent behind-the-scenes manipulations. It was not easy to maneuver James Angell, the Wistar men, or Vernon Kellogg, but Herbert's old friend Charlie did it. And, he was able to sidestep a demand by the Rockefeller Foundation to have representatives from mathematics and physics included in the committee. He kept it as one focused on the Concilium's forte, zoology.[32]

One reason for Davenport's triumph was his sensitivity to political trends in the science community. Another was economic. There was no money allocated for the committee. Consequently, its members were obliged to carry their own expenses. Also, Vernon Kellogg wanted to be close to the Rockefeller headquarters so he could consult its managers. As a result, only those with both time and money could attend the frequent New York City—not Washington, D.C.—meetings.[33]

Of the joint committee's eight members, with Vernon Kellogg acting as a technical advisor, six were associated with the previous AAAS group that had recommended the huge grant for Herbert. Most of the members were linked with the Concilium's agent in America, the American Museum of Natural History, and most had been trained in zoology. The two "representatives" of the council and its information plans, Robert Yerkes and Clark Wissler, were also friends of Herbert and the Concilium. Although Yerkes and Wissler wore two hats at the meetings, Charles Davenport had no trouble in guiding them in the correct direction. Council members with radically different views on information policy, such as McClung and Schramm, were not represented—except secondarily by Yerkes who half-heartedly voiced their opinions. He only made recommendations for a mild version of the old council expectations. He suggested that Herbert provide abstracts (as well as his numeric-classified indexes) and that he should cooperate with the council's other information initiatives.

As it had before, the term "abstracts" remained rather ill defined. While the agreement about abstracts allowed Yerkes to feel satisfied that he had met his council obligations, ambiguity led to some serious future differences between the Concilium and the council. "Cooperation" with other bibliographic efforts was also vaguely defined. Although the promise of cooperation helped Davenport dodge complaints from men like Burton Livingston that the Concilium endangered his *Botanical Abstracts* and other independent science publications, "cooperation" caused major conflicts in the future.[34]

The committee members noted Yerkes's comments then continued to forward convincing arguments in favor of generous support for Herbert. They stressed the importance of the recognition of the Concilium as the heart of any future international system. Stating that because the League of Nations was likely to settle in Switzerland, where it would establish an information research department and an international copyright office, and because Herbert was a respected information diplomat, Zurich was an ideal site for an American bibliographic investment. Davenport predicted that when the league-directed copyright office was established, two copies of every publication in the world would be in the league's library, making Switzerland an efficient base for bibliographic work in all fields of study.

In addition to the league-related arguments, Davenport's committee emphasized Herbert's "people skills." They made him the perfect man to coordinate the various future multination committees needed to make any Rockefeller-sponsored science information service truly international, they said. Significantly, the committee never mentioned Paul Otlet's international accomplishments or those of the Royal Society.[35]

That was surprising because Otlet was beginning a new phase in his career. With very generous support from the Belgian government, he was launching an international information project that had the potential to dwarf his previous operation in size and its capability for setting international standards in all information fields. The Belgian government was so generous because of its belief that Brussels, not Zurich or Berne, was to be the home of the League of Nations.[36]

The Rockefeller Managers Decide, then Dither

It took two months for the joint NCR-AAAS committee to compose its report and forward it to the Rockefeller's officers. The committee's final recommendation was essentially a restatement of the ambitious earlier AAAS proposal, but more financially generous. Over a million dollars was to go to the Concilium in the first year of the new program, then a half million during each of the next five years. In addition to money, Herbert was to have power. He was to be the director of the entire international system and was to organize the planned conferences in Europe and America to coordinate all world science information.

The Concilium was to remain in Zurich, with its first task being to rebuild itself as a zoological service. The committee hoped the Swiss government could be persuaded to sell its interest, thus making it easier for Herbert to turn the Concilium into a center for all fields of study while, of course, "cooperating" with the Council and the Rockefeller foundation. It was implicit that Herbert would be running what was truly an American operation, although a formally "international" one.

In summer 1920, Davenport cabled the good news of the joint committee's recommendation to Herbert. But Charlie soon had to inform his best friend of some not-so-agreeable decisions in America.

The Rockefeller managers had again decided to postpone a decision. They did not want the foundation to be seen as a heavy-handed American institution imposing on Europe's information turf, and they desired a triple-check on Field's potentials. They told Davenport that Herbert had to wait for a final word until an investigatory team returned from Europe. The foundation planned to send its secretary, Edwin R. Embree, to survey all the European information centers and to gauge the possible negative reactions in Europe to an expanded American-sponsored Concilium.[37]

News that was more depressing followed the words about the Rockefeller hesitation. Embree, who was regarded as a Concilium ally, could not make the journey. In his place Shepard Franz—a federal government clinical psychologist, an editor of the *Psychological Bulletin*, and worse, a council member—was asked to conduct the survey of England's and the continental nations' attitudes and willingness to cooperate in any Rockefeller initiative. Franz's appointment was not good news. Herbert worried about his future. As always, Davenport pitched in and had his colleagues, even Edward Mark, lobby the council and the Rockefeller foundation.[38]

Meanwhile, Herbert was only aware that someone suspected of being in the council's camp was to be the foundation's point man in Europe. His already intense anxiety about his finances reached a critical level. One reason was that his upper-middle-class background and his years living on his generous inheritance had set a very high lifestyle standard, one that seemed to be on the brink of collapse. Although he had a huge villa and an expensive car, Herbert fretted. He kept his worries from his children, however. He was a caring father, so he avoided burdening them with the woes of the Concilium or with his concerns about his precarious career. He also tried to buffer Nina from any strain—her operation and the pressures of the war had aged her, greatly so. But, there was one frustration he did share with her and his children.

The League of Nations, Herbert's Dreams, and the Concilium's Needs

As soon as Herbert disembarked from the *Imperator* he had inquired about the final stages of the debate in the United States over internationalism and the League of Nations. What he heard from his friends in the British government was not reassuring. By the time he arrived in Zurich, the final judgment had been made. The United States was not to join the League of Nations: the U.S. Senate, not wanting to allow an international body to determine the use of American troops or money, had killed the League of Nations treaty.

For Herbert, the decision meant a setback for peace, and a threat to the future of his grandest bibliographic dreams. Without the rich and powerful U.S. government's participation in the league it had little chance of a establishing a lasting peace, he thought. And, without the league's influence his goal of curbing information nationalism would be difficult to reach. He could not hide his disappointment, as a Quaker and as a bibliographer.

With emotion, he told the impressionable Noel and Elsie of his anger over the American politicians' rejection of his and Woodrow Wilson's righteous institution. Herbert may not have been fully aware that liberal internationalists in the United States were acting to ensure that America played some role in creating a new world order, despite the popular will. Nongovernmental organizations such as the National Research Council, were already joining league-sponsored initiatives on education, health, labor, and even science and culture.[39]

A Legacy for the Children, and Faith in the Zurich Center

Herbert remained despondent about the future of peace and reform, but he did not relent. He took his family on an excursion through the great battlefields of World War I with the goal of impressing upon his children the horrors caused by "nationalistic" wars. The children never forgot his talks or the trip through Germany and France. Perhaps that was because a time of great joy would suddenly be followed by the worst of family disasters.

Surprisingly good news came when Shepard Franz's European trip led to a report declaring the Concilium was the only choice for a central facility in Europe, or the world. Franz had met with most European information leaders and, except for Herbert, found them and their operations wanting. His report to the Rockefeller managers was good for the Concilium, but not for Paul Otlet, the Royal Society, or France's new information ambitions. Although the Concilium continued to face hard times, Franz concluded it was the only hope for an international system.

The new rooms and cabinets in Paul Otlet's expanded Brussels center did not compensate for what Franz perceived as a lack of purpose and control. He reported the Otlet operation was not, or ever had been, providing a valuable tool for scientists. He disliked its indexing methods and procedures and thought that it could never recover from the impact of the war, despite the generosity of the Belgian government. That generosity soon diminished when the League of Nations decided to locate in Switzerland, not Belgium.

Franz was not the only one to write disappointing reports on Otlet's work. Others found him a great theorist but not a proficient manager. It was only when Otlet gained a contract from the League of Nations and focused on cooperatively establishing standards, rather than on building his files, that the Rockefeller Foundation awarded him any grants. That was much later, in the early 1930s. The funding was small, and even then Otlet's international information bureau was unable to establish the degree of cooperation the league hoped for—although its operational control had been shifted to a man considered a better manager, the Efficiency-minded F. Donker Duyvis at The Hague.

Donker Duyvis had to devote much of his energy to revising the scientific schedules of the Dewey/UDC. They had become quite outdated, being without changes for more than a decade. Worse, there was still no acceptable edition of the classification in English and the era of cooperation between the UDC and the Dewey managers was ending. Meanwhile, the less elegant Dewey system, at first intended for an American market, began to spread to Europe.[40]

Franz had also predicted that Britain was not going to revive its old attempt to coordinate an international system. Anyway, the *International Catalogue* had never been timely or complete, he reported to the Rockefeller men. His prediction was proven correct a few months later. At another London conclave that Vernon Kellogg attended as the Rockefeller representative, he learned the *International* was dead, forever.

Franz did give a salute to Britain's *Zoological Record*. Yet, he added a coda to his compliments about its abstracts. He wrote that the *Record* was not timely in its publications and its management was not forward looking. In addition, Franz reported that Germany and Italy and other nations of Central and Eastern Europe were not information contenders. He never mentioned the Soviet government's emerging plan for a central scientific information bureau to aid economic recovery, a center that grew into the much-admired VINITI service in the 1950s.

Franz failed to point to other science information initiatives, even the league's, which might detract from his salute to the Concilium. The League of Nations' own information project seemed to have a bright future. The idea for it came from Otlet and La Fontaine's proposed International Committee on Intellectual Cooperation. Madame Curie was sure its efforts would someday lead to a Paris-based organization for world bibliographies.[41]

In 1920, Franz placed his faith elsewhere—in Zurich. His report to Kellogg went on to give more compliments to Herbert, who he had visited, and stated that if there were any European objections to the Concilium project being American, Herbert's diplomatic skills would quiet them. Then, word got back to Herbert that another recent league visitor to the Concilium sent a glowing report about its potential as a resource for the international medical community.[42]

The Rockefeller Foundation's recommitment to the second emergency bridge grant soon followed the unexpectedly flattering reports. It seemed to be another positive step toward the approval of Davenport's magnificent long-term Concilium proposal, and toward realizing the hopes that the Concilium would be the center for global science information.

Herbert's Fate

Then, the worst happened. Just as the foundation received Franz's positive evaluations, and was arranging for Vernon Kellogg to visit Zurich to make a final judgment, Herbert Field became one of the last victims of an international catastrophe, the pandemic influenza. It had killed millions around the world. A few days before turning fifty-three, Herbert died of a heart attack resulting from influenza, his obituary implied.[43]

14 A Concilium without Herbert Field: Nina Field and the Rockefeller's Great Decisions

In April 1921, Nina Field was left alone to raise her four children, including seven-year-old Letty who had troublesome health problems. Nina's parents were far away in England. Hamilton and the American Fields were on another continent. Responsibility, grief, and apprehension almost overwhelmed Nina. She had to arrange Herbert's funeral, notify people throughout Europe and America of his sudden death, and attempt to understand the family's financial matters. An initial burden was the heart-wrenching question: should she bury Herbert in Switzerland or have him cremated so his remains could be interred in his homeland? She chose cremation, despite Quaker traditions.[1]

She also had to step into the Concilium's affairs. Nina quickly found the situation dreadful. The Concilium was without money and leadership. The Swiss scientific society in charge of supervising the center was busy with more important matters. The only staff member at the center was Marie Ruhl. She was barely managing to pay bills, keep the offices heated, and to survive on her scanty and unsure salary.

Just days after Herbert's death, Nina believed the Concilium faced bankruptcy. She also feared for the family's savings, her only source of income. She immediately began a round of letters to America including desperate missives to Herbert's scientific friends. In return, Nina received many condolence notes. Laudatory obituaries appeared in science journals throughout the Western world. But, there was little practical help.

One of the responses was especially disappointing. Dated a month after Herbert's death, the note from the Rockefeller managers expressed their admiration and regrets but did not contain an offer of financial aid. The message stated only that the National Research Council and the Rockefeller Foundation remained interested in the Concilium and were going explore the situation.[2]

Unraveling the Concilium Accounts: A Brief Financial Respite

Fortunately, Nina's lawyer's preliminary examination suggested that Herbert had been able to leave her with more financial resources than he anticipated. She was not utterly destitute. The money came from the remainder of the old family inheritances that Herbert had been able to salvage, not from any generosity on his brother Hamilton's part. Hamilton had continued to spend the rest of his mother's funds and never kept his promise of sharing Lydia's funds with Herbert. Luckily for Nina, many of the investments that Herbert thought had become valueless, including the trust funds for the children, had rebounded. Soon, she and her attorney calculated she could afford to remain in Zurich and the children could continue to attend their select schools, perhaps for several months. Her attorney added a caution. She could stay only if there were no unexpected financial crises, and if the Concilium's affairs did not sap her personal funds.[3]

Although wary of investing any money in the Concilium, Nina hoped it would not die—it was her loving husband's Calling. She again urged her American contacts to remind the council, and the foundations, about their previous commitments. She wanted to persuade them to honor their promises of long-term support. She wished the Concilium could remain a Field family operation, but she soon realized she did not have the expertise needed to rescue what had never been even a financially break-even enterprise. Perhaps one of the boys would someday take up his father's crusade, she thought.

That, and the Concilium's and her economic future were not assured. Moreover, there was a pressing family decision she had to make: should she raise her children as expatriates? Nina had little time to ponder that because the results of her lawyer's final audit of the Concilium's books gave her a shock.

He told her he was still trying to unravel the accounts but it was clear to him that the situation was much worse than he first reported. By late summer 1921, the center's position was so grave that Nina once again wrote her husband's colleagues in America urging them to pressure the American foundations for an immediate decision. Soon after that, Nina received some other severe news. Her attorney discovered a complication that revealed how deeply the welfare of her and the children were involved with Concilium's finances.

She immediately contacted the council's representatives and the Concilium's supporters, such as Charles Davenport and George Parker, with dire news. The banks were again poised to take the Concilium's assets, and even the family's money. She believed that Herbert had pledged the majority of the family's assets as collateral for loans to support the Concilium. Nina's anxieties deepened as her attorney drew scenarios that were more ominous. He told her that Herbert had not been careful enough when he renegotiated the Concilium's legal structure and that her renewed fears about the family funds were well justified.

She sent out yet more pleas. In response to her letters, she received word of Vernon Kellogg's impending visit but, again, no assurances of larger and long-term support from America. Another small bridge grant might be provided, she was informed—but the word "might" was emphasized. Nina kept a degree of optimism, however. Herbert had told her of the Rockefeller interests in many activities that needed scientific information, so she believed the continued contacts by the foundation meant that the Concilium would receive aid.

Vernon Kellogg Begins a European Information Survey: Resurgent Protests and Supporting Roles

Herbert had been correct about the Rockefeller interest in information. The foundation's attention to the Concilium after Herbert's death was a result of more than the 1919–1920 negotiations with Davenport and the council. The Rockefeller managers had a long history of supporting medical education and a more recent commitment to the economic and social reconstruction of Europe. The Concilium's part in anatomical bibliography fit with the older Rockefeller medical initiatives and the Zurich center's place in general science information matched the foundation's desire to help reshape all parts of Europe's scientific infrastructure.

Nina did not appreciate something of great importance. It was the sense of honor among the foundation's leaders. They felt obligated to rally round the man who had become something of a friend, who they might have wronged, and whose family was in jeopardy. As a result, before the council and current Rockefeller man Vernon Kellogg began his European information survey in summer 1921, Nina received another check from the Rockefellers to help the Concilium continue for a few more months. However, the Concilium's and the family's welfare remained uncertain, awaiting Kellogg's judgment.

Kellogg began his trip with a positive attitude, but resurgent protests from within the council wearied him. Many voiced the belief that additional investments in the Concilium would be money thrown away. After all, the operation had never been free of debt. Some of the Rockefeller managers, despite Shepard Franz's conclusions, worried about tying the foundation to an effort that might entail ongoing sensitive negotiations with foreign governments.[4]

In addition, American nationalism was playing a role. Within the council there were again objections to diverting resources to a foreign operation, and those protests were becoming pointed and vigorous. One of those nationalists, Jacob Schramm, was arranging to have the council provide him an office and a small yearly subsidy so he could reinvigorate *Botanical Abstracts*. He and another man who had recently arrived at the council's offices with his own bibliographic goals were lobbying for much more. They wanted immediate funding for all of the proposed Research Information Service's programs.

Clarence J. West, a Special, Special Librarian

That new man, Clarence J. West, came with a mandate to put the National Research Council's information plans into operation. As did Schramm, West knew he had to compete for scarce resources. And, like Schramm, he began trying to convince the council's higher-ups, such as Robert Yerkes, who had the ear of the Rockefeller administrators, that supporting the Concilium was certainly not the most efficient use of information money.[5]

Clarence West was what Adolph Voge hoped to be: a man from a rather common background who became a high-status Special Librarian with a secure career. West came from a family of immigrant farmers that had settled in the Midwest. He was one of the few of his generation to finish high school and topped that by eventually entering the University of Michigan, where he opted for a major in chemistry. It took West much longer than Herbert Field to gain his PhD, not receiving it until he was twenty-six, in 1912. His following career was oriented to the application of the new knowledge produced by chemical research, as befitted a man who studied at one of the state universities that had pledged to the taxpayers to produce "useful knowledge."[6]

Clarence's first professional job was at a very prestigious organization—the Rockefeller Institute for Medical Research—headed by Simon Flexner in New York. West became interested in science information because of his assignments at the institute and because of his soon being appointed an associate editor of the important and growing American journal, *Chemical Abstracts*. He was not content with the routine chores of a journal editor and in 1918 he conducted and published one of the first statistical analyses of chemical literature.[7]

His chemical and bibliographic expertise became valuable to the federal government in World War I. As had Adolph Voge, he joined the army's chemical warfare section, receiving a commission as a major. The chemical corps had a great and urgent responsibility, to discover how to defend America's young men from the terror of World War I's version of biological warfare, poison gas. That entailed an offensive approach. The corps had to get ahead of the enemy and be ready with protections against new gases by developing its own. West became an important part of the discovery of chemical agents, and he became and remained one of the world's experts on poison gases. He helped to write a definitive book on the subject, one that continued to be a worldwide resource well after the war ended.[8]

West received a job offer as soon as peace was declared. It was one he could not resist. It turned him into a Special Librarian. The hard-to-refuse offer came from an innovative company founded in 1886 by two young MIT-associated chemists.

Arthur D. Little and Roger Griffin risked much when they set up a private, for-profit, chemical research firm. Despite the underdeveloped state of scientific knowledge in chemistry, they prospered. In fact, they are credited as being critical to changing the public view of chemists from being seen as merely pretentious "alchemists" to being true scientists.[9]

They made their first notable contribution in the field of papermaking. The nineteenth century's chemical revolution in paper processing had much to do with the technological revolution in libraries, offices, and the Concilium's technology. Little and Griffin's work contributed to the development of, among many other innovative products, the library cards Dewey and the Concilium relied upon, index cards, and the special cards for the electric tabulator. As West later did in the field of chemical warfare, Little and Griffin literally "wrote the book" on modern papermaking.

The two men went on to apply the new chemistry of the era to a long list of other industrial problems. The soon-to-be famous Arthur D. Little firm contributed much more to "information," including developing fireproof moving picture film, a product that was the basis for microfilm, part of an information technology revolution in the 1940s and 1950s. By the end of World War I, their Boston "invention" house was a prestigious corporation engaged in a stream of consulting and research projects. It was also bent on solving its own growing information problems. The company needed and sought an expert. They soon appointed Clarence West as the head of their new and innovative Information Department.

Many in the firm had met West when they worked together on the poison gas problems. They may also have encountered Adolph Voge, but West was their hands-down first choice. Clarence West considered the "information" position one with high potential for science, as well for as business information. But after only two years with the firm, he left for Washington, D.C.

The Council Moves Forward while Kellogg Stays Busy in Zurich

West had received another honor, a call from the National Research Council. In 1921, the council decided not to wait to implement the grand McClung Research Information Service plans. An energetic man was needed to direct the projects.

Clarence West began as assistant director of information operations, but was soon the "man in charge." He was a close ally of those within the council who were oriented to the needs of working, not theoretical scientists. Within a few years, he was placed at the top of the Research Information Service. He directed the service's various critical Mathematical Tables' projects, efforts that gave him experience in and perspective on international work. He and the council's librarian Callie Hull also published numerous bibliographies. Later, he became the pulp and paper manufacturers association's representative for indexing and abstracting matters.

His bibliographic focus began and remained a practical, not a theoretical or idealistic one. He, like others who studied at the state universities, wanted to provide information to those who worked in deadline-filled applied science. That was reflected in his achievements on one of his first assignments in Washington, to find ways to convince the Rockefeller and other foundations to help launch the McClung-Greenman-Schramm project for what became *Biological Abstracts*.

While West and the Research Information Service's group were organizing their campaign, Nina Field anxiously awaited word from the Rockefeller managers. Fortunately, the Rockefellers had assigned the task of making a decision about the Concilium to Vernon Kellogg, a man who had broader visions for science information than did some others in the council. It took longer than Nina expected, but Kellogg finally arrived in Zurich in late summer 1921. It was the first of two trips he would make to examine the Concilium's affairs and to craft new recommendations about its future.[10]

Kellogg had the power to make a life-or-death decision regarding the Concilium. He was the administrative head of the council and was the major Rockefeller advisor for scientific matters. He was also the point man for many other Rockefeller initiatives for European reconstruction. His stop in Zurich was part of a larger survey focused on postwar relief in Central Europe, including yet another assessment of all types of science information needs and related political sensitivities. That included making plans for allocating foundation money for the rebuilding of information systems for scientific fields ranging from biology and medicine through mathematics and physics.

Importantly for the Concilium's future, and despite Franz's earlier conclusions about Herbert's diplomatic skills, the Rockefeller Foundation continued to fear offending any European bibliographers—even the Swiss on the Concilium's supervisory board.

On his arrival in Zurich, Kellogg's examination found the Concilium's operation a depressing muddle. Its debts were far greater than its assets, it had a huge backlog of articles to index, and most of its workforce had not returned. Herbert Field, Kellogg remarked in a confidential letter, had been a terrible businessman. He had not kept good books, failed to collect past-due accounts, and lost track of inventory. Kellogg soon discovered a financial catch-22 that complicated any search for a solution to the Concilium's troubles.

The Concilium was a "Swiss" entity, with a convoluted legal situation, and Herbert had willed his nine-tenths of all the Concilium stock to the Swiss government on the promise it would pay off all debts. However, the Swiss economy was so badly hurt by the war it had no funds to cover the obligations. Thus, it now appeared that Herbert Field's personal estate would be responsible. Kellogg concluded that much of Nina's wealth was at immediate risk.

Luckily for Nina, the children, and Herbert's dream, Kellogg was a believer in the need for large-scale and global information services, and he was predisposed to help the Concilium and Herbert's family. He was determined to solve all the problems, and he was confident that any decision he made would be appreciated in America and Europe because, for one reason, he knew he had great influence among scientists.[11]

Vernon Kellogg Advocates Peace and Internationalism

Vernon Kellogg did have influence and respect. He was a noted entomologist, a zoologist, a well-known expert on evolution, and an esteemed emergency-aid administrator.

Furthermore, he was not one of the strict tooth-and-claw deterministic Social Darwinists. He believed that human intelligence could and should be used to solve problems.[12]

He was very sensitive to the condition of European postwar science and to the needs of those men who had their careers damaged because of World War I. Kellogg had a well-deserved reputation as a humanist who understood the great the suffering of Europe's population. He earned that partly through self-sacrifice during the war. He traveled behind enemy lines during the conflict as the right-hand man of his ex-student and friend, Herbert Hoover, who headed America's enormous aid effort. From then on, much of Kellogg's time was devoted to finding ways to use science to find long-term technical and political solutions to the world's food problems.[13]

Immediately after the armistice, Kellogg had toured Europe. He was tasked to help all of its nations rebuild their science capabilities and their libraries—and to do so in a way that signaled liberal America would not forsake the cause of world peace—no matter what the fate of the League of Nations treaty. So, his 1921 information survey was somewhat of a repeat performance, one for which he was well prepared.

Kellogg knew who to contact, what Europe's scientists desired, and he gauged the potentials for acceptance of a new international information system—one perhaps guided by America, but not dominated by it. He began his trip believing the Concilium should be a significant part, but not the whole, of a badly needed mutual international arrangement, one that could fulfill his high humanitarian goals. Significantly, to Kellogg, science information was not an end in itself: it was means to restore and heighten science's contribution to human welfare.

Kellogg and Documentalist Watson Davis Support a Library Technology Revolution

Kellogg had always supported a wide range of activities to help increase science's public image. Because scientists had yet to become high-status figures in the popular mind, he became a supporter of the efforts of men such as America's Watson Davis to make science accessible to the public and to raise it and its practitioners' reputations. Davis had assumed the leadership of the Scripps newspaper–sponsored James McKeen Cattell's nonprofit Science Service and broadened its mission from just encouraging newspapers to print favorable stories about science to one that would establish a more direct relationship with the "the people" and with influential decision makers. He made use of the famed Cosmos Club (where Herbert had stayed) to lobby the powerful in government and private industry to attend to the needs of his scientific sponsors.

Significantly, Watson Davis was becoming more than a publicist. He was establishing the American version of the European Documentalist movement. His emphasis was on the distribution of scientific literature at the lowest possible cost. In the late 1930s, he began a broad-based organization, the American Documentation Institute. It was the precursor of the American Society for Information Science.

Davis was more than an organizer. By the 1930s, he devised his own version of the earlier bibliographic hopes of Herbert Field and Paul Otlet—one whose outlines predated the Open Access movement by fifty years. Davis outlined a plan for a great and centralized science bibliographic service, one that inspired the grander schemes of England's "Red scientist" John Desmond Bernal. Davis and Bernal sought a cost-saving substitute for the disciplinary journal, an alternative that would democratically disseminate science information with the help of microfilm, the revolutionary technology of the time.[14]

But Davis first concentrated on the use of microfilm in libraries. Vernon Kellogg and the council would muster support for Davis's efforts to find a technological solution to the 1930s research library crisis. They achieved a funding miracle during the era of the Great Depression when, in 1935, they convinced the organization that managed the profits from confiscated German patents to donate $100,000 to establish a program for the development of microfilm for libraries. They regarded microfilm as "the" technology to solve the problems of diminishing space in libraries and the increasing cost of paper-based publication. Freemont Rider, an influential academic librarian, soon joined in their microfilm crusade.

Vernon Kellogg shared more than a sense of an information emergency with Davis. Both had faith that science could better the world because it dispelled prejudices and irrationality. They were not Reds, however. They were staunch liberals, as Herbert Field had been. Davis's writings and Kellogg's policy decisions reflected that. However, Davis did not confine his widely distributed science newsletters to comments on science or information. He had strong views on such things as the Scopes anti-evolutionary theory trial. And, like Kellogg, Davis was an internationalist. He supported information conferences and a newly constructed world language, Interlingua.[15]

Back to the Concilium in Zurich and Barriers to International Systems

Vernon Kellogg shared much with the likes of Davis and began his visit to Zurich in July 1921 with a favorable attitude toward the internationalist in, as well as the information visions of, Herbert Field. In addition, Kellogg had positive personal memories of Herbert. He had encountered him during his World War I food-aid trips. He appreciated that Herbert and his family, and his information dream, suffered because of his work with the Quaker aid groups. Kellogg also had first-hand knowledge of the Concilium. Because of his own scientific research, he had been aware of it since the 1890s. He had also heard many positive comments about the honorable way Herbert conducted himself in 1920 after Angell had questioned his integrity.

Of importance for Kellogg's evaluation of the Concilium's future was Franz's report, which counterbalanced the ongoing negative comments by council insiders such as McClung. In addition, Field's old friends kept up their lobbying on his behalf. After

Herbert returned to Europe in 1920, they sent a stream of letters to the council's office. For example, Ralph Tower, librarian of the American Museum of Natural History, had rethought his 1907–1908 negative evaluations of the Dewey/UDC card systems and made a strong recommendation that the council ensure the Concilium was subsidized forever, and at more than $700,000 a year.[16]

Kellogg responded favorably to such lobbying, but the Concilium was not the only contender for American postwar information funding. Kellogg was an admirer of the Royal Society's science information services and, despite reports such as Franz's, he had lingering hopes of resuscitating the *International Catalogue* and England's disciplinary publications. To complicate Kellogg's 1921 Zurich decision-making chore, the editor of England's *Zoological Record* asked for aid and Kellogg pledged that, at minimum, the council would protect it from any undue competition.[17]

Other complications arose as Kellogg was deliberating his make-or-break Concilium decision. He suddenly faced another surge of opposition. American biologists were making renewed and urgent demands for the support of a McClung-like system focused on quick publication of abstracts. They had one especially persuasive argument. They emphasized scientists needed a way to decide which full article to order before they ran out of scarce funds. That was extremely important, they said, for scientists in poor countries. Others repeated the old objections to the Dewey/UDC system for science classification.

Then, Burton Livingston went public about his objection to "one" science information center. He was now heatedly demanding support for older scientific society publications and the "typical" Small Science professor.

In addition, there was a touch of American chauvinism. Some resolute bibliographers in the United States reemphasized their demand that any center be based in the States because, they argued, it was the only place where "new ideas would not be rejected," as they usually were in Europe.[18]

Given the reenergized opposition and what he unearthed about the Concilium's near-death condition, Vernon Kellogg decided to reexamine the Davenport proposal, and to check with the Rockefeller Foundation about its policies. Nina was dismayed and frightened. Meanwhile, the Concilium was barely alive. To aid Nina, Kellogg asked Johannes Strohl, a professor in Zurich University's zoological department to act as an unpaid caretaker-manager.

Strohl agreed, happy with the idea of saving Herbert's grand dream. Luckily, Strohl had some experience. He also had Herbert's orientations to academic information. Before World War I, Strohl worked at the Concilium as a classifier and served as its vice-director, even remaining faithful after being dismissed because of financial difficulties in 1909. The German woman, Marie Ruhl, who had played such an important role for decades, agreed to continue as the technical director and the only working employee. Even with those arrangements Kellogg was unable to paint an optimistic picture of the center's future.[19]

Revolutionaries Detested by Herbert Defer to a Rockefeller Decision

Nina's concerns multiplied when Kellogg unexpectedly announced he had to leave Zurich because of an emergency. The Concilium, he said, had to wait—again. Kellogg put aside his Zurich work in early September 1921.

He had received an urgent request from Herbert Hoover. Hoover was heading what became another multimillion-dollar American government relief program, and Kellogg had already been helping to plan it. It was reported that starvation, even cannibalism, was sweeping the Soviet "workers' paradise." Hoover needed a first-hand account of the Russian situation from a trusted man who could judge what help was essential and whether the Soviet government would divert aid to further its own Red purposes.

The first years of the great Soviet experiment that Noel Field soon idealized were a disaster. The long and brutal civil war, the seizure of lands by the peasants, misguided Communist economic policies (including confiscatory taxation) and one of Russia's oft-recurring droughts, led to an appalling situation. Millions were starving as agricultural production plummeted. America's government and the nation's philanthropic organizations knew that the crisis was severe. However, they were not sure how much aid was immediately required, or if Russia should pay for any of it. The Red government claimed it had no resources for foreign food relief and did not admit that its latest land redistribution program played a major role in the drop in food production and in the country's severe decline in foreign trade and income.

Kellogg, established as an expert in food relief, was ordered to meet with American and British officials already in Russia; to contact the Soviet leaders; and, to discover as much as possible about conditions through direct observation. Importantly, he was to prepare reports that Hoover could use to convince the American Congress to spend millions more on grain shipments.[20]

Kellogg had been to Russia before, and he knew he was on a dangerous trip. The revolution's upheavals continued and the near barbaric civil war dragged on. Life remained precarious, even for visiting Communist stalwarts, "friendly" foreigners, or the League of Nations relief workers. Kellogg had just met an American the Soviets had imprisoned who told of frightening conditions within and without his jail. That man had received brutal treatment even though he had no link to Russia's counter-revolutionaries, British intelligence, or the Allied landings in Siberia that had generated so much fear among the Soviet leadership. The American's ordeal indicated to Kellogg that even official neutral visitors were at risk.

Kellogg later admitted he was afraid, especially because his excursion had another objective. It was to find out about the status of science and scientific men under the revolutionary government. He was to act as an intelligence agent and explore the possibility of devising a possible rescue operation for Russian scientists and intellectuals threatened by the new regime. That was something the Soviets would have interpreted as being rather hostile. He seriously considered a rescue plan, partly because of what he sensed would happen to biologists. Bolshevik dogma contained much that undermined the best of modern evolutionary science and the Communist party was already showing signs of purging those with "incorrect" views on genetics and heredity.[21]

In a letter to the council describing what was becoming an exhausting month-long tour, Kellogg pleaded that his scientific work be kept secret. He did not wish to suffer the same fate as the imprisoned man he just met. Like Herbert, Kellogg continued to keep his covert work hidden from view. In December 1921, when he returned to America to give testimony before a congressional committee on Russian relief, he did not reveal his intelligence efforts. Also, Kellogg did not mention his role in Herbert Hoover's behind-the-scenes bargain with the Soviets. Hoover had managed a quid pro quo. In exchange for American food relief, the Russians were to release Americans held in their jails, including the man who had headed the weak American undercover intelligence effort in Russia, Xenophon Kalamatiano.[22]

Hoover and Kellogg swayed Congress. The United States became a very generous donor of food and supplies, with multimillion-dollar programs continuing for years. Sympathetic liberals such as Kellogg and Stephen Duggan anticipated that the aid would lead to a softening of the Communist's draconian economic and social policies. However, while he awaited signs of liberalism in Russia, Kellogg began organizing a private relief effort for the Russian scientists, and for Soviet science. His group collected money to provide emergency aid to those who had already fled to Germany, and it sent tons of books and journals to Russian universities. Such work was not confined to Russia, however. Kellogg was part of the operation that was to send thousands of volumes and journals to devastated German universities.[23]

Kellogg gave hints about his views on evolving Soviet science and social policy, as well as his beliefs about eugenics and liberalism. He concluded his scholarly and temperate 1923 book, *Mind and Heredity*, with, "In Russia they assume that anyone can do anything, the only problem being to define the different things needed to be done and to assign by lot each person to a specified task. We do not believe in quite so simple a solution to social organization. I have tried to show one of the reasons why it isn't so simple. The reason is heredity in determining mental make-up."[24]

Armand Hammer, Famine Relief, and Later Connections to Science Information

There was another American in Russia who was involved in 1920s famine relief and, much later, post–World War II science information. In contrast to Kellogg, Armand Hammer always held an admiring view of Soviet policies and science. Vernon Kellogg may have met him in Moscow in the 1920s because the man's luxurious Moscow residence became the headquarters of the American relief effort.

Hammer, a young Columbia University medical school graduate, had different motives and different types of connections in America than Kellogg's. One reason for his attitude was his belief that he would become a multimillionaire by serving the

Communist system. Other reasons were that he, and especially his father, were important to the hard-core Communist movement in America; were Comintern agents; and, were central to the funding of Soviet spies in America such as Jacob Golos, who became linked to the Fields and their friends.[25]

Armand Hammer was a first-generation American whose Russian Jewish father, Yuly, earned a degree at Columbia before its medical school met Abraham Flexner's standards through lengthening its course to more than two repeated years. Yuly, once a bankrupt suspected of financial misdeeds, suddenly became quite wealthy because of his large New York City ethnic practice, his pharmaceutical and illegal liquor operations, and his partnership with a Soviet agent who was running the Allied American Drug and Chemical Corporation. Allied Chemical was a "front," and a very profitable one for the Hammers, as it received many exclusive trade concessions from the Soviet government. In return, the senior Hammer helped found the American Communist Party, supported public radicals such as John Reed and Big Bill Haywood (of the violent International Workers of the World) and funneled money to Soviet agents such as Golos.

In 1921, Yuly was imprisoned in America for the death of the wife of an attaché of the Russian Imperial government while performing what some regarded as an abortion. As his father was appealing his sentence, Armand went to Russia to escape any further legal problems and to salvage the affairs of Allied Chemical. As a late addition, he brought some medical supplies to Moscow. After being allotted a limousine and a lavish home and granted meetings with the Soviet higher-ups in Moscow (including Lenin) Armand and, soon, his father and brother, decided to become involved with food relief. In return for economic concessions from the Soviets, Armand arranged for the credit needed to divert shiploads of grain to Russia.

That was the beginning of an almost decade-long stay in the "land of the proletariat." Armand Hammer became an exclusive representative for many American firms, established a pencil factory, and acted as a go-between for much of the Soviet's foreign trade. Unfortunately, the Hammer family entangled itself in the struggles between Trotsky and Stalin in both Russia and America. Armand's connection with the American political commentator, Max Eastman, was especially troubling to Stalin.

Eastman had become an influential force within America's liberal establishment. The son of Congregational ministers, he studied philosophy at Columbia University under John Dewey. In 1911 he helped found *The Masses*, the influential left-wing magazine that became a near secular bible for young American radicals like Noel Field. Eastman made the grand tour of the Soviet experiment in the early 1920s. He returned with a Russian wife, but with a growing dislike of Soviet policies. After the takeover of his first magazine by a hardline Communist group, he became America's most influential voice for Trotsky. His writings in the *New Masses* (which was published beginning in 1926), and his editing of Trotsky's works, made him and those associated with him marked men. Eastman became one of the many among the intellectual left of the 1920s and 1930s that later turned against all communism. He became associated with a group of ex-Communists and Socialists that exposed Noel Field's Communist involvements.

Meanwhile, Armand Hammer, despite his family's early Soviet problems, became a millionaire, an international art dealer in America, and a liquor and oil magnate. If one believes the FBI's suspicions, for the rest of his life he was an important courier and power broker for Soviet intelligence. There also are reasons for believing that during and after World War II he was an important source for American intelligence.

Kellogg's Last Stand for the Concilium and the Struggle of Science Best Policy

In contrast to Hammer, Vernon Kellogg remained focused on his relief and information work. Although fatigued by his 1921 Russian expedition, and by his trip to the United States, Kellogg did not desert the Concilium. He returned to Zurich for a second time in December. It took the efforts of another auditor, and two weeks of Kellogg's own work, to decide that Herbert's center might possibly be revived. But a decision by the Rockefeller Foundation to provide long-term support now hinged on more than Kellogg's reputation, or his good will toward Nina.[26]

By 1922, the Concilium had become the center of a now obvious and deep conflict between the old and new guards in American science—and, between applied and theoretical workers. That long-building split over American science information policy surfaced in a very public way. The Livingston-AAAS and council representatives protested against what they suspected would be a positive recommendation by Kellogg, arguing the Concilium was a block to their plans. That made Kellogg's tasks an exercise in trying to be a mediator in strife-filled academic politics. The disagreements were deeper than the particular one over what became *Biological Abstracts*. Kellogg suffered under powerful cross-pressures.

After his initial visit to Zurich, the upsurge in complaints about the possible favoring of the Concilium was the result of more than parochial contests between a few nascent information scientists. The men from Wistar, Davenport's allies, and those who agreed with Burton Livingston that science's organization should remain as it had been for the last half-century, were motivated by their deep and differing opinions about science best-practice.

The 1920s was a time for decisions about the next phase in the development of American science, its universities, and science information. At the center of the debates were the council's Big Science advocates.[27]

Within the council, leaders such as McClung lobbied for what they considered the best form of information for Big Science. The lobbying worked, and by 1920 the council was committed to two things: abstracting services and information internationalism with the council playing a major if not dominant role.[28]

The abstracts were not to be the type Herbert's staff wrote when titles were inadequate. Nor were they to be like the terse descriptions written by Julius Victor Carus, the famed zoological bibliographer from Leipzig. The new summaries would be more than a few lines. They were to be one- or two-hundred-word descriptions of subject matter, methods, and conclusions. Such summaries could, many advocates claimed, replace all other forms of scientific bibliography. The abstract services would send out synopses of all the important literature in a field, and on a regular and timely schedule. The new abstract's advocates argued that the publications could efficiently survey the world's literature and could do so without translators, editors, or subsidies. The services could be so efficient (partly because of volunteers) that they would have low subscription rates allowing individual scientists in poor countries to afford the abstract journals.[29]

Of immediate importance to Vernon Kellogg's deliberations in 1921 and 1922 was the work of the abstracting advocates from America's Wistar Institute. They had more than ideas ready. They had practical goals and plans that included a direct threat to the Concilium's work.[30]

As Kellogg was at the final stages of his considerations of the Concilium's future, Jacob Schramm became reenergized and took the lead in the abstracts' cause. He was urging the council to finally convince the major foundations to give their limited funds for information to the proposed *International Biological Abstracts* that was to be located in Philadelphia and managed by the Wistar Institute. Clarence West joined the battle. He and Schramm promised a cure for the information disorder caused by the explosion of biological publications that was leading to some seventy-five thousand articles a year in the mid-1920s.[31]

Wistar's group had become unusually effective academic politicians. They convinced the council to give the help needed to form the Union of Biological Societies.[32] Next, Schramm made a direct hit on the fundamentals of Herbert Field's Concilium. He bolstered old arguments with "hard" facts. His group had prepared a survey and sent it to thousands of practicing biologists. As they had anticipated, the results showed that the overwhelming majority of scientists wanted book-form abstracts with, at most, simple alphabetic subject indexing. They did not use or want bibliographies with complex indexing; they did not want the card files; and, they did not want large cumulative files! That survey was something Kellogg could not ignore. (Studies of scientists' information-seeking behavior seventy years later echoed its findings.)[33]

The Wistar group had more ammunition. They proposed the use of *Chemical Abstracts'* successful "bundled" business model for their publications. Schramm's group emphasized that *Chemical*'s format had proven popular and financially viable since 1907. Those successes were among the reasons why the Council had selected Clarence West to fill its new information-specialist position. He had been at *Chemical*.[34]

While Schramm pushed his initiative, another and more general proposal was circulating around the council. Given the economic situation in Europe, the dislike of the Concilium in a few European circles, and the reports from Americans about frictions on the continent, it seemed even more likely that some group in the United States might have to take over the task of co-coordinating international science information. As the council awaited word on the fate of Paul Otlet's center, the of League of Nations' ambitions for an international science reference center, and the Royal Society's projects, thoughts were being put forward about convincing the foundations to support an International Bibliographic Office in the United States and, perhaps, an institute to teach bibliographic methods.

Charles Davenport and his AAAS committee soon provided a variation on that. They had taken that idea and wedded it to the Concilium's needs. They again advocated turning the Concilium into just such a comprehensive international information clearinghouse and laboratory, and to move it to the United States. Some council members, including Vernon Kellogg, had begun to entertain the concept.[35]

Protecting "Small Science": Burton Livingston and James McKeen Cattell

Meanwhile, Burton Livingston was becoming more active in representing the "average" scientist of the American Association for the Advancement of Science. The AAAS was not powerful, or wealthy, or intent on reforming American science. Most of its members liked things as they had been. They represented that "third voice," the academic who was not involved in either abstract research or large-scale applied science. After learning of the council's latest campaign for centralized information services, they reacted—even more heatedly than they had done when they realized the implications of Davenport's first 1919–1920 Concilium proposal. The secretary of the association, Burton Livingston, wrote many long protest letters to the Council. All contained objections against plans that would impose one information system on the many diverse fields of knowledge.[36]

Livingston's complaints about the council's policies were heated, but mild compared to those from the man who had been the driving force in the AAAS's growth since the 1890s, James McKeen Cattell. The son of a very wealthy family, Cattell devoted his life to the new field of psychology and, then, to science information. With American and foreign degrees in hand, he quickly became America's first noted quantitative psychologist. He also became a full professor at Columbia University when he was only thirty-one years old. He taught a generation of graduate students there, including Clark Wissler, and, at the same time, founded and edited a number of academic journals. He was the man who brought the journal, *Science,* back to life in the 1890s and then pushed it and the AAAS into public view.[37]

Cattell never pulled his punches. His straightforward manner, and years of friction between he and Nicholas Murray Butler, led Cattell to be dismissed from his post at Columbia University when he publicly voiced opposition to sending American troops to Europe in World War I. The dismissal did not end his career, however. While pursuing a successful lawsuit against the university, and while leading a crusade to establish academic tenure (and thus freedom from political pressures) for the American professoriate, he rose in the ranks of the AAAS. He became its president and he accepted positions such as that of the association's representative to a newly formed postwar international research coordination organization.

However, Cattell was a prominent critic of the National Research Council. From its birth, he contended with the council over funding issues and what he saw as its elitist and group-dictated approach to science and research. While on one of his European trips to secure a role for the AAAS in proposed League of Nations intellectual and scientific bodies, Cattell learned of the council's various information plans. He disliked them all and wrote of his displeasure to Vernon Kellogg.

Although he was not a socialist or radical, Cattell likened the council's plans to capitalists trying to take over unions, rather than cooperating with them to save the economy. Cattell also objected to the new plans for the Concilium. Although at times he served on the Concilium's American oversight board, he interpreted the Rockefeller role in the 1920s campaign for the Concilium as another attempt of elites to dominate science. He wanted the great foundations' limited funds distributed, as did Livingston, to the independent societies and their journals.[38]

Kellogg's Diplomatic and Compassionate Compromise

By late 1921, pressures from the Wistar group and professional leaders like Davenport, Livingston, and Cattell had become burdensome complicating factors, if not disruptors, of Vernon Kellogg's explorations of the Concilium's future. The deliberations over the center's rescue became tension filled. Kellogg was torn. Out of respect for Herbert Field and compassion for his family, he wanted to save the center. But out of loyalty to the council's possible strategic plans he wanted to avoid starving the other possible council information initiatives or alienating important scientists such as Livingston.

Kellogg settled on a compromise. He returned to Zurich in December 1921 with a watered-down version of Charles Davenport's ambitious $3 million proposal. Nina, Strohl, and the Swiss were disappointed and upset. Their emotional reactions compounded an already difficult situation.

The final negotiations were stressed and, at times, near to collapse. First, there were legal technicalities concerning the role of the Swiss government. They were finally overcome. The Swiss were to retain a place on the Concilium's supervisory body, but not to own the center. The new Concilium was to be a freestanding enterprise. Then came Kellogg's delicate task of avoiding the appearance of being an overbearing American, while at the same time making sure that Johannes Strohl, now recognized as the permanent leader of the center (with an assist by Marie Ruhl) followed good business practices.

Importantly, Kellogg had to gain promises that the two would act in cooperation with any of the council's future information programs. The exact nature of those programs remained to be defined, as did that powerful term, "cooperation."[39]

Kellogg counterbalanced his implicit demand that the Concilium follow the American lead by arranging for the Rockefeller grant to go directly to the Concilium, not the council. That gave Strohl some feeling of independence. After that, came a very embarrassing requirement for Strohl. There was to be a supervisory council / Rockefeller committee to work with the Swiss. It had the potential to micromanage, perhaps even damage, the Concilium.

Strohl was hesitant to yield any autonomy; but he acquiesced. He pledged he would: follow council business strategies; send frequent financial reports; allow an independent auditor to monitor the books; and, aggressively campaign for subventions within Europe. Strohl made additional commitments. He promised to make the Concilium self-sufficient within four to five years. He pledged he would never again (underscore "never") request additional funding from America after the termination of the five-year Rockefeller grant. He also reaffirmed that he respected the council's expectation that the Concilium was to be a part of an information system, not an independent center devoted only to Herbert's old biology products.

There was, however, no mention of the Concilium becoming the core of any international system, and Kellogg once more told Strohl that he was expected to go along with any forthcoming recommendations once the council's committees hammered out the final details of its strategy for international information. Strohl agreed, although he had little idea of what that might mean for the Concilium.

In early 1922, both Strohl and Kellogg were optimistic, although the final agreement had shortcomings. Strohl was able to hire a new full-time secretary-manager, Mr. E. Meyer, and to delegate details of the rebuilding to him and Marie Ruhl. There was another reason for optimism. With the predicted demise of the Royal Society's projects, the Concilium seemed to have an opportunity to obtain the funds many European nations had devoted to England's efforts. England's withdrawal could also mean more customers.[40]

Johannes Strohl Ushers in a Rocky New Beginning for Herbert's Commitments

Strohl's first duty was to bring the Concilium back into the existing international information system. He had to appear to be cooperative. He immediately began a stream of correspondence with Paul Otlet and Henri La Fontaine.[41] Strohl's initial task was to assure them of the Concilium's recognition of Otlet's leadership in standard setting despite Zurich's past disagreements with the Belgian center. Then, he had to help overcome an unexpected threat. The United States declared it would impose a tariff (a high one) on the importation of products like the Concilium's. When Strohl informed Herbert's friends at the American Museum of Natural History about it, they called on their political allies and averted a disaster. The government quickly reclassified the Zurich cards as nontaxable scientific items.

Despite Strohl's energetic start and his optimism about the Rockefeller agreement, there had been some things left unspoken between Strohl and Kellogg that soon led to disappointments. Strohl was committed to Herbert's form of numeric subject indexing, and he had an unshakeable loyalty to a bibliography based on a full reading of each document. A complete reading of an item by a trained subject specialist, Strohl believed, was the only way to keep faith with Herbert's dream of indexing ideas, not just documents. And, he believed a thorough examination by a professional was required so that all items relevant to a particular field could be noted, even if they received little space in a document. He was also determined to preserve the "memory" of science achievements through the cumulative card-file system.

Meanwhile, Kellogg was coming under increasing pressure to support the supposedly less expensive abstracts, ones without Herbert's detailed numeric or cumulative indexing.

Strohl delayed confronting one difficulty with "cooperation." Frictions between Otlet and the Concilium soon developed. Almost as soon as he took command of the Concilium, Strohl grew increasingly dissatisfied with Otlet's management of the Universal Decimal Classification. Strohl soon contemplated ending the time-consuming dependency on decisions in Brussels. He knew a break with Otlet would not be in harmony with the council's goal of avoiding any conflicts or competition with existing European organizations or products. However, he thought he might have no choice if Otlet created more difficulties as the Concilium restarted operations.[42]

The Council, Nina, and the Future of the Field Family

Meanwhile, in February 1922, before any frictions had surfaced, Kellogg returned to America and began the final negotiations with the council's committees and the Rockefeller administrators about the Concilium's grant. Bargaining with the council was not easy. Explaining the new Concilium plan to the Rockefeller managers was just as difficult. However, the hesitant foundation managers approved the funding arrangement and Herbert's center became one of the council's major projects of the decade.

The Rockefeller grant was, however, one-third of what had been proposed in early 1920 by Charles Davenport and his Columbia University allies. As well, the grant was for only five years, and if the Concilium's operations did not match expectations, funding could be terminated at any time. Yet the amount was significant, close to a million dollars, and it appeared adequate for the restoration of at least the original Zurich service.

Kellogg accomplished more in early 1922. He was able to channel considerable aid to Nina and the children, although neither he nor the foundation had a legal obligation to do so. Kellogg arranged for some of the foundation's grant to be used to compensate Nina for a large portion of the Concilium stock Herbert had willed to the Swiss scientific society. In addition, he gave Nina the remainder of Herbert's 1921–1922 salary. He also made sure that she was cleared of any Concilium debts. It was all a matter of "simple fairness," Kellogg reported to the council's board. The foundation's generosity to Nina amounted to approximately $100,000.[43]

Nina consulted her lawyer and then had Vernon Kellogg contact Herbert's stepbrother Edward Field in Brooklyn. He and Lilla were still at 104 Columbia Heights, but Edward had emerged from his emotional darkness of 1919. He agreed to act as Nina's attorney, to accept her council / Rockefeller funds, and to invest them. Edward informed Kellogg that he put Nina's money into bonds of the Dutch East Indies Corporation, which, he said, were yielding a very handsome rate of return. The budding left-winger, Noel Field, perhaps never realized that at least part of his future education and travels were funded through one of the most imperialist of enterprises.

Edward was not the only capitalist who helped secure Nina and the children's economic future. Nina later found she could rely on other members of the extended Field family. Herbert's favorite nephew Henry Field managed Nina and the children's money after Edward retired. He proved to be a rather skilled investor; as a result, Herbert's children would have a chance to live much as their father had—at least in a solid middle-class style.[44]

Strohl and the Rockefeller's Expectations

As Nina prepared for the next stage of her life, Johannes Strohl continued his rebuilding of the Concilium. He had a rather handsome salary, one higher than Herbert ever received. One-half came from the university because he remained on its faculty. Marie Ruhl received a raise. Both expected to act independently, as had Herbert. But the council soon pressed both of them to keep the several promises Kellogg had written into the Rockefeller agreement.

Kellogg soon had to assert one meaning of "cooperation." The Concilium was "encouraged" to work with those who were trying to establish America's *International Biological Abstracts,* and with the editor of London's *Zoological Record*—even though they were competitors that might lure away subscribers and benefactors. In addition, the council began micromanaging the Zurich center. Kellogg wrote Strohl that he expected the Concilium to resume full publication of its cards. He believed that immediate resumption was necessary to rebuild the customer base and to put the center on its way to self-sufficiency.

Kellogg was in a hurry. He had to show the foundation and the critics of the Concilium's methods that money was not going to be wasted. To kick-start the Concilium, the council itself paid for advertising broadsides. They appeared in U.S. science journals as early as March 1922. They announced the regeneration of Herbert's dream and the valuable role Zurich could play in American science. The advertisements emphasized that England's *International Catalogue* project was dead, so the Concilium's product would always remain uniquely valuable. The marketing once again emphasized one of the mandates of the council, to modernize science and its practices. The advertisements stressed that Efficiency (Taylorism) had to be applied to the field of science information and that the Concilium with its centralized international service was a prime example of such efficiency.[45]

In response, in spring 1922 a few subscribers returned; not many, but enough to temporarily satisfy the center's American and Swiss overseers. Strohl had also been at work writing articles aimed at England and Europe. He assured his readers that the Concilium was alive and about to resume all its activities. Strohl then took a rather courageous intellectual step. He did it perhaps at the urging of Kellogg and the Concilium's American zoologist friends. He began making the first revision of the Concilium's classification scheme in twenty years. He did it, he wrote, because the practice of zoology had changed so much since Herbert Field created the system. Zoology had turned away from comparative anatomy and hierarchical classification to questions of general biology and, importantly, experimentation. Most in need of recognition in the Concilium schedules were studies of heredity, cytology, and embryology. Strohl did not state it directly, but he brought the Concilium's classifications into the realm of the social and political hot-button issues of the time. There was another reason for the revision of the classification codes, he stated. The political upheavals following the Great War necessitated a revamping of the geographic codes.[46]

Strohl was also responding to the demands to put the Concilium on a sound financial basis. He began drawing up a new price schedule. When he issued it in mid-1924, it pleased the Rockefeller auditor. But Strohl's subscribers must have winced at the hefty subscription prices and his rather self-defeating demand that orders for old cards be for an entire file (at twice the cost per card of the current cards). He then announced that back issues of the book-form *Bibliographica Zoologica* would cost the equivalent of three hundred cards. Strohl priced the new *Conspectus* guidebooks, one for each of the main file's subdivisions, at the charge for twenty new cards.[47]

Old Problems Bite Back: Inventory and Book-Form Bibliographies

The Concilium's reincarnation then faced disappointments. Strohl believed that without copies of the card back-files, spanning 1896 through World War I, new subscribers might feel cheated. Certainly, the integrity of Herbert's system would be weakened if the entire file, some hundreds of thousands of cards, were unavailable. Strohl rushed a letter to Kellogg begging him to stop advertising the Concilium's rebirth until the depleted back-files were copied—perhaps by hand and at enormous cost per card. Kellogg and the council's supervisory group were not happy about that. There were letters to Switzerland demanding immediate sales of new cards.[48]

Strohl's decisions certainly frustrated Kellogg, but there were other concerns. Robert Yerkes had just informed him of some serious criticisms of the finalized Concilium grant. They came from men whom Kellogg thought had agreed to his compromise settlement. The first of the criticisms came from H. W. Wilson, who again advised that only book-form bibliographies were economically feasible. Wilson had done a thorough study of the possibility of his firm offering some specialized bibliographies in card form and found they not only cost sixteen times as much as the book form, but also that librarians disliked card files because they were expensive to maintain. Wilson wanted to help both the foundation and the Concilium and thus wrote to Kellogg that the Zurich enterprise was on the wrong bibliographic track. The grant, Herbert's friend said, was a waste of precious funding.[49]

After that came some stinging protests. The Wistar group contributed several more uncomplimentary suggestions about what should be done with the Concilium and its grant money. Although they were members of the oversight board of the Woods Hole, Massachusetts, marine research center, and had recently approved of the transfer of a full set of Concilium cards from Minnesota's Carleton College to the center, they hurled complaints against Strohl. Underlying all of the Wistar protests were two assumptions: that the council was the only interested stakeholder in the Zúrich center, and that it, not Strohl or the Swiss or the Rockefeller Foundation, should determine its policies.[50]

In the Name of "Pep and Push" Come Jibes and Complaints

Milton Greenman then sent rather detailed recommendations about operational matters. While mild in tone, they had the potential to undermine the Concilium and the Rockefeller commitment to information internationalism. He wanted the center to move to America where men of energy and "pep and push" could take charge of it. He wanted a radical simplification, if not elimination, of the Dewey/UDC coding, and demanded the cards be sent out every week. He insisted that the Concilium quickly gain five thousand (not five hundred) subscribers. Jacob Schramm, who was on the Rockefeller supervisory committee for the Concilium, then shrilly criticized Kellogg's earlier decisions. He also told Kellogg to move the center to America, whether or not Strohl approved, and he advised Kellogg to abandon the card system, completely and immediately. There also were protests over business particulars. There were complaints that Strohl was not collecting past-due bills quickly enough, and that his decision to give discounts to European countries having currency problems was unwise.[51]

Then came some attacks from usually friendly sources. Robert Yerkes hinted that he discovered some new information methods that made those of the Concilium look tired and outdated. Yerkes was excited about the visit he just made to an American company, West Publishing, which was doing revolutionary work in legal information.

Others pestered Kellogg with complaints about allowing Strohl to avoid composing abstracts until the old Concilium service was fully restored. Strohl was also the target for more jibes because, it was claimed, he was showing little sign that he was going to link, as promised, with international information efforts such as the League of Nations' emerging new organization for international intellectual cooperation and its committees on information.[52]

Kellogg's Concilium Efforts Waver in Light of the League's

Kellogg was tiring. He began to retreat from the Concilium. He failed to defend Strohl against the accusations of a lack of internationalism, even after Strohl signed off on a mid-1920s League of Nations–sponsored international recommendation to share and immediately disseminate abstracts that every author was to compose and submit with each article. In addition, Kellogg did not tell the Concilium's detractors that although he was becoming a central force in the league's information-related committees, along with Henri Bergson, Albert Einstein, and Bertrand Russell, he had forgotten to make sure that Strohl was invited to their meetings. However, Kellogg seems to have been instrumental in ensuring that Wistar's man, Jacob Schramm, became an official American representative to the International Committee on Intellectual Cooperation's new subcommittee on bibliography and Documentation.[53]

The link between Wistar and the international intellectual effort was important. The league now had ambitions to become a dominant force in the coordination of international bibliography. Kellogg, being a liberal internationalist, had to support the league. He had also come to believe that league-related information efforts would aid American science. Yet Kellogg again failed to ensure that Strohl played a role in the many bibliographic meetings sponsored by the league's committees, in a related grand Swiss international bibliographic project, or in a league-related bibliographic research institute soon founded in Paris.

Kellogg decided, however, to give Strohl additional chances to save the Zurich center. He reminded all the critics that the Concilium had a full five years to prove itself.[54]

At the same time that Strohl was coming under-fire, Nina Field was focusing on her children's' welfare. She had not cast off the Concilium, but it slid far down on her list of priorities. Anyway, she knew she could then do little to bolster Strohl's standing with Kellogg or the foundation.

II The Road to Intrigue, Revolution for the World before an Information Revolution

15 Information Science Infighting, Round 2—and a New Generation of Fields Comes of Age

In summer 1922, as Vernon Kellogg and Johannes Strohl labored to resuscitate the Concilium, Nina Field pondered her future. She hoped that she could convince Noel and Hermann to join in the Concilium's management after they completed their education, but her immediate problem was where to settle her family. She felt alone in Zurich and wanted help and companionship.

Nina surprised her own relatives when she announced she was not heading back to London. Rather, she would move her children to America, although it had never been their home and despite her being close to fifty years old with four young dependents.

She had not consulted the children. When they learned of their destination, they were angry. Noel protested, vehemently so. He told Nina that Zurich was his home, the only one. He was correct. At most, he had spent a few weeks in the United States. There was an additional reason for his protest. He did not wish to leave the girl he had loved since he was nine, his classmate and the family helper, Herta Vieser. Nina calmed him with promises that at the first opportunity she would bring Herta to visit the United States, and that she would finance Noel's visits to Zurich.[1]

Nina then stunned the American clan. Her passports listed Brooklyn as her hometown, but she did not feel it was necessary to be close to the Havilands or to the Fields who were managing her finances. Although Edward, Lilla, and Henry were there, and while there was a chance that Hamilton might allow her to use one of the townhomes, Brooklyn was not to be Nina and the children's destination. She decided to take her family to Harvard University's hometown, Cambridge, Massachusetts.

The immediate reason for selecting the Boston area was Noel's education. Hurried correspondence with Harvard's administration, and help from Herbert's Cambridge relatives, the George Parkers, had gained Noel admission to Harvard's freshman class in the fall. The Parkers had also indicated they would help finance the educations of Nina's other children.

As a grand farewell to Zurich, Nina allowed Noel to have a huge antiwar rally at the villa. He pledged to his friends and classmates that he would never give up the cause of peace, even while in the far-off, league-rejecting America. He more than kept that promise.

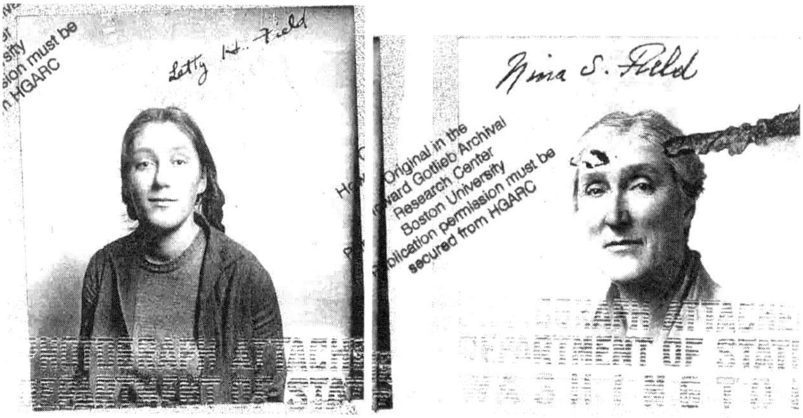

Figures 15.1 and 15.2
Letitia (Letty) and Nina Field, circa 1930, from Gotlieb Archive

In August 1922, after a brief stop in England, the Fields arrived in Massachusetts, just in time for the children to begin school. They first lived on Harvard's lovely historic faculty row in dear "Aunt and Uncle" Parker's house; then, there was a stop a few blocks away at 14 Sacramento Street. Finally, they settled into a small but comfortable home of their own at 21 Berkeley Street, a few doors from the Parkers. Nina purchased the home with the proceeds from the Concilium settlement. Later, Nina would move a short distance to 31 Concord Avenue.

All the homes were in an ideal neighborhood. The cream of America's professoriate surrounded the family. The Fields' neighbors were sophisticated, cosmopolitan, intellectually stimulating, and typically liberal internationalists. The local schools were nationally recognized. And, the houses were within walking distance of Harvard University with its wide range of cultural offerings.[2]

Nina Field's Endeavors to Provide the "Best," Especially for Noel, Her Needy Son

It took some time for Nina to calm down about the family's finances, but she finally realized that Vernon Kellogg had been generous. She also discovered that her situation was better than she thought. Because her own money and Herbert's old investments, including the trust funds for the children, were beginning to rebound from their low points in the postwar recession, she became confident that she could raise her family without too many material sacrifices.

She was not rich, however. Her funds were limited during the early 1920s. Unlike at the Parker's household, there was no maid at 21 Berkeley Street, and Nina did have, at times, to work. That was one reason why she drew a line for the children. Although none of them would have to take jobs during their school and college years, she told them they would have to strive toward careers. Nina's top priority was providing the education she believed they needed.

With the help of the Parkers, the children received the best schooling elite liberal America offered. That included the soon famous Shady Hill School and Cambridge's celebrated college-preparatory public high school. All the children except Noel went to the legendary Cambridge Latin High School, not to a private Quaker school or, of course, to a regular public school.[3]

After the first years in America, the financial situation improved. The family had more than enough income from investments to live a better-than-average middle-class life. Herbert's children never had to face the threat of dropping far down the social scale. Helped by the settlement of her father's estate in the later 1920s, some recovered German investments, and the expertise of Herbert's favorite nephew, Henry, the economic situation became even better for Nina in the 1930s. Although the Wall Street panic and the Great Depression threatened some of their investments, capitalism treated Nina and her children very, very well.

The Fields could even spend summers at Cape Cod and Woods Hole. "Uncle" George Parker was one of the founders of the marine institute at Woods Hole and always made the children feel part of it. For example, the bright and inquisitive Elsie found summer work as an assistant to the professors doing research there. Living in the Boston area and vacationing in such places led to important contacts for the children.

The family also frequented the old Field "summering" places in Maine. There often was enough money for frills, even during the depressed 1930s. Hermann spent one of his last college summers at an English folk-dance camp in Amherst, Massachusetts. There was also enough for something unusual: Nina put Elsie through Columbia University's expensive medical school. In addition, from 1928 through the 1930s, there were almost yearly family trips to Europe.

Herbert's wish, that Noel attend his alma mater, had been the motivation behind Nina's move to Cambridge in 1922. Although Noel could have lived on campus while Nina located elsewhere, she felt her eldest child wanted her nearby. Nina sensed that Noel was different and required constant special attention. She was correct. He was needy. Settling in the college town allowed Nina to look after the shy and withdrawn Noel as he tried to adjust to America.

Hamilton Continues to Bewilder the Family, Especially in Death

When Nina was in Zurich planning her children's future, Herbert's views on what had happened to his beloved Columbia Heights had made Brooklyn seem very unattractive. The picture Herbert painted of Hamilton Easter Field's new lifestyle convinced Nina that she could not rely upon him. She also worried about his friends' possibly negative influence upon her children. The fears about the Heights and Hamilton's clique had not ceased when Nina learned that Hamilton had unexpectedly died at the young age of forty-nine, almost one year from the day of Herbert's death in 1921.

Hamilton had also made an unanticipated decision. It was shocking to all the Fields. He chose not leave his properties to the family. He gave them to his protégé and artistic partner, the famed modernist sculptor Robert Laurent. He also made Laurent the executor for the complex and legally tangled estate. Hamilton's choice came despite all the discussions Herbert had with him over finances and family responsibilities in 1919 and 1920.

It seems that Hamilton had consciously decided to make a statement about his regard for his relatives and their traditions. The members of the extended Field clan received pittances in the will itself. Laurent was bequeathed the townhomes at 104 and 106 Columbia Heights (and probably number 110 as well), and the will gave him a controlling interest in the properties in Maine and New Hampshire. He most likely was the recipient of the proceeds from the auctioning of some of Hamilton's extensive art collection. The sales were held within months of "Easter's" death and brought in close to three-quarters of a million dollars.[4]

The will had left all of Hamilton's relatives bewildered, discontented, and alienated from Laurent. Nina was puzzled and concerned. If Hamilton had bequeathed his apparently sizeable fortune to the family, Nina would have had more options. She might have felt she could allow her children to live somewhat as Herbert and Hamilton had, as near permanent wanderers without practical-life burdens.

Fortunately, negotiations led to somewhat friendly relations between the Fields and the sculptor. Edward, Henry Field, and Nina reached a settlement with Laurent who, by 1923, had found that Hamilton had many unknown debts. Feeling somewhat sorry for Laurent, Nina and Edward agreed to accept partial payments (some $60,000) to settle what was owed them from loans to Hamilton and from Lydia's estate. Nina, although taking title to the Ogunquit property, allowed Laurent to use it if he paid the mortgage. Herbert's stepbrother Edward and the perpetually coal-stoking Lilla did not die penniless after all. By 1930, while they continued to run the old mansion at number 104 as a boarding house, they owned it.[5]

Although he had been part of America's radical artistic and intellectual set, Hamilton had not become a practicing revolutionary Communist, and his premature death from pneumonia in April 1922 ended any direct ideological influence on the impressionable Noel and Hermann. Nevertheless, he left his culturally radical mark on Nina and her children. For example, soon after returning to America, Noel attended some of the political "seminars" at the schoolhouse near Hamilton's old Ogunquit arts colony. The seminars reinforced the impressionable Noel's pacifism, internationalism, and disgust over America's rejection of the League of Nations. Later, after the strained relations between Hamilton's protégé Robert Laurent and the Fields were healed, Noel and Hermann, and even young Elsie, frequently visited the Maine farms and the arts colony—and met the artistic and political radicals who gathered there.

The accommodation with Laurent reinforced Nina's faith that her children would not be deprived. However, she knew they would never be able to go through life as had Hamilton: a social rebel without a job. So, she spent her income preparing them in every way for middle-class professional lives and a place in established intellectual circles. They were exposed to the "best" ideas by attending area lectures. They enjoyed Boston's finest music, theater, and arts; and they became linked to America's intellectuals.

Nina was careful with her money, but she indulged in some luxuries for her children. Nina went so far as to help Hermann buy a lovely farm-retreat in Shirley, Massachusetts, about forty miles from Boston. There were prior connections between the town and the Fields. The hundred-acre Valley Farm and its buildings dated from the eighteenth century, and Hermann declared he would keep its colonial character while turning it into a showplace for modern conservation methods, as well as a vacation spot for the family and a meeting place for creative thinkers. Nina accepted Hermann's youthful and lofty aspirations while she continued to warn her children that she would not tolerate, and certainly would not support, perennial "students."

Noel, Wanting More than Herbert's Liberalism, Suffers Rejection

Although Nina was not a radical, the Fields were drifting from being just peace-lovers to being left-wingers—despite father Herbert's earlier admonitions against Reds and extremism in general. After just a few years in Cambridge, Hermann, Elsie, even Nina, became more than sympathetic to reform causes of the left.

Noel went much further. Soon after reaching America, he became a devoted ideological adherent to communism and to its most virulent form, Bolshevism. Despite Herbert's warnings, he became a "true believer." He did so long before the Great Depression, the rise of fascism, the Popular Front movement, or the heart-wrenching Spanish Civil War. Noel was ideologically committed to Marxism well ahead of the time when the American Communist Party began to attract others than recent East European immigrants, and he supported "the cause" before it became fashionable for young American or British intellectuals to be "on the left."[6]

Being a youthful intellectual Red did not lead Noel to disregard his mother's career warning. He whizzed through Harvard, as had his father, and with academic distinction. He left when he was just twenty years old. His Swiss high school's courses had equaled the first two years of American college. But that was not the only reason for such a short tenure. Noel was brilliant and hardworking.

Noel exited Harvard with a different perspective on his life's mission than Herbert's, partly because he had taken a general studies (not science) program with an international studies concentration. Such programs had some new components, the social sciences. They reflected a belief that the methods of the natural sciences (not old moral philosophy) could be successfully applied to social problems. There were, the social scientists thought, "laws" to be found for society as well as the physical world. For idealists like Noel, once the laws were discovered there would be a short path to progress and equality. Although his professors were not Marxists, his exposure to the new social sciences reinforced his earlier hopes and deepened his devotion to the cause of peace.

Noel Field's devotion had a price. Although he received academic honors, he was not a fully rounded undergraduate. He was reticent, he still had a speech problem, and he was somewhat of an eccentric. He was a loner, and his few friends were socially unusual because they were from the radical fringe of the college and town. Noel and his Cambridge friends had gone further to the left during their college years than John Reed, a Harvard undergraduate of a decade earlier.[7] In 1924, Noel was not a regular Communist Party member, but as he entered into adult life he longed to become one of the "anointed."[8]

Noel rushed to fulfill his leftist and world-peace Callings. He decided he was not going to take any steps toward becoming a permanent student. He was not going to graduate school or to make the usual European cultural grand tour. Rather, he applied for a job in the American diplomatic corps in 1924. He had superb qualifications and recommendations. He was multilingual; he knew Europe; he had researched the League of Nations and peace proposals; he had a shining academic record; and he was known to rising stars in government circles, such as Herbert's friend Christian Herter. Noel anticipated an immediate and positive reply.

The reply was negative. The State Department told him that he was too young. He was placed on a waiting list and advised to resubmit an application in a few years. That stunned him. He would always resent the department's decision.

Noel's Commitment to His Own Callings: The Great Experiment and Democracy as Scientific Planning

The rejection did not end his devotion to his Callings. Also, it did not mean he would go back to academia, or have to go to work at a perhaps demeaning job and become a part of his idealized working class. He had some time, Nina had a bit of money to spare, and Noel was soon to receive the remainder of his trust fund. So, after again promising Nina that he would not remain idle, Noel changed his mind and decided to travel in Europe. However, it was not to be the typical Ivy League graduate's tour of the arts of the continent, nor were the important destinations Zurich and the Concilium.

Noel first went to Lugano, Switzerland, where his love Herta and her mother had resided. Then, as part of a YMCA mission, he observed the greatest "social science experiment" of all time, the Soviet Union. The Soviets guided the young Noel through an impressive tour of the rising proletarian paradise. He was overjoyed with what he was allowed to see. The tour reinforced all he had learned from reading Marxist literature and from the late-night discussions with his radical Cambridge and Ogunquit friends.

The apparent progress of a once-backward peasant land also validated Noel's belief in central planning by educated elites and the value of Efficiency. For Noel, the Great Experiment had all the advantages of modernization and none of its drawbacks, such as gross economic inequality. As a result, as other young, and not so young, intellectuals from the West later did, Noel accepted a new definition for *democracy*, one that had little to do with voting or popular will.[9]

Noel returned to Western Europe, making sure that he had time to stay in Geneva to see his father's great hope, the League of Nations, in action. At his mother's request, Noel made a stop in Zurich. He gauged and reported on the status of his father's information legacy and evaluated the chances of the Concilium again being a major player in science information.

Back at the Concilium: Strohl "Keeps the Faith" and Angers Kellogg

In the mid-1920s, the Concilium's situation was less than satisfactory. It was alive but not vigorous. Johannes Strohl had accomplished little and the positive relationship with Vernon Kellogg and the National Research Council was turning into a tense standoff between Strohl's standards for Herbert Field's "true" academic bibliography and the practical orientation of the Wistar Institute men.

The Concilium was in political trouble. Within a year of awarding the February 1922 grant, Kellogg began to question the wisdom of his decision to aid the center. He sensed that Strohl seemed intent on not cooperating with any other service and that he was obsessed with what Kellogg was beginning to evaluate as old and inefficient methods. Kellogg was also accumulating a growing list of new disparaging comments about Strohl, especially his refusal to be, as he had promised, Efficient. It appeared he was not willing to join the struggle to end duplication of bibliographic efforts; he seemed reluctant to fulfill the promise of composing useful abstracts; and, he appeared to be fabricating excuses in order to avoid preparing his cards from abstracts provided by others.

It was not just appearances. Strohl was wedded to the belief that only a reading of a complete text was professionally acceptable for indexing; and, that abstracts had to be composed, as did indexes, by bibliographic experts in each particular subject area. He was also insisting that he receive additional payments if the Zurich center was required to do any abstracts or indexing for others. Mimicking the words of his auditors, Strohl demanded that his "costs of production" be met.[10]

That was especially irritating to Kellogg who, in response to prodding by McClung's group and the international disciplinary societies, had begun a rather public campaign to support low-cost abstract journals. He now believed abstracts had a vital role because they served the isolated and library-poor scientists.[11]

Another of Strohl's seemingly inflexible policies dismayed Kellogg. Strohl believed that no limits should be imposed on what journals and subjects the Concilium surveyed, even if it meant duplication of the efforts of other bibliographic services. Therefore, he devoted valuable time to processing journals usually surveyed by other disciplines' indexers and abstractors. Of course, Strohl remained with an unshakeable belief that Herbert's sophisticated and costly Dewey/UDC classifications and cards were indispensable.

Kellogg was also becoming frustrated with Strohl's disinclination to make tours of all the European countries to gain political allies, subscribers, and subsidies. Worse to Kellogg were the financial results under Strohl's management. After a full year of operation, the Concilium's subscriptions were bringing in less than one-third of its yearly cost and future performance was predicted to be little better.[12]

Kellogg could not ignore the reports from Europe, or the newest complaints and demands in America. By early 1923, he was beginning to conclude that he might have been mistaken in his belief that the Concilium (or even Paul Otlet's Brussels center that he was monitoring) could ever become a significant part of any international system. More fundamental, he was questioning the possibility of any productive cooperative science information internationalism—even after the council aided new league-related groups for international science cooperation, such as the International Research Council and the International Committee on Intellectual Cooperation (ICIC).

One of the mandates of the international cooperation committee was to help form efficient bibliographic services and abstracting cooperatives in all major disciplines. That responsibility and many others remained unfulfilled. The committee, for example, found it difficult to reach out to the regular library community although it had some links to the library world's own international information institute. Librarians had formed the International Federation of Library Associations, which became connected to the league, but housed it in Geneva rather than in Paris, where the science information group was to be located. The ICIC also was unable to find a way to build a cooperative relationship with Otlet's Documentation organization.[13]

Kellogg was disheartened. He considered a very radical alternative. He began exploring the earlier suggestions to establish a separate International Institute of Scientific Bibliography in America. His concept now went beyond what Jacob Schramm and Milton Greenman had urged and beyond Charles Davenport's hopes to bring the Concilium to the United States. Kellogg was toying with nationalism although he knew that such ideas went against his and the council's commitment to world cooperation. Significantly, his alternative did not include the services in Zurich and Brussels. And, even if the American center had to be located in Europe, he wrote, it would depend on American money. So, such an institute would be under American guidance. Kellogg was thinking, as did Schramm, that only Americans could be effective and innovative bibliographers. Unfortunately, Kellogg's change of mind came just as Europeans finally began their own program for science bibliography.[14]

Paris, Not Zurich or Brussels: Nationalism or Internationalism?

Worrisome to Kellogg, (certainly disappointing to Strohl, Otlet, and Henri La Fontaine) was the league-approved institute to encourage, among other things, the development of international science information, which was being established at the Paris Palais Royal, rather than in Brussels, Geneva, America, or any neutral country. The

International Institute for Intellectual Cooperation (IIIC) had French, not American, funding. Unfortunately, it seemed to many to be an attempt by France to create a national organization under the guise of internationalism. However, France was the only league member that offered the needed funds for such an effort.

Kellogg and the council had not been asked to play a major role in shaping the Paris center. Strohl and Otlet were nowhere to be seen before or after the operation began. Worse for Kellogg was that—although amply subsidized by the French government (with a guarantee of a minimum of seven years of support) and begun with high expectations—the Palais institute could not overcome barriers to the creation of efficient international science information systems despite a special office devoted to that cause.[15]

From its beginnings, there were indications the Paris initiative could not succeed in fulfilling Kellogg's hope of having international abstract services in every scientific discipline. There was discord throughout Europe. Paul Otlet protested the new institute's founding because it threatened his Belgian work. When the Paris institute took over his small league-issued contract for information cooperation, he must have felt betrayed. He and La Fontaine had been the prime movers in making science information a League of Nations priority, and in convincing the league's librarians to use the UDC.

Then, the British refused to cooperate with the IIIC. Like most other major nations, including the United States, England did not contribute funds. Most regrettable, the French financial ties caused emotional responses about supposed national domination of bibliographic systems throughout Europe. The center's tendency to ignore older science bibliographers also caused difficulties, and its energetic support of the policies that excluded Germany from international science organizations was troublesome.

An attempt to avoid additional charges of national bias led to the appointment of a German scholar, Gerhart von Schulze-Gaevernitz, to head the IIIC's bibliographic/documentation office. Just as importantly, representatives of other major nations, such as Vernon Kellogg, were eventually asked to serve on its board of directors. To suggest that the office was oriented to the needs of all intellectuals, and that methodologically radical librarians did not dominate it, the term "documentation" was soon dropped from the office's official title.

None of that could compensate for the operation's deficiencies. The Palais Royal endeavor produced almost nothing of value for science information. Despite more than a decade's effort, it did not achieve, for example, that long-standing goal of establishing the coordination needed for worldwide abstracting services.[16]

Kellogg Extends Himself for the Concilium: Once More, Again

Vernon Kellogg may have sensed that was going to be the outcome as he was contemplating the American bibliographic center, but he still wished to aid "international" science information. Although he was increasingly alienated by Strohl's actions (and inactions) he continued to help him, although he knew that someone had to convince Strohl to change his behavior.

To avoid a conflict that might endanger the Concilium, and quiet another surge of criticisms, Kellogg searched for a diplomatic solution. It was a difficult challenge. He had to transform Strohl's ways, avoid offending the Concilium's American supporters, and appease the National Research Council members who remained upset about funding the Zurich center. Kellogg thought he discovered a means.

After a frustrating year dealing with an uncompromising Strohl, Kellogg arranged what he announced was a Washington, D.C. "conference" (sponsored by the AAAS and the council) to address international science information. He invited Strohl and his wife, wiring the travel funds for the couple. They left Zurich expecting to attend a disinterested discussion of science information. Strohl looked forward to the trip, especially because it meant he could visit his old friend, Nina Field. As Strohl was settling in at the Cosmos Club, his wife traveled to Cambridge to stay with Nina. He planned to join them later.

However, the substance of the March 1923 meeting was not what Strohl anticipated. It was not a free-ranging exploration of science information. Rather, it was a series of papers and speeches loaded against Herbert Field's information methods and Strohl's business policies. Kellogg arranged to emphasize two things. It was imperative that Strohl get on board the council's abstract initiative and that Strohl be Efficient. The most vocal participants were the men from the Wistar Institute. Their united message was simple. Strohl should switch to using and producing the long abstracts and downsize the Zurich work based on Herbert's overly complex, not-too-useful indexing.

Perhaps because of a language barrier, the Wistar message did not get across to Strohl—it certainly was not convincing. Strohl's final paper and his later reactions reflected his unwavering devotion to Herbert's deepest information beliefs. A true bibliographer, he again asserted, would never use abstracts for indexing, and a true scientist would never rely upon abstracts. At the conclusion of the conference, the most he promised was that he would explore the creation of common standards for abstracts and reconsider their use for indexing.[17]

Kellogg was displeased. Harmony had not been achieved. The Wistar group predicted the worst. Davenport's allies were protesting that Strohl had not been treated with adequate respect. After Strohl's return to Zurich, relations remained strained. Then, they deteriorated. The battles over writing and using abstracts escalated. And maddeningly for Kellogg, the Concilium's finances continued to be more than bothersome. In addition, he was hearing about a problem with the quality of the center's work.

Although well funded by the Rockefeller grant, Strohl felt pressured to take steps that lowered the quality of the Bibliographicum. To please some critics, he began to reduce the number of journals he indexed, although items of zoological interest were appearing in more journals as the field evolved. In response, very serious complaints began reaching the council. They hammered on one point: because the service was missing too many important articles, it was losing its usefulness. Then more bad news came for Kellogg's goal of cooperative internationalism. Strohl summarily broke off the relationship with Paul Otlet's UDC standards effort, protesting Otlet's failure to be timely and reasonable.[18]

Next, a report from the independent auditor of the Concilium arrived. It was devastating. The auditor predicted the center would never be self-supporting. Even with the few thousand dollars a year of anticipated subsidies from European scientific societies, he asserted, it would always need as much underwriting per year as the Rockefellers were now providing. His ominous prediction seemed credible as the Concilium's income continued to be far less than expenses. Strohl was also facing criticism over his failure to explore new money-saving techniques that might solve the Concilium's burdensome inventory problem. Among other things, critics stated, he ignored suggestions to use new plate-printing techniques that could eliminate the need to keep costly back-copies of thousands of cards.[19]

There were renewed demands from within America's National Research Council. Wistar's men were placing extreme pressure on Kellogg to have Rockefeller funds devoted to establishing their proposed biological abstracts journal. Kellogg could not placate them any longer. The Wistar group had plans for an operation that obviously would compete with the Concilium. And, it was one that appeared to some to be a purely nationalistic American operation.

The situation grew tense. The McClung group made it clear they would present its plan to the Rockefeller foundation themselves if Kellogg continued to hesitate.[20]

When Strohl learned of the intensified campaign for what became *Biological Abstracts*, he was infuriated. He rushed protest letters to Kellogg. Then, he began to enlist Charlie Davenport and Herbert's other Harvard friends in an emotion-laden campaign against the Wistar Institute's ambitions. Because of his limited English, parts of Strohl's letters are difficult to interpret, but he clearly stated a warning that funding the *Abstracts* would seriously endanger the Concilium.

Strohl went further: he protested that what Kellogg was about to do would breach the previous gentlemanly agreements ensuring the Concilium's full protection from competition during its five-year grant period. He also warned that the welfare of every international bibliography was in danger because of the new "nationalistic" council initiative.[21]

Strohl did not stop with reminders about earlier promises. He went on the attack. In his attempt to save the Concilium, he angrily criticized American science practice and its professional ethics. Strohl now emphatically stated that abstracts were "secondary," while indexes were "primary," information. He even asserted that abstracts were for lazy people, not real scholars.

A Very Temporary Truce, then Wistar and Philadelphia Win

Davenport and the Concilium's AAAS friends voiced similar criticisms, although in much less severe tones. Davenport's protests infuriated the Wistar group. They immediately began another round of letters of complaint to Kellogg. To avoid a serious conflict between the groups, Kellogg arranged another conclave.

This time, Kellogg worked some diplomatic magic. At a New York City conference, the AAAS and council-related participants concluded that both the *Abstracts* and Concilium might live together. They agreed to give Strohl and European bibliographers another chance while the *Abstracts* plan moved forward. To implement the recommendations, Kellogg left the meeting with a mandate to organize a grand science bibliography conference in Europe. Its goal was to revive the dream of a coordinated and cooperative international information system, but with the American *Abstracts* to play the major role. Kellogg was also encouraged to provide advice to Strohl before the European meeting.[22]

However, the seeming harmony following the conference did not lead to a save for the Concilium, or for information internationalism. Kellogg was unable to set up the great European forum, the French remained in charge of the league's science information agency, and Kellogg could not find someone who could provide practical advice for rescuing the Concilium. The future of the Concilium and of European-centered international bibliography seemed bleak. In contrast, the Wistar Institute's bibliographic fortunes rose in Philadelphia.

Just as Strohl began yet another campaign in defense of Herbert's methods and cards, the near opposite arguments of McClung, Clarence West, and Schramm were reaching the top council leaders and the Rockefeller managers. They were persuaded to give serious consideration to providing the Wistar group with an unprecedented level of funding; and, as they examined the *Abstracts'* possibilities, they lowered their estimate of the Concilium's worth.

Greenman, McClung, Schramm, and West had become a team of skilled grantsmen. They dusted-off their reports and let the foundation know of their quantitative surveys of biology's information needs. The results, they reiterated, provided convincing evidence that few working biologists wanted indexes, that most desired abstracts, and, importantly, that some four thousand people were ready to subscribe to the proposed *International Biological Abstracts*. They presented evidence of the Wistar Institute's proven experience in publications management, and they described how modern and professional its operations had always been.

Then they exhibited details of their plan for establishing a worldwide network of respected volunteers to compose abstracts and to do so without pay. They emphasized that all indexing, if any was ever needed, could be done from those abstracts. There would be no major multi-language problem, they intimated. However, they did not mention that some Europeans thought the Philadelphia plan was one that would make the league's efforts, and those of the French-sponsored institute, unnecessary.

The Wistar's promoters also presented a persuasive economic plan. With the abstracts being composed by unpaid volunteers, who had journals already in hand; with any indexing done from the abstracts; and with the use of modern printing technologies, their cost per information item, they promised, would be a fraction of those of other services, including Strohl's. When they seemed to prove that editorial costs would be minimal, they had a strong argument.

They won their case. In late 1924 the Rockefeller administrators committed $3 million over ten years expecting to be subsidizing only the *Abstracts'* administrative costs as the journal quickly worked its way to profitability. In 1924, there was no indication the subsidy would soon grow to $7 million.[23]

News of the Wistar's 1924 victory did not close the debates over the Concilium. It certainly did not silence Strohl or his allies in the United States. With only two years before the termination of the Concilium's five-year Rockefeller grant, Strohl and poor Marie Ruhl became frantic. They began another emotional letter and telegram campaign that, tragically, became self-defeating. They soon completely alienated the seemingly ever-tolerant Vernon Kellogg.

Strohl was direct in approach and in words. He sent lists of European scholars who supported Zurich's indexed bibliographies and who all, he claimed, had little complimentary to say about abstracts. Quite soon, Strohl put aside the etiquette of the old era of science.

He did something unacceptable to Vernon Kellogg. He bypassed the council and wrote directly to George Vincent at the Rockefeller Foundation. Strohl's letters to Vincent were accusatory. It was the council's fault if the Concilium was not able to raise money in Europe; it would be the council's responsibility if the service failed; and, the foundation must act on its own, not follow the council's advice. Strohl, expectedly, demanded the renewal of the subsidy for the benefit of pure and truly professional science.[24]

Besides contacting Vincent, Strohl made sure that Davenport, Edward Mark, and George Parker knew of the Concilium's impending doom. He warned them of the consequences of the termination of the five-year grant, and of the damage *Biological Abstracts* could do to the Bibliographicum's markets. He also called on Nina who, in turn, began sending pleas to the members of the AAAS to lobby Vincent for continued support for the Concilium. Strohl was so distraught he was even planning to write to Stephen Duggan, who had never been a great admirer of the Concilium, to obtain a grant from the Carnegie's International Institute for Education.[25]

Kellogg Stays Calm and Conciliatory in the Face of Strohl's "Storm"

In contrast, Vernon Kellogg remained a calm professional and again tried to convince Johannes Strohl that if he at last became part of a cooperative abstracting effort the Concilium could possibly receive more help. Then, Kellogg suddenly made a rather strange decision.

He followed up his advice to Strohl by arranging to have Jacob Schramm of America's *Biological Abstracts,* not Otlet or any representative of European bibliographic organization, begin drawing up another plan for international cooperation.

Strohl then made a serious miscalculation. On hearing of Schramm being picked to spearhead the new plan he sent a number of impassioned and demanding letters to Kellogg. In response, although burdened with other work, and with a limited budget, Kellogg traveled to Switzerland to attempt to appease Strohl. The diplomatic Kellogg bet that a face-to-face discussion would lead to a satisfactory compromise. He was wrong. Strohl stood his high professional, "real science," ground.[26]

The trip was unsuccessful. The Concilium lost forever one of its best and most influential friends. Strohl has an attitude, wrote Kellogg, who left Zurich resentful of Strohl and Ruhl's beliefs that the council "owes" the Concilium. Strohl also felt alienated. He thought the Americans were not paying respect to the intellectual approach of European scientists, or to the Swiss government's mandate to continue Herbert's indexing service—of which, he asserted, it was very proud. He also resented what he was interpreting as America's attempt to dominate all international science information.

Strohl played his own version of a nationalistic card. He wrote to Vincent and Kellogg that the Swiss government would never agree to become, as the Schramm cooperative abstract plan implied, a mere appendage of an American operation and just another of the many abstracting houses. That did not set well with Kellogg. He thought he was only trying to achieve the abstracting goals the league's intellectual cooperation committees had finally agreed upon.[27]

Meanwhile, the Wistar group was making the situation very difficult. In early 1925, McClung published an article with complimentary comments about England's *Zoological Record* but not quite nice things about the Concilium. In response, Strohl hurried another heated letter to America. It had a tinge of paranoia. Not only were the people allied with the council not supporting the Concilium, they were trying to destroy it, he declared. Kellogg, resuming his role as a conciliator, managed to have McClung state that he had never meant to say any unkind things, and that Strohl had misinterpreted his article because of language differences. Kellogg reassured Strohl that the council would not set the Concilium adrift.

Despite that, the situation became unmanageable. Strohl was saying even harsher things about American methods. Of the most significance was a comment he wrote about the many errors in the abstracts sent to him, which proved that no one should ever use abstracts for professional indexing. Strohl also complained that it was the editor of the *Zoological Record,* not he, who had been uncooperative.[28]

Another conciliator appeared but failed in his mission. A rush trip by Charles Davenport to Zurich neither comforted Strohl nor convinced him to be cooperative. Davenport's assurances that he would end any rumored attempts by American biologists, including McClung, to develop a new classification scheme for biology or zoology did not lead Strohl to decide, as Davenport was now urging, to cooperate with the joint cooperative abstracting efforts. Nor did Strohl respond to Davenport's suggestions about using new technologies to reduce the Concilium's costs. Strohl gave an unenthusiastic response to the suggestion that he use photocopying or addressograph technologies to allow the reduction in the storage and reproduction expenses of the back-files.[29]

Strohl again went on the offensive and insulted more people. In one not-too-guarded public rejoinder to criticisms he argued the Concilium's economic problems were not the result of the quality of his work but to attitudes of librarians and individual scientists. He hinted that American universities were also to blame. Although saluting American and Swiss research libraries (which now held most of the subscriptions) he tried to undercut the power of librarians. He pleaded for the establishment of separate bibliographic budgets in science departments because only true scientists could appreciate the need for the Bibliographicum.[30]

Alienating Kellogg, Outfoxing Charlie: The Concilium Suffers in More than Spirit

By mid-1925, Kellogg, already fed-up with Strohl's attitude, sought an end to the relationship. He wanted to get on with what he had come to feel was a worthy project with a potential for enormous benefits. He now viewed *Biological Abstracts* as "the" tool for the solution of critical world health, demographic, and agricultural problems, and he had regained his hopes for the future of international cooperation.

Kellogg was completing a work extolling international cooperation in science, "Isolation or Cooperation in Research," and he wanted to have his actions match his statements As a result, after the failure of his many pleas to Strohl for cooperation with the Wistar group in Philadelphia, Kellogg responded to one of Strohl's angry letters with a surprisingly undiplomatic one of his own.

He wrote that the council no longer wanted any relationship with the Concilium, although the council might allow Zurich to be a part of the international information system being developed in Philadelphia and Washington. When Strohl protested, Kellogg was so angry he topped his declaration with the appointment of three not very friendly men to a new supervisory body for the Concilium. This time he out-foxed Charlie Davenport. The members were all associated with *Biological Abstracts*; none was from Herbert Field's group of long-time supporters.[31]

Strohl seemed incapable of appreciating the danger he was facing. He kept making decisions that alienated his remaining American friends. He also continued to keep a distance between the Concilium and other information projects in Europe.

In early 1927, as the initial grant period was ending, the Rockefeller managers notified Strohl they, too, were cutting the ties to the Concilium. They did not emphasize it, but not only had they tired of the Concilium's failure to meet its goals, they had just committed an enormous amount to an even-more international cause. The

Carnegie and Rockefeller foundations would commit tens of millions of dollars to build the League of Nation's library and to ensure it could acquire what became one of the largest and finest book collections in the world. Rockefeller and the Carnegie funds remained essential to that library, even financing the shipping of its volumes to the safety of Abraham Flexner's Institute for Advanced Study at Princeton when World War II began.

By the end of 1927, the Rockefeller managers were treating the Concilium as somewhat like a black sheep in the family. They wanted to forget their Zurich misadventure, so much so that none of the Rockefeller histories ever gave significant attention to what was once the foundation's greatest information investment.[32]

Even Charles Davenport backed away from Strohl. In 1926, after receiving one of Strohl's demanding letters, Davenport informed him that Cold Spring Harbor was terminating its subscription. Too few of its researchers used the file; and, he said, it was only his friendship with Herbert that had led him to spend thousands of dollars on the system over the years.[33]

Amazingly, despite such a rejection, Strohl did not back down. Again becoming frantic, he responded to Davenport's announcement by playing more "guilt cards." Didn't Davenport remember that Herbert had stated in his will that the Concilium had to be maintained? Didn't Davenport recognize the Concilium was an American institution and responsibility, and that many of Europe's true scholars depended on its continuation?

Strohl also initiated a new advertising campaign, even posting full-page announcements in "enemy" territory. In one of the first 1926 issues of the new *Biological Abstracts,* he advertised that the Concilium stood ready to provide more than new cards. He offered the public a complete set of the some 520,000 cards produced since 1896 for $25,300, or subsets at a higher charge per card. He seemed to expect a number of institutions to be willing to spend such a large amount.

The advertising did not help, and criticism by the Concilium's old nemesis, England, added to Strohl's woes. Articles appeared stating that the Concilium's system was of little use, that the classifications were outdated, and that the price was too high.

Some friends remained, however. Davenport mellowed and returned to the fold. The Harvard / American Museum of Natural History / Columbia University clique had become outsiders in the council's information program, but they now returned to lobby Kellogg. Others, such as the Smithsonian's board, announced it, too, supported renewed aid to Strohl.[34]

That accomplished little. There was no response from the foundations, and the AAAS-Harvard based supporters own new emergency fund-raising campaign failed. They had conducted a drive, but they had to tell Strohl that American scientists had little interest in the Concilium. The most they accomplished was to obtain a small one-time $2,000 grant from the old Boston-based Elizabeth Thompson Foundation (probably because of George Parker being on its board of directors), and a tiny donation from the naturalists' society.[35]

Herbert's old admirers were discouraged and weary. Worse, many began to think that Herbert's once-revolutionary approach and methods had become outdated. To the disappointment of the Davenport group, it was learned that one very influential friend of Herbert's, Thomas Hunt Morgan, had written Kellogg that the files at the American Museum of Natural History had always been a nuisance and not worth continuing. He was not a lone voice at the museum. By the mid-1920s, official correspondence with the Concilium complained of the cumbersome size of the file and an unbearable cost for filing and maintaining ("updating") what was intimated to be the too-many cards it was receiving each year (15,000). In the mid-1930s, the card cabinets took a space sixty-feet long by six-feet high and the librarians wanted that precious space for their expanding book catalogs. By mid-decade, the museum even stopped using the Dewey/UDC for classification.[36]

The Zurich center continued on in spite of all the depressing news from America. In the mid-1920s, it was sending out hundreds of thousands of cards a year. However, compared to its growth two decades before, the Concilium was declining in every way. Moreover, it was out of the information mainstream. Strohl was finally invited to attend some of the international bibliographic meetings in Paris in the late 1920s. Yet, the Concilium was unable to send a representative to the important 1937 World Congress of Universal Documentation conference, which was sponsored by the league's organization for intellectual cooperation, at the International Exposition of Arts and Technology in Paris.[37]

Some Americans had again stepped in to help in the later 1920s, but organizations such as the American Society of Naturalists could contribute little. The assistance from the Swiss government remained minimal. The Concilium continued to be in debt, and it lost its innovative spirit.

New Economic Realities and the "Files" Problem

Meanwhile, other scientific publications were providing more timely and comprehensive bibliographic services. Vendors, such as *Chemical Abstracts*, *Engineering Index*, and the *Zoological Record* improved and expanded their offerings. A worse outcome occurred for Strohl because America's early version of Big Science information seemed a model operation: *Biological Abstracts* became a science-information poster child with the publication of its first volumes in 1926.[38]

Some of the problems in Zurich were the result of Strohl's decisions, but not all. The post–World War I Concilium had to work within a cluster of unfavorable economic and political conditions. The long-term economic impact of World War I meant that few European nations' libraries and universities could afford to resume their subscriptions—just when support in America was declining. The subscriber's list dwindled to less than one hundred. With fewer subscribers, prices had to be raised to cover expenses. Few institutions could afford a subscription and maintenance cost of $2,000 a year for just one of the main databases. By the end of the 1920s, the annual charge for the zoological file was four to five times that of a book-form abstract offering. By the mid-1930s, while the world's economy remained battered by the Great Depression, the average price of an in-print Concilium card in constant dollar terms was close to 75 percent more that it had been immediately before World War I.[39]

Capitalism Fails the World, and Science Information, Too

Of course, the Great Depression was a major factor in the Concilium's decline. Just as the European nations were recovering from the war, industry, agriculture, and commerce plunged into chaos. The world financial situation was so desperate that a call by America's Smithsonian Institution for a rebirth of the *International Catalogue of Scientific Literature* was treated as a near fantasy. As well, international science faced more than bibliographic troubles. The economic collapse was causing a host of problems. For example, Charles Davenport and his geneticist colleagues spent valuable time trying to find the pittances needed to rescue their specialized discipline's international meetings and to continue the flow of scientific information across national borders. The International Federation of Library Associations pleaded with the German scientific publishing industry to lower its prices so that science education could keep up with new findings. As Herbert Field had feared, German houses such as Springer had regained a preeminent position.[40]

The depression gave the Concilium a near fatal blow. Strohl was overwrought, and he was preparing to do something that was embarrassing to Nina and Charles Davenport. In 1929, he told them he was about to write directly to President Herbert Hoover for financial help, hoping to play upon Hoover's Swiss heritage. Davenport and the Fields hurriedly dissuaded him from that. Strohl fretted for a year or so. Then, he decided to try an end-around move. He wrote to Nina that he was arranging for the Swiss government to ask the Americans for help. She was receptive to that idea, even suggesting that he directly contact her sons.[41] That brought Noel Field, who had finally secured a position at the State Department back in 1926, into the life of the Concilium once more.

Noel to the Rescue?

Although Noel was extremely busy at his job, in 1931 he immediately responded to Strohl's letters. The correspondence from Zurich told of the impending request by the Swiss consul for American government aid. Noel knew he could not play any official role, but he was determined to save his father's lifework. He researched the recent history of the Concilium and then arranged a meeting with Jacob Schramm of *Biological Abstracts*. Schramm's apparent unwillingness to help surprised and saddened Noel. Noel did not know that Schramm and his colleagues had made it known to Davenport and others that Strohl's uncooperative behavior since the mid-1920s led them to write off any further attempts at collaboration.

Strohl, in fact, had refused more offers to join the cooperative abstracts effort, despite Schramm's success at recruiting scientists around the world to contribute to *Biological Abstracts* without compensation.[42] Noel did not give up. For two more years he wrote to Kellogg, other council members, the Davenport group, many of the most important remaining American Concilium subscribers, and to his contacts at the Rockefeller Foundation. He even lobbied Herbert's old friends at the State Department. All provided advice, but no financial help, even when they learned that Marie Ruhl had died. Without an infusion of money, she could not be replaced.[43]

Noel began to realize that Strohl's hard-headedness concerning the use of abstracts was the cause of much of the opposition to the Concilium. He also began to appreciate that Zurich was not keeping up with the modern trends in science information. Noel's correspondence with Strohl was confirming that he was dealing with an obstinate and overly emotional man. So, in late 1933, Noel concluded there was only one way to deal with the Concilium. The Rockefeller managers should give Strohl a firm "either/or"! If he did cooperate with the abstractors he should be granted some money; if he did not he would certainly never deserve an American subvention.

Once more, Strohl refused to bend. Even if he had, it was unlikely that he would have received any significant help. America's science information infrastructure was enduring its own crisis; and the nation's philanthropic foundations had their own financial difficulties.[44]

16 The Information Consequences of "Capitalism's Disaster" and the Shift to Applied Science Information

The science information system in the United States and the economic foundations of U.S. universities were eroding as the nation struggled to overcome the crash of 1929 and the following Great Depression.[1] By 1931, the American abstract and indexing services the National Research Council had nurtured were near death. No one was willing, perhaps able, to revive them. At the same time, the economic pressures on businesses encouraged a search for efficiency in dealing with applied-science information. Sci-tech librarians in England and the United States were energized and began to take a more prominent role in information organization.

A profound disappointment was the fate of *Biological Abstracts*, the Rockefeller Foundation and council favorite. The biological journal had been unable to merge with the International Biological Union, and it was among the first American science information service to announce it was facing bankruptcy—after receiving some $7 million of grants including a subsidy for a not-quite-successful attempt to build a viable cumulative index.[2]

Despite the Wistar group's earlier arguments that scientists did not need such expensive things as indexes, the *Abstracts* leaders had to recognize that history did play a part in scientific work, even applied science. The *Abstracts* mid-1930s index project, however, proved to be a huge economic burden just as the service plunged into financial crisis. The service's great friend, the National Research Council, made a feeble attempt to rescue the *Abstracts* but a 1934 idea for raising a huge endowment of $20 million went nowhere. The council's crusade to gain further cost-saving cooperation among biological journals accomplished little.

The great American *Chemical Abstracts* was the next victim of what Noel Field's leftist friends were calling "capitalism's disaster." It barely survived after a painful reorganization and reduction of services. *Psychological Abstracts* and *Social Science Abstracts*, even *Engineering Index* and its new card notification system, soon joined the disabled list.[3] America's universities and faculties had their own hard times. There was less for research, high unemployment, and starvation-level library budgets.

There was no one to rescue them. Although Franklin Roosevelt's administration was beginning to pump funds into major sectors of the economy, science was not high on the Democratic agenda. As well, the nation's great foundations were under financial stress. By 1935, the Rockefeller Foundation had thrown in the towel by announcing it would not provide any more subsidies to science journals, with one distasteful exception. It offered a small grant to allow *Biological Abstracts* to wind up its affairs. The Rockefeller information interests turned to supporting humanities libraries and the now hard-pressed Library of Congress. While science languished, the foundation made large grants to preserve manuscript collections in famous European libraries and to microfilm their rare classic works and library catalogs. The foundation's leaders wanted American and other libraries to have copies to aid general education, and they wanted to ensure the preservation of the classic works if European nations again went to war. That was one reason for their support of the development of microfilming technologies in the 1930s.

The National Research Council did attempt to shore up all of science information, but the council was feeble and penniless. It held an emergency conference on conditions in the mid-1930s, and another a few years later. Neither contributed anything of substance. The findings of the council's Oswald Veblen were correct in being pessimistic.[4]

Jacob Schramm's Heroic Struggles in the Name of "Big Science," Johannes Strohl's Last Requests

It seemed that America's march toward Big Science information systems was over. Attempts to obtain special government subsidies for operations such as *Biological Abstracts,* including some from the science-friendly and socialistic head of the Department of Agriculture, Henry Wallace, produced little. The *Abstracts* barely avoided bankruptcy. It managed to survive only through one of the government's unemployment relief agencies. It replaced much of its staff with the federal government's make-work National Youth Program's students. It saved additional expenses by relying on what it said it distrusted, author-composed and mainly self-edited abstracts, not abstracts written by impartial scientists. All of Charles McClung's ambitions for the application of advanced technologies, such as microfilm and electric tabulators, remained only desires.

As they fought against the Rockefeller advice to close down, Jacob Schramm and his *Abstracts* board valiantly searched for other solutions. In the mid-1930s, they made some desperate attempts. Ten thousand letters were sent to recruit new subscribers; Schramm and his staff worked without salary; and many others in the Philadelphia center were fired. When possible, they were replaced by more "relief" workers. But Schramm did not succeed.

Soon, the journal had to suspend publication; almost all the board and higher-level staff were replaced; and, in 1938, Jacob Schramm was given an academic version of a corporate "golden-parachute." It was not a lavish one. He was granted a special professorship at the University of Pennsylvania and a small windowless office.[5]

Johannes Strohl may have gloated a bit over the *Abstracts'* problems, but he had no reason to feel superior. The Concilium limped through the 1930s until 1940 when World War II finished it. The few subsidies Strohl obtained, such as the one from the International Union of Biological Sciences, had always been too little and too late.[6] Strohl endured a final insult from the Americans. After he sold the Concilium's building to pay some of the center's debts, he begged the Rockefeller managers for funds needed for the final shutdown. They refused him. Then, he placed advertisements in American scientific journals offering the Concilium's collections and furniture for sale. There were no takers. There were more embarrassments. Strohl put another run of advertisements in American and British journals pleading for buyers for the remaining stock of cards and cabinets. With the embarrassing end of the Concilium and the devastation of the war, Strohl must have died a very unhappy man in October 1943.[7]

The Decline of Paul Otlet's Information Internationalism Crusade in Brussels

An additional disappointing scenario for information progress occurred when the larger and more politically influential Otlet-Fontaine data center and the UDC projects experienced hard times, as did the Paris-based intellectual cooperation project. The great Brussels information center's decline started in the mid-1920s. That came as a surprise to many because the Belgian government had made a major commitment in 1919 when the Otlet-Fontaine information bureau was central to the initiative to convince the League of Nations to make Belgium its permanent home.

For a time, Otlet's operation seemed to be recovering from World War I. It was processing 1,500 requests a year, it built bibliographies for an increasing number of subjects, and it expanded its number of file cards to 11 million. But the league's decision to locate in Geneva led the Belgian government to end its large subsidies in the mid-1920s. It even asked Otlet to move from his sumptuous quarters. As well as suffering from reduced support from the Belgian government, the center had to endure cuts in support from the league and to have many of its responsibilities co-opted by Paris' Institut International de Cooperation Intellectual. Even a valiant effort by the American Library Association raised little money for Otlet's "international" catalog. Just as depressing was the realization that there were few new users of the Universal Decimal Classification system. In 1951 only some 1,600 libraries worldwide were using the UDC. (Dewey's system already had many thousands of users). Ironically, England then had the third highest number of UDC libraries. There were very few in the United

States. The only major American users of the UDC were the Meteorological Society, some electrical engineering journals, and a physics organization linked to the CIA.[8]

By the later 1920s, Otlet's work began to take on the look of an attempt to encourage voluntary cooperation for international standards rather than a cutting-edge database and information research center. There was a new name reflecting that, the International Institute for Documentation. Within a few years, the name was again changed. It became the International Federation for Documentation and, then, the International Federation for Information and Documentation (FID). The name alterations indicated Otlet's decreasing attention to his great file and his shift to a focus on theory and reprography.

Paul Otlet's hopes for his other interests also suffered. His and Henri Fontaine's lobbying had led the League of Nations to create that high-level policy-recommending organization, the lofty commission for International Intellectual Cooperation. Sadly, the Belgians were never invited to play a significant role in its work, although the committee's interests had become wide-ranging. Education, art, publishing, student-faculty exchanges, interlibrary book loans, education in international relations, even labor unions came within its domain.[9]

However, interest did not mean accomplishment. Although it gained the support of world-renowned scientists such as Albert Einstein, the diplomatic skills of the National Research Council's Vernon Kellogg, the aid of Stephen Duggan, and the approval of the Council on Foreign Relations, the international committee accomplished much less than anticipated, especially in the domain of science information. In addition to an overly idealistic orientation, nationalistic concerns had turned it into what many continued to regard to be a vehicle for French cultural advancement.[10]

Universal Decimal Classification—a British Voice from Its Past and a Hint of Its Future—in the Race for Applied Science Methods

There was a paradoxical twist in the history of science-information internationalism. The last great attempt at constructing a UDC/numerically classified massive card file took place in the land of Herbert Field's protagonists of thirty years before. In the 1920s, as Melvil Dewey's group showed declining interest in coordinating with Otlet's work, a young English scientist-bibliographer, S. C. Bradford, put new life into the Universal Decimal Classification, even convincing the publishers of *Science Abstracts* and, then, other British journals, to begin using UDC numbers. He would also successfully lobby to have an English edition of the UDC manuals ready by 1944.

However, neither Bradford nor the inheritor of Otlet's UDC and standards-setting initiatives, Donker Duyvis, overcame all the economic and psychological hurdles to the application of a universal classification system. One reason was that too much had already been invested in the Dewey and Library of Congress schemes.[11]

Bradford's efforts went beyond picking up the torch for numeric classification. He attempted to reestablish an Otlet-like card file and to begin a new version of cooperative international indexing. He built a two-million-entry file of numerically classified science information. However, despite his hopes for his London information center, and its ties to Otlet's operation, two things defeated his plans: a lack of users and the Great Depression. His cards even disappeared after World War II.[12]

Bradford's UDC work had a significantly different goal than Otlet's. His interest in the UDC was not primarily academic. Bradford was central to an extension of England's post–World War I search for technical progress and for efficiency in handling applied science and technological information. Driven by the pressures of the Great Depression, there was a demand for immediacy in information delivery that was more intense than even the Wistar's. Speed was also a major focus of England's Association of Special Libraries and Information Bureaux (ASLIB) and the U.S. Special Library's organization during the 1930s. Like Bradford, they became supporters of the National Research Council's drive for efficiency in abstracting. Bradford often voiced his special dislike of the waste caused by multiple abstracts of articles, and he showed little tolerance for information specialists who squandered time on "theory" while users awaited practical results.[13]

A second UDC exception came a generation later. In the early 1960s, as what the American liberal pundit (and Field acquaintance) Walter Lippmann termed the "Cold War" was dividing the world, the Soviet Union, also rushing to organize sci-tech information, declared the UDC mandatory for its and its satellites' science indexing and classification centers. The UDC's numerical classifications seemed the only way to serve the multilingual Communist bloc.[14]

Adolph Voge Returns to the Library of Congress as a Special Librarian in the Name of Efficiency

In the 1930s, as Strohl, Nina, and Noel were attempting to save the Concilium, they gave no thought of calling for help from the man who spent so many years trying to inject Herbert's center with an entrepreneurial spirit. Strohl and the Fields knew of Adolph Voge's availability, but Herbert's earlier cautions about him made them wary of their one-time friend. If the Fields had known the details of Voge's post–World War I career and behavior, they would have been even less willing to trust him.

After leaving the Concilium for the second time in 1912, Voge tolerated many disappointments as he sought to become a well-paid and respected librarian-researcher-entrepreneur, and as he attempted to become what was later termed an "information broker." All his efforts to establish a commercial Concilium-like information service for chemistry withered. The next defeat was his failure to secure a position in the Library of Congress's prestigious forerunner to the Congressional Research Service.

That organization was becoming central to allowing American policymakers to be progressive and Efficient and, hopefully, to base their policies on scientific research, not political pressures. After that disappointment, Voge's trip to the Paris Peace Conference in 1919 led to personal entanglements rather than recognition as an international information expert.

Voge's most profound frustration came just a few months after he was offered the prestigious job at DuPont's innovative information center. At the beginning of 1920, the postwar recession led the company to dissolve its new intelligence unit. It was just one of the dozens of American centers that were abandoned in the wait for better economic times. Voge was without work at age forty-five. He could not find a good position and he became pessimistic. Although he marketed himself as both a librarian and one of those new "chemical engineers," he could not get a permanent position during the 1920s. He became desperate. Then, fortune seemed to bless him.

Voge had planned a great chemical-entrepreneurial venture but he lacked capital. During all the years he worked at the Library of Congress, he conducted experiments at home on sources of caffeine. He knew there was a growing demand for the chemical, and that anyone who could reduce its cost of production would become wealthy. He was in search of alternatives to the increasingly expensive materials needed for its extraction. He thought he found a class of plants that could be easily cultivated and processed. The problem was to find where the best plant grew and determine where it might be commercially produced. At one time, he believed the plant was located in Bolivia. Using his tiny savings he sailed there to trek through its jungles. It was a costly and dangerous six months—with no payoff.

Then, his research led him to believe that the plant might be located deeper in the Amazon Basin. He did not have any funds, so he began a search for a partner. It took him years, but he finally located a man who was willing to grubstake him. Adolph again left for South America. After hiring a native guide, he and his partner trudged through the Amazon jungle for half a year, going to places where "no Anglo-Saxon" had ever been.[15]

On his return to the United States (just two months before Herbert Field's death) Adolph Voge posted a long letter to the great national librarian Herbert Putnam. He declared he had found the correct type of plant and that he was forming an international corporation to exploit it. He said the process was going slowly, and in the interim he needed a reference for a Special Librarian position. Although it had been years since Voge worked for the Library of Congress, Putnam agreed to compose a recommendation if an employer sought him out.

Voge rejoined the ranks of the Special Librarians in late 1921, but not as a prestigious one. For the next two decades he held a series of short-lived positions with chemical companies and printing-research organizations in New York City and Philadelphia. In between those jobs he worked as a chemical engineer on mining expeditions in Peru,

Chile, and Bolivia. Life was not good to him. His caffeine project died. He was able to make only a modest living for himself and his new French-born Scottish wife. He could not afford to purchase a home.

When he was near sixty, Voge returned to Washington, D.C. Fearing that he had little chance of the Library of Congress rehiring him, he wrote to the man charged with bringing efficiency to the American government and the nation, Lyman Briggs of the National Bureau of Standards. Voge did his best to convince Briggs to make him the chief of the bureau's library. He also applied for a similar position at the government's massive printing office. Both agencies quickly responded with a "no thank you."

Voge was humbled. He was able to obtain only a relatively junior position at the Library of Congress, and he almost lost that. As before, he could not contain his ego or his temper. Just after he began the job Voge wrote a letter of protest about his immediate boss. He brashly sent it directly to the new head of the library, Archibald MacLeish, the famous humanist, left-leaning poet, and soon to be central influence in World War II's intelligence activities. Voge did not receive the sympathetic reply he expected. MacLeish responded, in essence, with a warning to "put up or shut up," telling Voge either to resign or stop sending such complaints. Voge complied. He had few options. He kept invisible and spent hours at an isolated desk in the library's stacks conducting eugenics-oriented research into his Quaker family's history while worrying he would be dismissed from the library.

A New Generation of "Specials" Contends with Military Projects and "Terms"

World War II saved Adolph Voge's job. MacLeish was turning the library into an information center for the war effort, including establishing the super-secret research branch for the Office of Strategic Services. Voge did not seem to fit the needs of the spies, but the call for experienced researchers led to his faults being forgiven. He was assigned to a hothouse version of a center for applied science information, the United States Navy's research program at the library. In an example of "what comes around, goes around," the project somewhat reincarnated Voge's old hopes for an applied science research information service. It was large and innovative, and it helped lay the basis for all of the library's expanded technology and science information programs of the 1940s and 1950s. Those programs and similar efforts at new military information centers led to the development of revolutionary and cost-saving information techniques for Big and applied science.

Voge died before the flowering of those new *sci-tech* information programs. He passed away in February 1945, and was buried in Arlington National Cemetery. Ironically, by the early 1950s, Voge's special field of electrical engineering was the most prominent user of the UDC for classifying articles in its journals. Of 404 journals worldwide that used the UDC, 20 percent were in that applied electrical field while less than 1 percent were in any biological or zoological area.[16]

There were two young men working on Library of Congress military projects who soon played a major and revolutionary role in creating a new wave of science information methods. The two who became new types of Special Librarians were Mortimer Taube and C. D. Gull. They were not quite aware of it at the time, but their careers and methods would become bridges to the era of for-profit science information and would generate a new round of debates over "best practice and organization" in science information.

Taube became a leader in developing methods that skirted the need for high-cost, labor-intensive, and time-consuming indexing. His and similar "term" methods did not impose any concepts, such as UDC categories, or subject headings, on documents. Their new type of indexing used only the individual major words ("terms") within an article. The "term" advocates believed that the older methods, with their focus on one or a few items of interest, favored the particular perspectives of an indexer, thus hiding too much information. By relying on the original words in a document, and allowing user-determined combinations of "terms" for a search, the retrieval of information would be maximized and costs minimized. The professional reaction to "terms" and the rejection of the indexing methods developed over the two previous generations were intense and sparked a replay of the debate between those oriented to creating a "science" of information (Herbert Field) and those focused on speed and cost savings.

The Ideological Consequences of "Applied Science": Evolution, Eugenics, and the Dilemmas of a New Generation

That replay of debates between those using old and new science information methods lay in the future. In the 1920s and 1930s, there was a more profound struggle, an ideological one, caused by an early attempt to "apply" science to everyday life. The science in question was biology and the application had implications for the fundamentals of religion and morality. The struggles engaged the attention of Adolph Voge, many of Herbert Field's closest friends, the public, and, significantly, Herbert's children. The new findings emerging from biology and zoology became entwined with older human evolution issues and caused conflicts that reached deep into society, as well as philosophies of history.

In the early twentieth century, the explorations of the role of genetics in human life were tagged with names such as "human evolution." Later, as evolution studies merged with a broader movement attempting to apply science to improve world health and social well-being, the name became "eugenics." The term then had no malevolent connotations. Additionally, at the time, the use of family histories to judge if Charles Darwin's theses applied to the human world was acceptable practice. Family biographies, such as those collected by Charles Davenport, held the promise of weighing the relative influences of social-environmental factors against genetic determinants. Intelligence tests, like those administered by Robert Yerkes, also seemed a way to evaluate nature versus nurture arguments.[17]

There was a compelling need for answers. A growing sensitivity to mental and public health problems was driving a merger of science and social policy. The eugenics movement began as liberal-progressive explorations of possible solutions to the apparently increasing numbers of the insane and mentally retarded. However, the implications of accepting genetic and evolutionary premises began to cause a rift in the ranks of liberalism. What was emerging from the research on human heredity was, for some, beneficial; for others it seemed to be devaluing human life. For still others like Noel Field, the results of family, zoological, and biological research undermined hopes for humane social reform.

To the Concilium's scientific friends, such as Henry Osborn, Charles Davenport, and Robert Yerkes, their research had liberating potentials—despite their work pointing to nature being more powerful than environment in determining human behavior.

There were opposite and heated opinions within academia about genetic theory. Among those who feared and disputed the new scientific findings was the foreign scholar Columbia University hired when Nicholas Murray Butler was pushing all of its departments into world-class research. That physicist-turned-social-scientist, Franz Boas, is now regarded as the father of modern American anthropology. To many at the university, hiring Boas was a mistake because his career centered on debunking eugenics research, much of which was being done by his colleagues. Boas did not stop with his anti-eugenics crusade. Liberals admired him for exposing and chastising the American anthropologists who were secretly working for the Office of Naval Intelligence in its search for German agents and submarine bases in Central America.[18]

Within a few years after his arrival at Columbia, Boas became a public figure and a beacon for the American political and ideological left. Even Hamilton Easter Field and his friends were acquaintances of Boas. Young idealists such as Noel Field revered him because Boas was taking the lead in shifting anthropology and psychology to an environmentalist stance, one that appealed to the younger generation of intellectuals and, of course, to Marxists. As a result, Boas became the perhaps unwitting spokesman for many radical political organizations throughout the 1920s and 1930s.[19]

In contrast, many of Herbert Field's scientific friends considered Boas, at best, an irritant—and in the 1920s and early 1930s the views of such old "liberal" physical anthropologists and biologists held sway in academia.

While only a few American scientists became advocates for the use of drastic measures to protect the nation from a decline in abilities and "moral character," even intellectuals hailed as Progressive reformers joined in a movement that approved of measures ranging from immigration restriction to birth control, sterilization, and even abortion. The list included one woman who seemed at the time to be on the left, Margaret Sanger. She supported birth control (some say abortion) and the new left-leaning American Civil Liberties Union. She did much that shaped the later career of Herbert's daughter Elsie, just as Boas's work reinforced Noel's predilections.[20]

As the emerging scientific findings on evolution mixed with social policy, a more formal eugenics movement arose and some of Herbert Field's friends moved beyond their laboratory work and began relating their research findings to social policies.[21] Charles Davenport, more than others of his generation, focused on the questions of human evolution, especially those concerning intellectual capabilities and social behavior. He also became the most politically conservative member of the old Harvard zoological group. Along with another of Herbert Field's supporters, Henry Osborn, Davenport became an influential international spokesman for the segregation of defectives, for immigration restriction, and, at times, for sterilization. In the 1920s, these men and Clark Wissler were even giving lectures to the United States Army on eugenics and immigration policy.[22]

Osborn and Davenport were very outspoken. In 1934, Osborn publicly declared the Great Depression was a result of the decline and slovenly behavior of the American population. In the same work, he condemned what Noel so admired, the Soviet scientific system. He denounced it because of its refusal to let "science be science" and recognize the "laws" of genetic inheritance. Davenport was also blunt. He became a leading American spokesman against the Soviet political-economic system and its suppression of scientists whose findings confronted the accepted Marxist doctrine that learned behavior was genetically passed on.[23]

Even George Howard Parker, Nina's kindly benefactor, became a noted commentator on evolution and an advocate for controls over human "breeding." Others within the Fields' social-intellectual sphere joined in. The influential relative of the Flexners, Carey Thomas, who headed the Quaker's famed elite college for women, Bryn Mawr, was in the ranks of those who supported limitations on the creation of new human beings.[24]

Ideological Convictions and Conflicts: Noel Field, Herbert's Friends, and Scopes Trial Perspectives

In contrast, while accepting birth limitations, Noel became a firm believer in Boas's "nurture" and was increasingly alienated from his father's associates. For a young left-wing liberal like Noel Field, the evolution-related work of his father's colleagues, such as Davenport and Yerkes, had to be classified as unprogressive. To him and his fellow radicals of the 1920s and 1930s, such men were as backward as the traditional religious groups the Red revolutionaries had so cruelly tried to destroy.

Noel, however, was not confrontational. As with the issues of birth control and abortion in Elsie Field's life, Noel chose to avoid the conflict within him between the "facts" of science, his Quakerism, and his new secular ideology of communism that soon turned against the fundamentals of genetics. Additionally, Noel avoided publically attacking his father's colleagues.

One example of Noel's tendency to push aside intellectual difficulties was his reaction to one of the most famous battles between the new sciences and religion in America, the Scopes anti-evolution trial. The mid-1920s conflict was over allowing the teaching of Darwinian evolution in a public school. A young teacher, John Scopes, disobeyed a Southern school system's rules and taught that humans were a part of the chain of evolution, rather than being special creations of God. He faced criminal charges.[25]

Liberals and radicals came to his support, but not for the same reasons. With the help of the press, especially the radical press and the Marxist Comintern, the "left" made the trial an international event—a battle between science and "irrational" religion, as well as a crusade for freedom of speech. Leftists temporarily put aside the incompatibility between genetic science and Communist doctrines as they waged their Scopes battles.

The Scopes affair involved the progressive liberals of Herbert Field's circle, the ones Noel was beginning to see as ideologically backward. Important to the defense in the Scopes case were Nicholas Murray Butler, the scientists at his Columbia University, and the men at the American Museum of Natural History. There were good reasons for Scopes's allies calling on their help. Their connection with Scopes had a long history. For example, Columbia University's graduates contributed to the textbook Scopes used in his classrooms.

The Columbia contingent came to his support by hosting the young teacher, writing popular articles in behalf of Darwinian evolution, and acting as advisors to the Quaker and Unitarian-supported American Civil Liberties Union that was subsidizing Scopes's defense. The union hired the famous agnostic lawyer Clarence Darrow to lead its case, but he was unable to do more than have Scopes's eventual conviction set aside on a technicality.

While Butler and his scientific fellows had clear-cut and consistent messages about "best practices" for science and academic freedom, for Noel's ideological soul mates there was an unacknowledged significant conflict between their commitment to science and their devotion to environmentalism. They seemed not to notice, for example, the approval of eugenics in the text Scopes had used. Noel was an example of the left's reaction. He dodged the underlying contradictions. As he did in so many instances, he pushed uncomfortable things to the back of his mind when the Soviets dogmatically turned against the fundamentals of genetic science and purged, "with prejudice," their leading geneticists.

Noel was becoming increasingly committed to the twentieth century's most dogmatic, hierarchical, and "church-like" philosophy. As in the conflict between evolution and Marxist perfectibility, Noel would repress the contradictions between his underlying Quaker-Progressive liberalism and Marxist totalitarianism. He would never let himself see the "dark side" of his ideological devotion.

17 The 1930s Ideological Journey of the Fields and Their Liberal Friends

In 1924, when Noel returned to the United States after his trip to the Soviet Union, he was alternatively elated and depressed. His tour reinforced his already ultraliberal beliefs and his faith in the Great Socialist Experiment. Although he was still somewhat of a Quaker and was devoted to nonviolence, the on-demand free abortions provided by the new Soviet government did not diminish his faith in communism. At the same time, being without an offer from the State Department, he was deeply anxious about his future. He was unemployed and had little desire to return to school.

Nina did not allow him to waste time, and Noel agreed to go to work. Although he was a champion of the "working class," he did not seek a blue-collar job. Rather, he took a short course at the radically tinged Boston School for Social Work and then found a position as a state mental health worker. Barely twenty-one, the state allowed him to act as a prison psychiatric counselor. He would not stay in that job for long, however. He was focused on other things: a diplomatic position, peace, and the contributions he thought he could make to leftist causes.

Noel was more than an armchair, ideological leftist. He joined organizations such as the Fellowship of Youth for Peace and became somewhat of an activist. He founded another world-peace club, supported an antiwar group, and joined numerous protests sponsored by radicals. He did not confine his activities to what the then just-left-of-center Nina approved of, such as the peaceful antiwar demonstrations by Swarthmore's liberal Quaker students, or the work of Cambridge's Red Women's International League for Peace and Freedom, which had the backing of the millionaire Jane Addams and her more radical friend, Anna Louise Strong.

Direct Ties to the Party and the Peaceniks: The Clark Family and Frederick J. Libby

Noel's interests and his circle of radical friends were expanding. For example, the Fields were becoming close to the Alfred and Susan Ainslie Clark family of Cambridge. It had links to New England's radical labor unions and allied organizations. Noel later stated that his more direct ties to American Communists came in the mid-1920s when, with the aid of the Clarks, party members led organized protests over working-class issues such as the Sacco and Vanzetti death-sentence case.[1]

Noel also conducted additional academic research when he became associated with an organization run by a Puritan-turned-Quaker, Frederick J. Libby. For a time, Noel served as the man's personal assistant and speechwriter. Libby's National Council for the Prevention of War was a large and well-funded "peacenik" group of the 1920s and 1930s. It held demonstrations and lobbied for arbitration of disputes, for arms limitation, and even for its own version of American isolationism. The council's controversial policies led private anti-Communist organizations to put it on their watchlists.[2]

Noel must have been aware that he could endanger his chances of becoming part of the American diplomatic corps by having Libby's group issue his first book. At age twenty-two, Noel published *Banishing War through Arbitration: A Brief Sketch of Post-War Arbitration Treaties*. He based it on a paper he wrote while at Harvard, and on what Herbert had taught him about European peace movements. Libby's council released the first of its several editions just before Noel faced the State Department's final assessment.[3]

Ramifications, Now and Later: Noel's Stammer and His Idealism

In 1925, when Noel took the State Department's examinations, he excelled on his written tests. Then, he became bitter. His face-to-face interviews led the department to conclude that while he was brilliant, his speech difficulties (like Herbert, he stammered), and his general demeanor made him unfit for an overseas posting as a diplomat.

Despite a letter to "State" protesting Noel's appointment because of his and his family's "well-known" left-wing ties, the department offered him an analyst's position in Washington to begin in 1926. Perhaps his father's friendship with many in the department helped overcome any reservations about Noel. His radical activities were probably discounted as youthful enthusiasms. Noel accepted the offer of a desk job, but resentment over the department's refusal to assign him to a diplomatic station pushed him further into what he later called the "other" part of his life.[4]

Noel later admitted that by 1926, when he entered the State Department's training school, he was leading something of a double life. Along with his visible behavior, he had what underground political workers called an "illegal" existence. Although he was not a regular formal member of the Communist Party, he longed to become one and he maintained contacts with local party members. However, because his beliefs had an abstract and unrealistic tinge, and because he was someone who refused to wait for party orders before acting, he had to remain outside of the ordinary party organizations. Something else about Noel was important. He was not brave or physically strong. That, and his particular situation in Washington, limited his contributions to the "cause." Noel left it to others to man dangerous picket lines and to fight "scabs." His early contributions were mainly intellectual, including writing articles under assumed names.[5]

Noel was a romantic idealist and had blind spots—and he was gullible. He uncritically accepted the radicals' benign interpretations of the actions of America's violent International Workers of the World's unionists, the anarchists who shot businessmen, and those who sent bombs to government officials. Even though Noel proclaimed himself a pacifist, such actions were, for him, those of heroes, not terrorists. He also failed to realize that agents of the Soviet government and its subsidiary, the Communist International (Comintern), orchestrated much of the radical propaganda and organized many of the supposedly spontaneous protests in America.[6]

Moreover, Noel was unable to allow himself to see the brutality of the Communist's management policies. He never recognized that the Soviets had turned Melvil Dewey and Fredrick Taylor's Efficiency into an inhumane management system that made work routines at Henry Ford's mass production factories seem kindly. Noel and his family also somehow managed to reconcile Marxist abstractions with the Soviet policies that led to the 1920s and 1930s mass starvation in the Ukraine, the seizure of thousands of small farms, the horrendous purge trials of the 1930s, the millions deported to Russia's interior, and the Hitler-Stalin Pact of 1939.[7]

In the 1920s, Noel's family was beginning to match his radicalism and romanticism. Like Noel, they filtered out the unpleasant and contradictory in the era's leftist beliefs. Even in the later 1930s, after Nina and the other children toured Russia, they failed to notice the repressive nature of the Soviet state. They were so enthralled by their idealistic views they did not support the 1930s movement led by many respected American liberals, such as John Dewey, who protested Soviet totalitarianism. The Fields held fast to their faith well beyond the 1930s. They remained blind to Communist excesses, even after their American hero, Henry Wallace, apologized to the nation for his failure to recognize the brutal nature of Soviet post–World War II expansionism.[8]

Of all the Fields, Noel remained the most idealistic. After he began his job at the State Department he wrote his younger brother Hermann that his job was just a way to support his "important" career. His letter to Hermann stated, "my real life begins in my free hours . . . studying, writing little essays, dreaming about greater ones, trying to understand the reality behind the mask of things." He recommended new Soviet films to Hermann and Elsie, and he proudly wrote that he had recently sat through three consecutive showings of the *End of St. Petersburg* (and he stated he wanted to see it again). His description of the film gave no sign that he could sort out propaganda from historical fact, or that he had taken heed of the writings of the famous anarchist Emma Goldman who was already warning of tyranny in Russia. Significantly, he never imagined that his enthusiasm for the Great Socialist Experiment would soon be interpreted by hardened Communist operatives as a weakness, and as an indicator that he might be a hated Trotskyite. That term, "Trotskyite," was reserved for traitors who refused to adhere to Stalin's version of true Communist policies.[9]

The Young Fields Come of Age: Modern Marriage for Noel, Adventure for Elsie

Noel was a romantic in other ways. He hurried into marriage. In May 1925, the girl-shy, twenty-one year old wed Herta Vieser, the German-Swiss girl who had loved Noel since he was a boy. Nina, fulfilling the promise she made to Noel when the family left Zurich in 1922, arranged to have Herta, and then her mother and sister, brought to Cambridge. Herta helped around the house and then took an outside job. She lived with Nina primarily to stay close to Noel, she later told her friends. Personal letters reveal that Herta and Noel had been more than just close friends while they were in Switzerland.[10]

Noel's remaining veneer of traditional Quakerism had worn thin. Foregoing years of formal courtship followed by recognition of their bond in the Meeting House, Noel and Herta married in a rushed ceremony before a Boston justice of the peace after quietly obtaining special government immigration permission for Herta. As much of their life would be, the first day of the marriage was chaotic. Although the wedding took only moments, Herta was unable to return to her job quickly enough and she was fired the same afternoon. The marriage ensured Herta's American residency, and the couple did make some amends to the family. They allowed Nina, in conjunction with Herta's mother Katherina, to hold a quiet Quaker-like reception at the family's Cambridge home.

One key family member may not have been at the reception. Elsie Field had just graduated from high school and was preparing to enter one of the elite Seven Sister women's colleges. Nina had promised her a graduation reward, a trip abroad on her own. As had Noel, Elsie did not follow the usual pattern of the upper-middle-class New England student. Instead of a trip to sample European culture, she chose an around-the-world junket. It included a stop in Zurich to visit with Johannes Strohl, but her target destination was China. Although Nina knew China was a country in disarray, including the continuing battles between the Communist and Nationalist factions, she agreed with Elsie's choice, perhaps because Hermann had made a safe trip there a year or two before.

The summer adventure changed Elsie's life. Since childhood, she wanted to be a veterinarian but the sight of the poverty and suffering in even the safe parts of China led her to decide upon a career aiding humans. The trip also validated the left-wing interpretation of the world her brother was espousing.[11]

On returning home, Elsie entered Cambridge's prestigious Radcliffe College, where she took a pre-medical as well as a liberal arts undergraduate course. Important to her life at Radcliffe was joining another network of elite liberal families. Elsie was more gregarious than Noel and made many friends, including young ladies connected to the nation's leading Unitarian families, such as the Wheelers of Baltimore. The Wheelers had made that city's Pratt Public Library one of best in the nation, and they were an important part of the city's progressive establishment.

Such friendships did not dilute Elsie's radical political commitments, however. She openly lauded the "socialist" ideas of Bertrand Russell, as one unfriendly listener reported to the FBI. She also joined the Sacco-Vanzetti protests in Cambridge. Like Noel, she did not realize how much money and effort the Comintern put into creating the political rallies against the decision to execute the two immigrant anarchists.[12]

The Odd but Successful Couple: Quaker Pacifism and the Hiss / Duggan Connections

When Noel and Herta moved to Washington in 1926, they soon became known as a rather eccentric hypochondriac couple that gobbled patent medicines and had "political" tendencies. There were health problems. Herta had to undergo a thyroid operation and Noel seems to have inherited many of his father's defects, including recurrent headaches. They were so severe that at a critical point in his career he had to request a fortnight's sick leave from the State Department. There were also social impediments. Their lifestyle and political beliefs did not mix well with some of the department's up-and-coming young members, such as George F. Kennan. Unlike many of his colleagues, Noel and Herta had no car, no servants, and only a small, disheveled apartment with many cats—although Noel's salary was quite generous and he had a trust fund awaiting him. He was soon receiving more than three times the average American income. One reason for that was a bureaucratic move he made in the summer of 1930. He resigned from the regular State Department service and rejoined as a professional. [13]

Noel frequently left Herta alone. She did not work, was childless, and was bored. She felt somewhat out of place in Washington, but she joined in furthering Noel's professional life. The national press mentioned him when the couple attended prestigious Washington social events including the Soviet ambassador's gala presentation of the Chekhov play *Uncle Vanya*.

There were marriage difficulties. Noel told one friend that Herta had turned out to be not his type. The frictions led to an agreement between Noel and Herta. They would take semiannual vacations from each other to preserve harmony in their cluttered apartment. That agreement, and his deepening attachment to an alternative lifestyle, might explain the "Ellen" who was listed as his wife on his trips back from Europe in the 1930s and early 1940s. There were rumors that Noel and Herta had become believers in "free love."[14]

As he did at Harvard, Noel gained a reputation in Washington as a loner who seemed to make an effort to be different from others. However, he and Herta did not remain completely isolated. Their Quaker connections, his Harvard background, and their left-wing activities brought them into contact with many other elite ultraliberals, including Alger Hiss and Laurence Duggan, Stephen's son. Those two young men would rise to the top of post–World War II's liberal establishment. Noel and Herta may even have met one of the heirs of the Vanderbilt fortune, Frederick Vanderbilt Field.

He was becoming an expert on Asia. He was also becoming a radical and a recruiter for the Communist cause. Unfortunately for him, his relatives would soon disown him because of his deepening radicalism.

Noel's eccentricities did not immediately harm his career. The department first assigned him to the Swiss and League of Nations desks where he compiled numerous reports that filtered up to America's working diplomats. They were impressed. Noel focused on those subjects for several years and gained a standing as a solid analyst. He also studied peace initiatives. Then, a part of his dream came true. The department selected him to be a member of America's disarmament teams that attended numerous European conferences during the 1930s.

At the same time, however, Noel was gaining a reputation as being sympathetic to the Soviet Union, to Communist causes, and left-wingers in the United States. By the late 1930s, some in the State Department presumed he was a true Communist. At times, Noel even came close to abandoning his Quaker pacifism, partly because he was embittered by the threats to his investments due to the 1929 market crash. In 1932, he did throw stones at the tanks Douglas MacArthur's troops were using to rid Washington, D.C., of the unemployed Bonus Marchers, the World War I veterans who had squatted in abandoned buildings in the city and demanded payments promised by the government. Noel then wrote more anonymous articles for radical magazines.[15]

Noel continued to ignore traditional social norms. He was not hesitant to talk of the birth-control measures he and Herta were using. He instructed his younger brother on the best ways to deal with premarital sex and "ladies problems." Noel seems to have made himself familiar with the then ultrafashionable Freudian psychosexual theories. The young Fields were also open to discussions of homosexuality and lesbianism, weighing the pros and cons of their friends' experiences.[16]

A Rising Star of "State": Learning Secrets, Contacting Agents

Despite Noel's unconventionality, the State Department did not dismiss him. He was very good at his analyst's assignments. He received excellent job ratings and he was respected enough to be asked to help write speeches for the Secretaries of State during the 1930s. Importantly, he was conquering his stammer and was learning how to dress and to comport himself as a diplomat. He even grew a distinguished mustache. As his career progressed his sphere of expertise broadened, and he became central to the department's study of Western European affairs. Then, he was given more responsibilities for the critical issue of the era, disarmament. He was also gaining access to sensitive information. By the mid-1930s, he was the head of the technical committee advising America's disarmament negotiator, Norman Davis. Noel was privy to reports on armaments and he witnessed Japan's startling demands to be recognized as a world-class naval power. Noel wrote many advisory policy papers and came in professional contact with more of the rising stars of Washington and New York, such as his father's old contact, Allen Dulles.[17]

Despite that, Noel remained resentful. He was repeatedly denied a permanent foreign posting. His trips to disarmament meetings in London and Geneva were not fulfilling his long-held dream of becoming an important voice in the league's deliberations. He and Herta were also becoming increasingly alienated from American culture. She wanted to return to Europe where she felt comfortable, although her sister, Traudy, had permanently moved to America.[18]

More than a stalled career caused Noel's deepening frustration. As the Great Depression ravaged the economies of the United States and Europe, his discontents multiplied. The unemployment and social unrest intensified Noel's estrangement from the capitalist system. He soon found an outlet for his frustrations, a truly "illegal" one.[19]

By at least the early 1930s, Noel had contacts with agents of the Soviet Union and its extensions, the Comintern and the American Communist Party. Noel became part of one circle of Washington government employees supplying information about American policies and programs. Later, several Russian defectors and American ex-party members, including those who pinpointed Alger Hiss as one of the most important American sources for the Communists, testified about Noel's aid to the Communist cause before World War II.[20]

The Soviets had gained an inexpensive windfall in Noel and his colleagues. While America's code breakers labored to read the messages of foreign diplomats to learn of such things as their disarmament policies, Noel's circle gave out some of the State Department's most precious secrets—for free.

Yet the inner ranks of Soviet intelligence declared Noel too immature. Its agents in America informed Moscow that Noel, while a valuable source, was too naive to be allowed full open membership in the party, or a central role as an agent. That meant he remained unaware of some of the seamier aspects of Soviet information gathering. For example, he never learned of the cadre of Soviet agents that was planning to steal the secrets of America's atomic and radar projects, or of the Russian travel-agency group in New York City that helped to arrange the assassination of Stalin's nemesis, Leon Trotsky.[21]

Moving On: Noel, the League, and New Adventures

Noel remained at the State Department until 1936. After attending a critical naval disarmament conference, and handling the most confidential papers of the lead American negotiator, Norman Davis, Noel accepted a job at the League of Nations in Switzerland. Some believed he sought the position at the prodding of Earl Browder. Browder was the American Communist Party's new leader and an agent of Soviet intelligence—as was Noel and Nina's new close friend Hede Massing. Browder, it was claimed, wanted someone on the inside of the league. Hede Massing believed that Noel took the job so that he would not feel guilty about passing secrets to her. A more acceptable explanation is that Noel wanted to serve the league since his youth.[22]

There were other motivations for taking the position at the league. Herta wanted to return to Europe, and Noel's bitterness about the State Department's refusal to give him a diplomatic posting had reached an emotional level. Although he now had better social skills, and although he had proven himself an expert analyst, the department had once more refused him. He was thirty-two, without children and with a stalled career. Also, he was sensing that the department was concerned about his leftist viewpoints. His "Pink and Red" opinions had finally led to inquiries. He began to be anxious about being dismissed.[23]

Noel continued to have some influential friends at State, however. When he told them he was considering a job with the league, one that was being vacated by a acquaintance at the department, there was mention of a significant salary increase and, perhaps, being put in charge of the State Department's vital German section if he would stay in Washington. He did not pursue that option. Noel headed for Switzerland and was soon telling reporters that he was "formerly with the State Department."

However, Noel hoped the department would treat his move as only a leave of absence. He wanted a job ready in case he decided to return to America. He kept in close contact with the department throughout the mid- and late 1930s, making frequent visits and informing it of his activities. In turn, many in the department continued to see him as one of their own.[24]

Hermann "Ties the Knot" to the Radical Clark Family

Meanwhile, Hermann, the younger of the Field boys, had followed in his uncle's footsteps. As had Hamilton Easter Field, Hermann opted to attend Harvard with architecture as a career goal. He was an outstanding student (receiving honors) although not quite the amazing academic star his brother had been. However, he was a social success. Unlike Noel, he had some time to adjust to American life before entering college; he had a wide circle of friends (not just radicals), and, he was engaged in enough activities to keep him from dwelling, as Noel did, on any personal injustices he encountered.

Hermann was an outstanding writer, singer, and musician. He was also quite a ladies' man. In addition, he exhibited an interest in politics. He was a leader in Harvard's Liberal Club where he expressed some radical views, but he had not become a Communist. Like Noel, Hermann was a romantic, and he was after more than just training for an occupation. He wanted a skill that would contribute to the social good. The way he spent one of his summer vacations reflected his idealism. Shortly after beginning college, he traveled to China and worked in a hospital.[25]

Hermann had also drifted from the old Quaker social ways. Just as he was graduating cum laude from Harvard's general arts program in June 1932, he married his school sweetheart, Jean Clark. The unusual marriage arrangement (Jean was still in college and not a Quaker) did not alienate Nina and the responsibilities of marriage did not end Hermann's career ambitions.

Hermann's new father-in-law came from a well-to-do family that had owned a jewelry business in nearby Lynn, Massachusetts, and then a banking business in Boston. Alfred Sinclair Clark became a Harvard student in 1896 and was a classmate of the founder of the U.S. Army's (and the nation's) first radio-spying agency. Upon graduation, Alfred's father helped him become the treasurer of his Boston brokerage firm. Alfred did very well for six years until a bankruptcy led to him become a not-well-paid newspaper reporter. With that, he began to question the results of capitalism and what he called "unthinking conservatism." He rebounded, soon becoming a rather famous and well-paid feature writer for the *Boston Post*.[26]

Alfred was temperate in expressing his left-leaning political views. His wife, Susan Ainslie, was not. Her respectable upper-middle-class family had moved from the Midwest to Brooklyn where she attended the exclusive Packer Institute for females in Herbert Field's Brooklyn Heights. Helped by a scholarship, she graduated from one of the Seven Sister women's colleges, Wellesley, in 1903. She returned to New York to work with the Progressive reformer Lillian Wald. After that, she helped the crusading Muckrakers who were publishing sensational exposés of the results of capitalism. Although in her twenties, the young Ainslie became associated with many famous reformers such as Florence Kelley, Ida Tarbel, Lincoln Steffens, and even Louis Brandeis. Ainslie did research and writing for the progressive and vocal Consumer's League, including a work that made her famous among later feminists. Because of *Making Both Ends Meet,* a 1911 book she co-wrote with Edith Wyatt about the condition of women workers in New York City, the name "Sue Ainslie" became nationally recognized in reform circles. Her book and her other writings included demands for paid vacations for women workers for two days a month (to help with "lady" problems) as well as a surprising salute to those advocating the use of Frederick Taylor's science of Efficiency to determine standards of fair working conditions.[27]

After almost a decade in reform organizations, Susan Ainslie married Alfred in 1909, when she was thirty. Marriage and a move to Boston did not end her progressivism. She became associated with more radical organizations, such as the Women's Trade Union League (WTUL). The WTUL took a leadership role in many of the very violent New York garment district strikes. Because of that, Susan Ainslie Clark came to the notice of the government, as did a fellow supporter of the league, Eleanor Roosevelt.[28] In Boston, Susan continued her research on women workers for the moderate National Civic Federation, but her radicalism was deepening.

In 1911, the massive strike of textile workers in Lawrence, Massachusetts, was taken over by the ultraradical Bill Haywood and his fierce IWW union. That led to hunger and violence. Many soon condemned the strike and the IWW, but Susan Ainslie Clark did not. She published a broadside that included a none-too-veiled salute to the IWW and its strategy of violent confrontation. That was a dangerous statement for her to make. Bill Haywood was under investigation for an assassination and was soon convicted of sedition. In response to the conviction, he jumped bail and fled to the Soviet Union.

Susan Ainslie Clark withdrew from public view after she had the first of her four children in 1912, but not entirely. Then, as she quietly moved further "left" and into Communism, she got involved in the development of what became one of the most respected and influential elite progressive grammar schools in America, one critical to the futures of Nina Field's children and their lives "on the Left."[29]

Educational Incubators for Intellectual Radicals

There had long been discontent among the more liberal Harvard University professors over Cambridge's public schools, but it was not until the best of them (the one named after Louis Agassiz), was to be closed for reconditioning that something was done. A man noted as "the" liberal of the era had returned to Harvard and soon led the search for an alternative.

William Ernest Hocking was a romantic theologian who reconciled modern science (including evolution), pragmatism, and higher purpose for himself and other progressive Unitarians. He, his poet wife, and several other faculty members who did not wish to pay the high fees demanded by local private schools, decided to begin a cooperative coeducational teaching venture for their own children. In spring 1915, with faculty wives as instructors, the Hockings began a school on their porch. They called it the Cooperative Open-Air School. It was in the open air. Students wore winter clothes, even sleeping bags, to keep warm.

By 1925, the name of the school was changed to the Shady Hill School and tuitions were increased, but it remained an innovative college-prep school aimed the children of the local intellectuals. The students, including Nina Field's and Susan Clark's children, learned all the traditional subjects, although the teaching approach was very, very progressive.[30] As the school's attendance grew, instruction was turned over to a professional staff under Katherine Taylor. Her work attracted other progressive teachers, such as Henry Wadsworth Longfellow's granddaughter, Anne Longfellow Thorp, and a fascinating woman, Carmelita Chase Hinton. Hinton became a famous progressive educator and a noted fellow traveler with New England's Communist Party. She was also the mother of some of America's most prominent radicals of the 1940s and 1950s.[31]

In 1924, Carmelita Hinton was on her own after her husband Sebastian committed suicide in a Massachusetts sanatorium while being treated for his long-recurring depressions. She suddenly had the responsibility of raising three young children, Jean, William, and Joan. She quickly sensed a great opportunity for herself and her children when her friends from Bryn Mawr, Hull House, and other reformist organizations told her of the forward-looking educational work at the Shady Hill School. Hinton's application for employment was quickly accepted. She hired on as a second-grade teacher, but laid plans for innovations in high school education.

While taking her students on adventure-filled, *Treasure of the Sierra Madre*–like trips to the Mexican mountains and on other "international" summer learning excursions (including one to Germany) she honed her educational ideas. By 1933, she had the idea for an unstructured, cooperative, sensually oriented, and "natural" high school education. In 1934, after a trip to rural Putney, Vermont, for an international peace meeting, Carmelita Hinton thought she had found a site for her democratic "educational beacon." The next year, Carmelita opened what became an ultraliberal, if not decidedly Left stronghold, the Putney School. It attracted students from the intellectual elite, including the son of the Johns Hopkins professor, Owen Lattimore, and Alger Hiss's son. There was something of a barrier to making Putney a true "people's" school in the 1930s, however. Its tuition equaled half a year's income of a skilled working-class man.

At the same time, Hinton's heroes, William Kilpatrick, John Dewey, and Eleanor Roosevelt, were turning their visions for a modern women's college into a reality. They were about to open the curricular-unstructured and totally secularized version of a Seven Sister school, Vermont's Bennington College. Unfortunately, Bennington charged higher tuition than Putney. Somehow, Carmelita Hinton was able to send her two daughters to that new college.[32]

Shady Hill and the Field Children's Exceptional Lifelong Friends

Before then, in 1922, although Cambridge's Shady Hill School remained in primitive buildings, and while it still allowed some parents to work off tuition, it was fast becoming a school for intellectual, liberal, and economic elites. Alfred North Whitehead sent his son there. Its students felt special, formed lifetime bonds, and usually went on to Boston Latin and then to the Seven Sister and Ivy League colleges. It was no surprise that Susan Ainslie Clark and Nina Field enrolled their children in Shady Hill.

Although Shady Hill was not yet the ultra-expensive "prep" school it became ($50,000 tuition a year by the early twenty-first century) enrollment at the Hill was somewhat of a financial burden for Nina. But it seemed worth it. Hermann and Letty had loved the unusual school and they met the "best" of classmates there. Besides the Clark children (Jean, Margot, Joy, and Alan), there was May, the daughter of George Sarton, the famed historian of science, recipient of Carnegie funds, and colleague of Paul Otlet. May Sarton was an important link to the artistic Left for the Fields. She became a friend of the progressive and leftist East Coast intellectuals and actors of the 1930s. Some of them, such as Burgess Meredith and Will Geer, were later entangled in the postwar Hollywood anti-Communist traumas.

Attending Shady Hill drew the Clark and Field families together. In 1933, after Susan Clark was widowed, and her rather upscale standard of living was endangered, she became more radical and even closer to the Fields. Fortunately, her children received scholarships to Shady Hill and, then, college. Susan and her daughters went on to become American Communist Party stalwarts with Margot running the party bookstore in Cambridge, Joy serving as an underground organizer, Jean helping the cause in Switzerland and California's Mill Valley, and Susan supporting a Communist cell for intellectuals in Boston.

There was another Shady Hill chum whose later life became entangled with Communism and with the leader of America's atomic bomb program. Jean Tatlock (daughter of a famous professor of English) was Robert Oppenheimer's San Francisco lover and a suspect in the investigation of leaks to the Russians about America's greatest World War II secrets. [33]

Hermann, Holistic Architecture, and the Drift toward the Far Left

While the new bride Jean Clark Field was completing her final two years at Vassar and earning a Phi Beta Kappa distinction, Hermann Field was taking Harvard's graduate architectural courses. He could afford it because of Nina's generosity and his own trust fund. Nina helped pay tuition and provided a home for him, but she let it be known that she would not support him for too long.[34] Hermann recognized his mother's limits, but stood by his version of an architectural career. He sought one that would fulfill vital social needs rather than monetary goals. He could not put a name to it then, but it later became known as holistic architecture. Hermann worked hard and soon received word that he was in line for a fellowship at a European school of his choice. Although he did not intend to spend his life as a student, he was willing to give another two years to education. The offer became more attractive when the organization that had snubbed his father and Johannes Strohl, the Institute for International Education (IIE), informed Hermann it could also arrange a fellowship for his bride, Jean Clark. (At the time Edward R. Murrow was assistant director of the IIE, which had been founded by Nicholas Murray Butler, Elihu Root, and Stephen Duggan.)

Before the fellowships began, Hermann decided that one of the best places to learn more about holistic architecture was Soviet Russia. He had heard many exciting things about the new Communist architects, especially about their town designs. He and Jean, accompanied by Joy and Margot Clark, took a summer trip there in 1934. Hermann and Jean looked forward to the distinction of being among the first Americans to attend the Soviet's new Duggan-linked American-Russian Institute program at Moscow University.

Hermann and the Clarks were part of a group of some two hundred, but few were students—most were leftist stalwarts who rushed to see the results of rule by "the workers" as soon as allowed by America's formal recognition of the Soviet regime.[35] During his summer 1934 Moscow visit, where he likely encountered Stephen Duggan who was there establishing another exchange program, Hermann fell deeply in love with Russia's "planned" economy and society, partly because he made arrangements for the trip through an agency that skillfully showed tourists only Soviet successes. A few years later, its connections to Soviet intelligence became public knowledge: New York City's World Tourist had always been part of the American and Soviet Communist Party's undercover apparatus.[36]

Unknown to outsiders, World Tourist was a major source of financial support for America's Communists and a channel for Soviet influence and espionage. Its manager, Jacob Golos, was also the controller for many spies in America, and he was one of the many American Communists taking direct orders from Moscow. Profits from World Tourist subsidized much of the American party's illegal work, just as had profits from Armand Hammer's Allied Chemical Company, income of the American-Rumanian Film Company, and counterfeit U.S. currency smuggled into the country. Some of those funds were used to finance a year's trip to Russia by an American Communist whose life would become dangerously entangled with the Field and Clark families: the young Joseph Figueiredo was assigned to study "methods" in the Soviet capitol in the mid-1930s.[37]

Hermann knew nothing of that, and he thought the summer trip to Russia was an invaluable experience. Although he was displeased about being charged for a course that was unexpectedly canceled, he was so enthused after his trip that he considered taking a job in the Soviet Union with its economic planning organization instead of attending the famed Swiss Technical Institute in Zurich, where Duggan had arranged fellowships for both Hermann and Jean.[38]

18 Intrigue's Beginnings, in Switzerland, England, and Cambridge

Immediately after their 1934 Russian tour, Hermann and Jean Field traveled to his Swiss hometown. Their student fellowships at the Swiss Institute were in the same city, Zurich, and in the same university. He studied architecture; she majored in German literature. Ironically, the Institute of International Education (IIE), which had recently refused to help Johannes Strohl and the Concilium, sponsored both awards. How the couple was able to obtain grants for the same institution (co-founded by Stephen Duggan who then hired Edward R. Murrow to help him) at the same time, and from the same agency, was never detailed, but Herbert Field's old social and diplomatic connections, and Noel Field's close friendship with Stephen Duggan's son Laurence may have played a part in securing the young couple such "plum" assignments.

The Swiss Institute, founded as a high school, had developed into a world-renowned center for science and technology. Einstein had been a student there. In contrast, when Hermann looked in on the Concilium, he saw little promise for his father's center ever again being "world class." Hermann never mentioned meeting with Strohl and his wife.[1]

Hermann and Jean were pleased with their European experiences. They mixed with stimulating students and intellectuals, including young Swiss Communists. Hermann had at least one long discussion with leading German Communist Party officials who visited Zurich. He reported to Nina how exciting that was.

During the summers, as part of the fellowship agreements, he and Jean worked as apprentices in Italy and Germany. In 1936, at the end of their fellowships, and as Noel was on his way to Geneva, the couple disliked the idea of leaving Switzerland. Jean was still laboring on her dissertation and Hermann loved being "home" again. However, after completing his studies, Hermann could not avoid taking a job. He was unable to find one in Switzerland, but he did not return to America. His Swiss mentor secured him employment as an architect in England. Hermann's job was as a planner for the new manufacturing complex and scientifically planned "garden town" for Roche, the Swiss pharmaceutical company. He was happy, but not Jean. The marriage became stressed.

Hermann the Ideologue, Jean the Radical: Shifts and an Eventual Split

While in Zurich, Hermann had developed into somewhat of an ideologue. His commitment to Marxism was so intellectual and abstract (perhaps fanatical) that Communist agents such as Paul Massing, while trusting Jean with covert tasks, refused to assign Hermann to serious undercover operations.[2]

Hermann had revealed his shift to the Left in an emotional letter to his mother. Nina, already sympathetic, partly because of her hatred of fascist policies, agreed with Hermann's views. Hermann protested the lies being told in high-finance capitalist Switzerland against the Soviet Five Year Plan for industrialization; he lambasted the liberals who had delayed the "ultimate revolution" in Russia in the 1910s; and he denounced those who had initiated the failed American recovery polices of the 1930s, although Roosevelt's New Deal had been moving to the Left. However, he and Jean (who was becoming more radical than Hermann) appeared to be an ideal young couple and did not gain reputations as eccentrics or radicals while in England.[3]

Despite socializing with party members, Hermann was not involved in Communist underground activities in the mid-1930s. Jean, however, became entangled—even before she and Hermann began a separation that eventually led to the unheard of for Quakers, a divorce. Hermann's deeper leftist involvements only came after his marriage into an English family. Jean's came when she became entwined with Soviet agents and a radical Jewish immigrant in Switzerland.[4]

During his first years in England, Hermann devoted himself to architectural-social planning. However, he fit the pattern of liberal young idealists of the 1930s, and he must have encountered the writings of the British Marxist scientist and information theorist J. Desmond Bernal. Hermann probably met him because of Pablo Picasso's friendship with his uncle Hamilton Easter Field, the Havilands, and Bernal. Picasso was so close to Bernal that he painted a great mural on the scientist's apartment wall.

Bernal protested what he saw as grave problems resulting from the way modern universities and science had developed under liberalism. His publications on the social function of science emphasized planning, central direction of science, the need for morally dedicated scientists, and a radical change in the nature of science communications. He admired and supported the Great Soviet Experiment, perhaps because during his many visits to the "workers' paradise" he, like Boas's student Margaret Mead, saw only what his ideological commitments allowed.[5]

Hermann also continued to be unable to "see" everything around him. Although he would frequently travel to Switzerland for vacations, and to visit with Noel, he did not realize how much of a Marxist his brother had become. Hermann was unaware the Swiss Communist Party had secretly granted Noel a special status, or that he was passing information to his new Soviet controllers. All Hermann saw in the 1930s was a successful man in the grand peace organization, the League of Nations.[6]

Cultivating Friends of the Communist Party: Noel and the Fields, at Home and Far Afield

While Hermann and Jean were moving toward a separation and divorce, Noel and Herta were establishing themselves in Geneva. They were happy and thought they had found their niche. The job in Switzerland was a step up in every way. The salary was generous; more than three times that of an American office worker in a period when the American dollar was very strong. Partly because of that, Noel and Herta decided to end their Spartan lifestyle. Their apartment and their chalet outside the city were sumptuous, and they aided Noel's attempt to show that he had the makings of an ambassador. He felt he now had the ability to "meet and greet" as was expected of policymaking diplomats.[7]

At first, however, his league assignment was writing general papers on a range of what he considered mostly insignificant subjects. One assignment, to handle disarmament issues, provided some excitement and a chance to pass on information to Paul Massing, one of his Communist contacts.

The stimulating culture of Geneva compensated for confinement to an in-office job. Herta felt at home in Europe and did not express bitterness about having a childless marriage. She and Noel had a full social life, and they were always in contact with family. Her mother, Nina, and Hermann frequently visited, and they all took pleasant auto trips through neighboring countries. The Communist agent, Hede Massing (Paul Massing's wife), joined them on some of those excursions. In addition, Noel earned enough to afford frequent trips to the United States.[8]

Nina was a frequent and active traveler. When in Switzerland, she contacted her old Swiss, American, and British friends and ingratiated herself with Noel's new political circle, especially those of the Far Left. On one of her trips to Switzerland, Nina consented to take forbidden Communist literature into Germany in order to help one of Noel's Comintern allies. She also secretly contacted a German official who could supply Soviet intelligence with information on the state of the German economy.

Then, at age sixty-three, she made a grand tour of Russia and returned as a Communist. The skilled public-relations experts of Intourist, and its American branch, World Tourists, arranged the trip. Like Hermann, Nina had no idea of the real work of the American travel agency. Besides making money for the American Communist Party, it was a front for distributing false passports to Comintern agents and, later, the Spanish Civil War's International Brigade fighters.

Nina was unaware of the deep involvement of her travel agency in espionage, but she must have known of the man who ran the American branch, Jacob Golos. He was a colleague and controller of one of her and Noel's closest and most remarkable friends in the mid-1930s, Hede Massing. Golos was also linked to one of Noel's party contacts, another spymaster, J. Peters. Some of Golos's activities became public knowledge during a trial in 1939. However, it was not until the late 1940s, when his ex-lover, Elizabeth Bentley, began informing on the old Soviet networks in America, that the depth of his espionage work became public knowledge. Golos's reach even extended to operations by newspaper men such as the famous Winston Burdett at the *Brooklyn Eagle* (later of Edward R. Murrow's CBS radio program).[9]

Like Jean Field, Elizabeth Bentley was a Vassar graduate influenced by the liberalism at that Seven Sister college, and at Columbia University where she did some graduate work. After completing her studies, Bentley worked for the American Communist Party and Soviet intelligence. At the end of World War II, alienated by the actions of her comrades, she revealed some of Golos's underground activities. Only later did government investigators discover that Golos was also using the secretary of the liberal columnist Walter Lippmann (an influential acquaintance of Herbert Field), as a source of "inside" political information. Much later came revelations about Golos and the Keeneys, the American "librarian spies" who provided the Soviets information on classified work in the Library of Congress and America's World War II intelligence agencies.[10]

The Massings, the Duggans, and the New Left

Nina's friendship with a woman who also had espionage ties to Jacob Golos was of direct significance for the later intrigues of Noel and Hermann Field. When Noel began passing information to the Soviets in the mid-1930s, one of his intermediaries used the name Hede Massing. Nina had first encountered her on one of her visits to Switzerland. Nina and Laurence Duggan then helped Hede and her husband at the time, the Marxist theoretician Paul Massing, to immigrate to the United States.[11]

Hede, a Vienna-born long-time Communist activist had spent a few years in America when she was younger, even visiting Mill Valley, California (one of the U.S. communities where party supporters felt comfortable), while on a countrywide get-together with prominent leftists and while gaining American citizenship through a marriage. She met the other Fields when she returned to America in the 1930s. A writer for the American Communist–backed *Daily Worker* arranged a meeting with Noel in 1934. After the introduction, Hede went on to build an even stronger personal relationship with Nina and her children.[12]

Hede had an impressive Communist resume. There were years of undercover work in Germany as a reporter, a friendship with Richard Sorge who would lead the main Soviet spy ring in Japan, and marriage to Gerhardt Eisler who managed espionage agents in the United States and China, and who later became an East German official. Then, Hede married a martyr to the cause, the German leftist scholar-author Paul Massing.[13]

Paul Massing provided Hede with an entrée into the American liberal elite. The couple became the darlings of New York's intellectuals. The Germans had imprisoned Paul in the 1930s because of his communist writings. On his release, his articles detailed for the first time the horrors of Germany's new labor camps—and brought him fame. There was more in his background that impressed the American liberals and the Fields. Paul was one of the members of the famed Frankfurt school of philosophers who were the academic-philosophical torchbearers for theoretical Marxism. Because of that, Massing became one of those favored by the Stephen Duggan group that was rescuing academics from fascist persecution. Stephen's son Laurence, although assigned to Central and South American affairs at the State Department, had stepped in at a critical moment to aid Paul's exit from Germany.[14]

That was not an isolated case. As soon as Hitler came to power, Duggan and other liberals began forming organizations devoted to saving European intellectuals, especially Jews and leftists, who they feared would be silenced and, perhaps, punished. Duggan and his allies raised money to finance escapes, and they used their influence to ensure those brought out of fascist countries could continue their work. Massing, for example, eventually was given a coveted position at Nicholas Murray Butler's Columbia University where others of the "Frankfurt School" of Marxist philosophy had settled. Another Jewish-Marxist refugee, Herbert Marcuse, would join them. He later became known as the philosophical father of the 1960s New Left.

While at Columbia's Institute of Social Research, Paul Massing forged ties with other émigré scholars, such as Paul Lazarsfeld. Lazarsfeld did much to shape mainstream, not leftist, postwar American social science while he worked closely with the nation's intelligence agencies. Strangely, Massing went on to marry Lazarsfeld's ex-wife after his own separation from Hede.

Massing did not become as integrated into American life as others of the Frankfurt School. Marcuse worked for American intelligence during World War II and Franz Neumann was a top-level strategic analyst. He was employed at the secret offices at the Library of Congress during the war. Massing had a somewhat too-checkered background for the leaders of the intelligence agency running the library's center for the Office of Strategic Services.[15]

Noel and the Agents, Even Those in Hollywood

In the mid-1930s, however, both Massings were acceptable to America's Eastern elite, especially those already on the left. Hede's charm and Paul's status as a major intellectual soon led to a very close personal relationship with all of the Fields, especially Nina. The Massings stayed with Nina and Hermann at the Shirley, Massachusetts, farm; they had dinner with Noel and Herta at their Washington apartment where Hede met other possible intelligence contacts; and all the Fields attended left-wing political meetings with the Massings.

Noel's contacts with Soviet agents and sympathizers went even deeper in the 1930s, and he knew more about their work than did other family members. He even had a Soviet codename: "Ernst." His special relationship with Hede Massing, his later rescue of the Abraham Lincoln Brigade's men in France (including the Hollywood screen writer Alvah Bessie), and his work with European left-wing Spanish Civil War refugees, brought Noel in touch with a wide range of Comintern agents in Europe and America. They included Otto Katz and Hanns Eisler.

Katz formed the later-to-be infamous Anti-Nazi League in Hollywood in the 1930s. It raised funds and recruited influential supporters who helped shape a positive image of the Soviet experiment through films and radio broadcasts. As an ex-husband of the film star Marlene Dietrich, and friend of Peter Lorre, Katz gained entry into the entertainment industry's liberal circles and recruited numerous famous writers, actors, and directors to the Communist cause. Some of them, like Alvah Bessie, became members of the Hollywood ultraliberal factions. Those who were already members of the American Communist Party formed the initial core of Katz's supporters. Then, intellectuals such as Dorothy Parker and David Ogden Stewart became loyal torchbearers.

Hanns Eisler, a famed European music composer, was another of the Communist intellectuals brought to America by rescue groups such as that supported by Stephen Duggan. He settled in Hollywood to write award-winning scores, to collaborate with Bertolt Brecht, and to pen anthems for the struggle against fascism in Spain and America. Later, he fled to the East and to a dark fate similar to Noel's.[16] Katz and Eisler's 1930s work was one of the reasons for the early Hollywood anti-Communist investigations, and for the later Hollywood Ten affair.

Nina Mourns Letty: Looking for Companionship and a Cause

While Katz and Eisler were influencing the movie industry, the relationship between Nina and Hede Massing became one of dependency. Nina had become emotionally needy. Her family, at times, thought she was off-balance. The house in Cambridge was becoming an empty, emotional burden. She wanted someone to be close to and she desired a meaningful cause.

Nina's loneliness had become intolerable, partly because of young and demonstrative Letty's sudden death just as she was to graduate from Boston Latin. Nina had great hopes for her affectionate daughter. She had done so well academically that Radcliffe accepted her while she was a junior and she was about to leave for Geneva for a year of study. Her death came after she, Nina, and Hermann returned from a three-month European trip in 1930. It had included another stop in Switzerland and a visit with the Strohls. Back in America, Letty began having severe pains. When her copious menstrual bleeding wouldn't stop, the doctor rushed her to surgery. He performed an emergency operation. Although recovering in a renowned Cambridge hospital, she succumbed to a massive infection.[17]

Despite her Quaker stoicism, Nina was devastated. She was left with what seemed a vacant life. She attempted to fill the void through the frequent trips to Europe and by caring for Hermann while he remained at home. But, she required more. Nina sought attentive people. She found two, the Massings. She had helped Paul enter the United

States by posting a financial guarantee for him, and then frequently gave him money. Especially because of their anti-fascist credentials, the Massings gave Nina a sense of purpose. For Hede and Paul, Nina was a valuable secondary intelligence contact and a woman with riches.[18]

Nina's older daughter was not available to fill her mother's needs. Elsie had a life of her own. Unlike her grandmother Lydia—who, after her mother's death, waited to marry until her widowed father found a new wife—Elsie did not let family obligation stand in the way of her future; she married before Nina adjusted to Letty's death. It was a difficult decision for Elsie as it also meant eventually leaving Cambridge. In 1931, she wed a young Harvard statistics student, Joseph Doob, who soon began moving to academic posts across the country. Leaving Cambridge did not mean that Elsie abandoned her modernism (through at least the 1950s, she signed her name as Elsie Field, not Elsie Doob), her left-of-center beliefs, or her contacts with the elite of America's liberal culture. The marriage to Doob also indirectly linked the Field clan to some of America's greatest secrets.[19]

Joseph Doob was a precocious New Yorker who, like the son of another famous man connected to the Fields, Franz Boas, went to Felix Adler's New York City avant-garde Ethical Culture School rather than to a local public school. The students at Adler's "experiment" included Robert Oppenheimer, the son of a very rich Jewish merchant. After Adler's school, Doob enrolled at Harvard where he studied statistics with Marshall Stone, a respected advisor to America's code breakers and code makers during the 1930s, World War II, and after.

Herbert Field's old protagonist, the National Research Council, recognized young Doob's abilities. One of the Council's responsibilities was to act as an intellectual talent spotter and mentor to the brightest young scholars who might further American science. The council rewarded Doob a coveted National Research Council / Carnegie postgraduate fellowship at Columbia University where Elsie completed the MD training she had begun at Harvard. As she had done at Radcliffe and Harvard, she excelled at Columbia's medical school.

By the mid-1930s, Joseph was teaching at the University of Illinois and gaining a reputation as the man who developed modern mathematical statistics. Doob had a broad perspective and many interests. His commitment to liberal causes helped Elsie overcome her reluctance to put her medical degree and feminist ideals aside. She chose to be a housewife and only a part-time physician.[20]

At Illinois in the late 1930s, Doob helped train a number of young men who became major contributors to the U.S. Navy's most secret cryptanalytic efforts and who went on to be leaders in America's ultra-secret code breaking National Security Agency. During World War II, Doob served as a consultant to the navy, always insisting that he work on such things as mine detection, not cryptanalysis. That may be true because some believe his possible links to American Communism made the intelligence agencies wary of him. Given the World War II intelligence agencies efforts to seek all those who were statistically skilled, not taking on Doob needs an explanation.[21]

Elsie's leaving Cambridge in the early1930s was hard on Nina. Fortunately, Nina had money. As she grew older, she could afford to hire a servant. She could also afford to help Spanish Civil War veterans who managed to reach the United States and Canada.

Meanwhile, Nina's relations with the Massings became so close that she shared in the purchase of a farm-retreat some ninety miles from New York City in, appropriately, Quakertown, Pennsylvania. The area was part of a literary and arts colony that later became significant to Hermann's life. Nina hoped that by living at the farm she would have friends to care for her in her old age, a conduit to the cultural offering of New York City, contact with the many intellectuals drawn by Paul Massing's writings, and another place to house returning veterans. Unfortunately, Nina was too old and frail to enjoy much time at the farm and personal and ideological conflicts developed between two increasingly strong-headed women.[22]

Noel and Herta Are Anointed, the Massings Are Unfaithful: Precarious Situations in a World of Secrets

Before Nina and Hede became estranged, Noel, still in Switzerland had kept in touch with his family, his friends, and with his Communist contacts. Staying in touch did not mean he was fully open with them, however. He frequently sailed back to America, at times leaving puzzling historical traces. In 1936, he supposedly told his Communist handlers that he would not give them any more important information while he was still an employee of the U.S. government, and he later told others that one of the reasons for his taking the League of Nations position was to allow him to be a guilt-free provider of secrets. That led the Massings and others to think he had resigned from the State Department.

There are indications that Noel was keeping something vital from his comrades. On his return trips to America, he listed himself as an employee of the department and his place of business as Washington, D.C. There was also that problem of his wife being listed as "Ellen," not Herta, on the ship manifests, even in 1941.[23]

Noel certainly kept the reason for one of the most important trips of his and Herta's lives a secret. He told his league colleagues his 1938 excursion to Russia was just a summer vacation, but there was much more to the visit. Before they left for Moscow, Noel and Herta met with Hede and Paul Massing in Geneva and discussed important Communist Party matters. Hede put Noel into contact with new "controllers" who were to be his contacts after his return to Switzerland. One of those controllers used the name "Krivitsky." In 1941, after defecting to the United States to escape Stalin's purges, he was found dead in his locked hotel room.[24]

When Noel and Herta arrived in Moscow in 1938 and took rooms in a luxury hotel, he, and perhaps Herta, received a long-dreamed-of honor. It was secret but formal recognition as Communists—and as secret agents. Noel treasured the acknowledgment. Although he did not take a regular Soviet Party membership, remaining as a secret American/Swiss comrade, he began to repay the award many times over, even after the Soviet government executed one of his controllers. Noel knew of that mysterious death because European and American newspapers reported it, but he "kept the faith." He continued his clandestine work, never revealing his ties to the assassinated controller or Krivitsky, and never protesting their fate.[25]

During the later 1930s, Noel intensified his efforts for the Red cause. In contrast, the Massings were angered by the lack of recognition from the Soviets. They were beginning to question Soviet policies in general and, worse, they feared the Stalinist purges of even the party faithful. Although Jean Field had told the Massings they were overreacting to the purges, they became afraid for their lives. They had good reasons for that. The Soviets believed they were deviationists and had summoned them to Moscow in 1938 for "examination." Even Noel then seemed to view the Massings as traitors. The Massings were on the verge of being "corrected with prejudice." Hede felt lucky to leave Russia alive. It seems that she was saved only because of the American citizenship she had gained more than a decade before.[26]

After returning to America, Hede became more than a suspected policy deviationist. She made a formal break with the party, although not with Marxist ideals. Nina was appalled. She and the Massings became estranged. Hede's dissatisfactions with Soviet behavior deepened soon after World War II. She became a witness against the Soviet's American spy operations, even telling about Noel's involvements.

Noel, however, remained a Communist Party man. But now he kept his ideological commitments to himself. After his return from Moscow in 1938, he no longer paraded his leftist views. He even began to lie to protect his reputation to allow for the possibility of returning to the State Department, and, importantly, to increase his chances for a new and important league assignment. In 1939, he went so far as to lie under oath.

His friends at the State Department, who were advocating for his return, had notified him of a crisis. They had immediately written to him after he was accused of being a Communist agent during testimony before a U.S. congressional committee. Noel swiftly composed a reply and sent a sworn affidavit to America. It denied any connection with communist activities of any sort! His friends submitted it to the department. That helped protect his chances of returning to "State" if and when he wanted to. Then, world events pushed aside his thoughts of a high-status insider job in the United States. The league suddenly presented Noel with an exciting opportunity.[27]

Although it was obvious that Noel was frustrated by the league's inability to prevent conflicts, he was a man-on-the-move in the organization. He was even selected to represent it at the New York World's Fair. The league then recommended him for a very important position. It involved delicate negotiations with Germany, many other countries, and with the explosive issues of Palestine and Zionism. The job was a chance to be a true diplomat, something he had always desired.

In February 1939, Noel accepted a well-paying and very high-status position with a new organization created through the efforts of U.S. president Franklin Roosevelt. Roosevelt was an internationalist who seemed determined to involve the United States in Europe's affairs despite the strong opposition in the country to foreign entanglements.[28]

As a cooperative effort of over thirty nations, the new agency's first task was to help European political refugees, especially Jews from Austria and Germany. Noel quickly informed the State Department of his new posting, and then he began to coordinate his work with the efforts of Stephen Duggan, who had been advocating for the creation of such governmental agencies to help refugee intellectuals.

But, just as Noel was beginning his Intergovernmental Refugee Committee's work, he was given another assignment. The league did not realize it, but the task was a good fit with Noel's Communist sympathies. The League of Nations had taken on the delicate and dangerous task of disarming, repatriating, and resettling foreigners who fought on the leftist side in the bloody three-year Spanish Civil War. The league asked Noel to be the operational head of the project. He quickly accepted and arranged to have Herta assist him.

Noel and the Spanish Cause, Conflicts within the Ranks

The couple hurried to the French border to find thousands of members of the "volunteer" legions stranded and in need of the most basic supplies. The Fields, of course, had known that many of the refugees had been part of the Soviet-directed Comintern contingents and the American versions, commonly called the Abraham Lincoln Brigade. Complicating the war effort, many men and women from Eastern Europe and Germany among the recruits had run afoul of the Soviet agents assigned to the international units. Stalin's representatives wanted to eliminate the ideologically unfit within the ranks of the Spanish fighters. Stalin had instructed his agents to weed out (even by execution) any who seemed to deviate from his policies, especially Trotskyites who placed international revolution above Russia's welfare.

Not all of those who supposedly advocated Trotskyite internationalism, or were supporting their home nation's desires, had been eliminated. As the civil war was ending, many remained on Soviet watchlists while heading for internment in France, and an uncertain future. During the harrowing escape from Spain, a few realized they were poised between the dangers of being caught by fascists or being "disciplined" by the Soviets. They, even hardline Communists, would seek help from those like Noel who might provide aid.[29]

After witnessing the human suffering at the Spanish border, Noel and Herta made a dangerous trip to Barcelona. It was a great adventure. While negotiating for supplies, they endured shelling and machine gun fire. They then rushed back to the border camps where their focus was on the starving and ill refugees. Women and children, as well as the brigades' members, were reduced to eating bark and grass. Noel and Herta supplied food and medical care, and they arranged for travel permits and transportation to the refugee camps inside France. They filled other roles. They obtained special travel and residency permits for those ex-fighters they thought were ideologically correct enough to deserve being at the top of the lists for settlement in countries besides France. Mexico, with its new radical government, was a favored destination. Noel favored Switzerland for those who he thought had potential as post-war leaders.[30]

Herta took increasingly important roles in the work and made some of the critical administrative decisions. She also was responsible for a change in the family. It was her pleading that led to the informal adoption of an adolescent girl who was serving as a "Left" nurse. Her German Communist physician father worried that he could never escape the Stalinists and he wanted her saved. That girl, Erica, became, in some ways, the child Noel and Herta never had.

Knowledge of Erica's and her father's plight did not weaken Noel's faith in the Communist ideals, however. He and Herta even discounted the stories that began emerging about Stalin's agents' treatment of fellow Communists inside the refugee and internment camps in France. The stories about mass executions of civilians, Socialists and the clergy by the Communists during the civil war were also ignored.[31]

Noel and Herta were hearing increasing numbers of such reports as they labored at the border posts longer than the league had expected. Arguments over where the refuges could settle had delayed the repatriation program and their own return to Switzerland. Some of the brigades' Communists could not be sent home because their native countries would punish them for cooperating with the Soviets. Others stayed on because the Germans occupied their homelands—and leftists did not live well or long in German or fascist-controlled territories. They were stranded in France although that government wanted to be rid of refugees. The Soviet Union was taking in only the ideologically pure. Many free nations, such as the United States and England, had immigration quotas that were already filled. Even the Americans who had fought in Spain had difficulties arranging a return home, partially because they had disobeyed America's neutrality laws. Some had used false passports obtained from Jacob Golos's and J. Peters's organizations.

With their energies focused on the fighters from Eastern and Central Europe, Noel and Herta tried to overcome all the obstacles stalling their repatriation and resettlement efforts. The Fields came to know and help hundreds of European Communists. Some of those ex-soldiers spent World War II in detention camps forging networks as they awaited a return to their homes; some fought the fascists by serving in underground units in occupied Europe; and some managed to find refuge in neutral countries such as Mexico or Switzerland, where they became connected to local Communist parties.

All hoped that when the war ended they could play leadership roles in turning their home nations into Communist "paradises." They also hoped they could contribute to rectifying the mistakes President Wilson and his naive social-scientific Inquiry experts had made when they had redone Europe's post–World War I political geography. There were many borders to redraw and many ethnic populations to reshuffle. Noel's new friends wanted to set the policies.[32]

Noel's Perpetual Job Search: From "Modern Saint" to Security Risk

By early 1939, Noel was again nervous about his future. Even before Germany invaded France, Noel's league refugee projects were ending and the league's continuation was itself in question. Noel worried that he would be out of work and, because of resurfacing rumors of his leftist activities, that his reputation in America was tarnished. He also feared the league had learned of his favoring the members of the Far Left in the Spanish camps, so even if it continued on he thought it might dismiss him. He also became angry at the league when it seemed poised to expel the Soviet Union because of its conflicts with Finland.[33]

He began seeking a new position. He made a quick trip back to the United States, perhaps to see if he could salvage his diplomatic career. He listed himself as a State Department employee on the *Queen Mary*'s passenger list—something that would have aroused suspicions in Russia. Although an international crisis was at hand, and while the State Department was in need of people, Noel did not receive positive responses to his inquiries. He explored the possibility of other types of work. As before, he gave no thought to joining the ranks of the working class. He let his contacts know he wanted something in the diplomatic or aid fields. It did not take long for his network to provide some leads. But Noel was not going to take just any position. He had a list of requirements that included some ideological items.[34]

He ignored tentative offers from international relief agencies that he considered to be politically immature. Two organizations that had offered Noel a position in the late 1930s, the Red Cross and the Young Women's Christian Association, had a policy of strict neutrality. That meant they were ready to provide aid to any refugee or population in need. When they seemed interested in him again in 1939, Noel turned them down. He did not want to help any country or individual on the Right. In contrast, he enthusiastically responded to a possibility his friend Alger Hiss had developed. It was not with or likely to be in direct support of the Left, but it was a position that might prove a source of valuable information for Noel's comrades.[35]

In October 1940, the American High Commissioner for the Philippines sought an executive assistant and asked Hiss to recommend a qualified man. Hiss's brother, Donald, who had been acting as an advisor to the Philippine government, began a search. (Donald was also a source for Soviet intelligence with the codename "Junior.") Noel was his first choice. With the help of a department leading light and friend of Noel, Laurence Duggan (Stephen's son and friend of Hede Massing), Hiss overrode some objections based on Noel's supposed left-leaning tendencies. Noel was placed at the top of the recommended list. He would not get the job, however.[36]

At first, Adolf Berle, the department's specially appointed security officer, approved the recommendation. Although he was informed that Noel and some of his closest friends in Washington, including Duggan and the Hiss brothers, had Communist leanings, Berle believed such charges were, at worst, exaggerations. He came to that conclusion despite receiving negative memorandums from J. Edgar Hoover of the FBI.

Berle was a Harvard and Columbia graduate, an internationalist liberal, and had some experience in intelligence matters. He had worked with army intelligence on the Russian problem during the Paris postwar negotiations. He became a friend of Christian Herter while doing the work. Perhaps because of Herter's evaluation of him, Berle decided that Noel was politically safe. He was on the verge of giving Noel a glowing recommendation, especially because of the other highly complementary letters of support Hiss forwarded to him.[37]

All those recommendations stated that the Communist charges against Noel were false. They went on to salute his capabilities. One of Noel's former supervisors called him "the most brilliant man" ever in his department. The high-ranking diplomat, Norman Davis, described him as a "modern saint." The department's soon-to-be highly influential Lewellyn Thompson called the accusations that Noel was a Communist absurd. Alger Hiss then wrote a summary recommendation. It stated that Noel's professional reputation was excellent. Then, he tried to increase Noel's chances for the job. Hiss, to avoid any taint of bias, claimed he had never met Noel in person. Noel also bent the truth by submitting another notarized letter. He once more asserted that he had never been a Communist and had never been engaged in any radical activity![38]

Berle was ready to give permission to proceed but, unfortunately for Noel, several of the letters of recommendation described his managerial and administrative skills as less than satisfactory. There were also some pointed mentions about Herta's German background. Would a wife who perhaps retained her German citizenship be appropriate for such a position after the European war had begun and Germany had taken the threatening Japan as an ally?

That tipped the scales, in an ironic way. The vulnerable Philippines was not a place for a diplomatic family that carried any security baggage, Berle decided. For Berle the Right, not the Left, connections of the Fields were the dangers.

At the same time, Berle changed his mind about Laurence Duggan. He warned him that his political leanings were jeopardizing his career. Some of his colleagues, such as Spruille Braden, were becoming convinced that Duggan was surpassing the ideological by doing things such as removing critical documents about Communist agents in South America from the State Department's files. Berle's encounter with the problems concerning Noel and his friends contributed to a later change in his attitude. After World War II, he became a leader in the movement to eliminate Communist influence within the U.S. government.[39]

Finally, for Noel: The East Coast Liberal Unitarians and the Stephen Fritchman Connection

With Berle's rejection, Noel became extremely anxious. However, he received an offer from Boston that revived his hopes. Noel did not realize it immediately, but it was for a near-perfect job, one that that fit with his Callings and his background. Noel hesitated for a time, wondering whether a position as an on-site administrator of the European relief work of the Boston- and New York–based Unitarian Service Committee could provide him an opportunity to further both his peace and Red causes.

There was also hesitation in Boston. Noel had failed to inform the church-based Unitarian group that he had adopted atheism, but some in the church knew of his leftist leanings. While atheism might have damaged his chances for the job, Noel's ideological preferences were certainly acceptable to the organization's current leaders in Boston. After weighing the possibility of any political repercussions, they decided to squelch opposition to Noel. His ideological bent was, in fact, a rather good fit with the East Coast liberal Unitarian community's beliefs.[40]

Unitarianism, especially on the East Coast, had become increasingly liberal during the nineteenth century as it was spiced with the Transcendentalist's romanticism, perfectionism, utopianism, and even William James's pragmatism. From there, it was easy to accommodate to science and socialistic ideas of the twentieth century, as well as to drift further from Biblical foundations. Much of that, of course, was compatible with the Fields' liberal Quakerism. In turn, Noel's shift to atheism welcomed the world-centered view of "modern" Unitarian teachings.

When he came to believe the organization would tolerate him behaving as a not-quite-pure neutral, the job seemed near ideal. The offer was also attractive because Unitarians were very influential in American politics and culture. The powerful Harvard University was their institution and their organization was an embodiment of American liberalism. In addition, the various Unitarian units cooperated with other liberals' efforts. Quaker-Unitarian groups had become central to all liberal causes. Unitarians were also leading supporters of internationalism, and they had long been involved in the kind of European aid that was acceptable to Noel.[41] The denomination's organizations had cooperated in providing assistance to endangered Czech as well as Spanish political refugees.

In the late 1930s, the Unitarians began to rescue left-wing intellectuals. When their representatives traveled to Europe carrying thousands of dollars in donations to help already established Quaker relief efforts, they also brought a larger fund gathered by

Herbert Field's old nemesis, Nicholas Murray Butler of Columbia University. After the first rather informal Unitarian efforts, prodding from a left-leaning faction in the church led the Boston-based hierarchy to form a larger, permanent, and, significantly, independent relief unit in 1940, the Unitarian Service Committee (USC).[42]

The very radical Unitarian clergyman Stephen Fritchman was one of those who would convince the USC's leader, Robert Dexter, to trust Noel with extremely sensitive work for the committee. Fritchman became the editor of the Unitarian's magazine *The Christian Register*, a member of several radical-left organizations, a friend of the Franz Boas family, and a harsh and vocal critic of the U.S. government's social and foreign policies. The friends of the Field family, the Clarks, had formed close bonds with Fritchman as they participated in ultraliberal causes in New England during the 1930s. Also, for years, Fritchman had "summered" at his coastal retreat in Maine where he may well have encountered the Fields and Clarks as they vacationed.[43]

After World War II, Fritchman endured a congressional investigation stimulated by an ideologically based political struggle in the Unitarian Church. Then, taking over an already ultraliberal pastorate, he became a well-known Korean War peacenik in Los Angeles. While in California, he was a friend of the Hollywood Ten and maintained close contact with Hermann Field's ex-wife Jean Clark, her new husband, and her sister, Margot. Those ties had been so close since the 1930s that friends believed Jean Clark had named her son Stephen after Fritchman.[44]

That was in the future. In 1940, the very liberal Robert Dexter became the head of the Unitarian committee checking Noel's background. Dexter, a confirmed anti-fascist, agreed with Fritchman and thought Noel had perfect credentials. Noel was a product of the Unitarian's American hometown, Boston; he came from a good middle-class family; he had experience in refugee problems in Spain and France; he had been a diplomat; he had ties to the Quakers who had previously supported the committee's relief efforts; and, of great importance, he was a Harvard man. Some contacts at the State Department gave Noel strong recommendations. Charles Joy, who had been in Europe establishing centers for the USC in Portugal and France had met Noel and found that he had a solid reputation among leaders of established relief agencies. In addition to all that, Noel's mother Nina was one of the first donors to the USC.[45]

Just as Noel was hired, rumors began recirculating within the denomination that he had gone too far to the Left. The organization's liberal managers dismissed the complaints as unfounded rumors. As well, they considered Noel's recently acquired disability, blindness in one eye, as irrelevant to his job. Whether or not he told the committee that he still had some links to the State Department is unknown. It is likely that he did not, because many in the organization feared having a government man representing the USC as it fulfilled its advertised role as a purely neutral non-governmental organization. However, a few, including Robert Dexter, may have known and approved the continued association.[46]

When Noel learned that Herta would also be hired, he accepted the offer from the Unitarians, despite his reservations about the USC's commitment to anti-fascism as he defined it. Its leaders had publically stated they were following their traditional beliefs and would aid refugees no matter what their political background. Noel, however, correctly sensed the organization's "neutrality" was eroding and rushed to Vichy France where hundreds of thousands of refugees, including Spanish Civil War fighters, were stranded.

"Neutral" Aid Relief: Edward Barsky, the Anti-Fascists, and Varian Fry

A sign of the Unitarian's intent to be less than neutral was a tie the USC soon made to a man and an organization already seen as on the Far Left by the U.S. government's intelligence agencies. Dr. Edward K. Barsky had been associated with several Spanish Civil War and refugee relief efforts. He would merge them into the Joint American Anti-Fascist Refugee Committee (JAFRC) in 1942. Stephen Fritchman and other leftists were on its local board of directors and some of the other Unitarian leaders had lent their names to Barsky-aligned organizations. The JAFRC had American Communist Party connections, ones to international communism and, certainly, links to the leftists in Spain. One of those in its inner circles was Hede Massing's ex-husband and Soviet espionage agent, Gerhardt Eisler. Barsky's groups had long been publicly criticized by liberals and Socialists as Communist front organizations that were aiding only the members of the International Brigades.[47] The predecessors of the "Joint" committee had been denied certifications, and their overtures to relief agencies such as the Red Cross and Quaker groups had been rejected. Even the "Joint" committee did not receive a full certification by the U.S. government until late 1944. Barsky did not have a secure legal/political position or bureaucracy in place to conduct foreign relief in many areas. So, he sought a "front" organization. He asked the Unitarians to administer the "Joint" organization's money to support relief, immigration, and medical aid for selected Spanish war veterans in France. The USC accepted Barsky's request in 1941 soon after Noel took on the work in Vichy France.[48]

Barsky had some earlier contact with the Unitarians, and with Noel Field and Stephen Fritchman, through his previous relief organizations and, somewhat, through the Emergency Rescue Committee (ERC). Intellectuals, including Eleanor Roosevelt, Stephen Duggan, and Hamilton Easter Field's old friend, John Dos Passos, supported the ERC. When they had learned that France was ordered by the occupying Germans to send refugees within its borders back to Germany, or to the nations it occupied, leading New Yorkers quickly stepped up with funds and political influence. The ERC's young Varian Fry rushed to the unoccupied France to provide aid to a list of endangered émigré intellectuals. Along with a rather disobedient American diplomat, Harry Bingham, and some help from the USC, Fry quickly built much-less-than-legal networks that aided intellectuals such as Pablo Casals and Marc Chagall and, then, some thousands more. Varian Fry's admirable but frequently illegal efforts soon led the French, the U.S. government, and other aid organizations in France to demand his return to America. He managed to delay that for year.[49]

After being hired by the USC in October 1940, Noel continued the USC's operations with Fry. Fry and Noel worked in Marseilles while a future USC director and Noel admirer, Charles Joy, ran the station in Lisbon. As did Noel, Fry and Joy helped the Left, but did their best to appear to be neutral aid workers so they could operate in all national territories. They had another practical reason for appearing neutral. The Unitarians were soon to become one of the major recipients of the U.S. government's relief and rescue funds.[50]

The Unitarians and the Office of Strategic Services

Critical to Noel's future was the public image of the Unitarian Service Committee's chief operating officer, Robert Dexter, as a "neutral." However, Dexter and other leaders of the denomination's liberal core had determined to focus their work on helping anti-fascists, especially those of political, intellectual, and artistic merit. Dexter was not going to enforce strict neutrality, despite what USC's publicity proclaimed. His actions would include helping the USC to circumvent the U.S. government's rules against allowing Communist refugees headed for permanent residency in Canada or Mexico to travel through the United States. Dexter made another non-neutral commitment: he would aid America's newly formed intelligence agency, the Office of Strategic Services (OSS). Dexter was not alone in working for the OSS (and British Intelligence). All the USC's top men (and women) in Europe became sources and operatives for the OSS.[51]

Noel was glad to work for Dexter, although he and Herta's income was a pittance compared to his previous salaries. They began with pay of one-tenth of what Noel earned at the League of Nations. They soon received an increase, which gave them a combined pay of just one-third of the League of Nations salary.[52] There had been other than ideological reasons for accepting the Unitarian's ill-paying offer, however. An important one was that Herta had again let Noel know she had no desire to live in America.

Noel later told his Communist colleagues that he saw the Unitarian Service Committee as a tool for building a system for what he called a secret Red Aid cadre. Using Unitarian money and the USC's credentials as a neutral organization, he hoped to extend his work in providing funds, border crossing documents, and medical aid to exiled Communists and Spanish War refugees with the correct political credentials, not just Jews fleeing Nazi persecution. His intentions soon turned into action. Among many others, Noel gave special help to the Communist refugee leaders Laszlo Rajk of Hungary and Paul Merker of Germany.[53]

Noel and Herta held posts in two of the most infamous centers of intrigue, Marseilles, France, and, at times, Lisbon, Portugal. Both were where the displaced gathered hoping to find refuge and transport. They were also cities where intelligence agents of all countries played spy upon spy.

The Unitarian Service Committee did not care to notice that Noel was granting significant aid to those with the truest ideological commitments and who had potential to be the correct sort of leaders when Europe was liberated.[54]

19 New Loves in a Family of Agents, and Science Information in a New World War

Noel was not the only Field bonding with those who became important in postwar Central and Eastern Europe. It had taken little to convince his brother Hermann to abandon his post as an architect in England and, in 1939, perform an exciting task for a British relief agency. The convincing came from a charming lady who soon became Hermann's second wife.

Kate Thornycroft entered Hermann's life shortly after he and Jean Clark decided on a trial separation. Then, the truly unexpected for Quakers happened: a formal separation in 1938. Hermann believed that Jean had liaisons with several men, including Hermann's Polish Communist friend "Lolek," and one with Paul Massing. How modern and liberated the Fields were is indicated by Jean's younger sister, Joy, also having what she described as an "affair" with the Pole.[1]

Jean had encountered another man on the Swiss ski slopes. He was a Swiss Russian Jew of humble status, but Jean, although a Seven Sister graduate, could not resist him. She began an extramarital affair. By early 1939, she indicated she wanted to stay in Switzerland for more reasons than finishing her doctoral studies. She was giving strong hints she would seek a divorce from Hermann.[2]

Jean's newest love, Sali Liebermann, was from a relatively poor immigrant Jewish family that fled Russia because of the pogroms and his father's Communist activities. Lieberman's education had gone no further than a vocational school for interior and theatrical decorators. That Swiss school was practically oriented and taught him how to decorate store windows in case he could not find work in theaters or the film industry. Lieberman was not physically impressive. He was short, he wore thick glasses, he was stooped, and he walked with a strange gait. However, he was exciting.

Like his father, Lieberman was an activist for the Red cause and, most likely, a Communist party member. He had been arrested several times. In contrast, Hermann seemed stuck in a rather boring job and lifestyle. Also, Hermann avoided heeding all of Noel's earlier advice about sex and a happy marriage. Sali Lieberman, however, was charismatic, hot-blooded, and dynamic. He was attentive, although he was busy working in films and in local theaters—even helping to make illegal underground propaganda films for the extreme Left.

Another member of the Clark family was attracted to such social opposites. Jean's sister Joy would later find her true love in a Portuguese Jewish American labor agitator and party "muscle man" of little education, Joe Figueiredo. She followed him from the textile strikes in New England to California in the early 1950s to help wage another left-wing crusade and to guide Harry Bridges's radical International Longshore and Warehouse Union (ILWU) into a new era.[3]

Stuck in Switzerland, Jean and Sali's Strange Adventures

In Switzerland, although Sali Lieberman and Jean Clark had a bohemian streak, they bent to convention, and perhaps the need for passports. They married in 1940 as soon as Jean's divorce was finalized. The couple's sex life continued to be satisfying, and they soon had two children. While Lieberman was closely connected to Switzerland's Communists, he had always shown interest in America and Americans, and had learned some English. That had led him to people like Jean. He also had a drive to know and socialize with many in the intellectual world of the European left. The electrifying intellectual-political life in Switzerland was one of the reasons Jean stayed with Lieberman in Zurich, although she had obtained her PhD in 1939.

However, when war seemed imminent she had prepared to leave and obtained the required papers. Then, despite her social position in America, she could not borrow enough money to pay her passage. Lieberman could not get the papers he needed to travel or, later, a release from the Swiss army. As a result, throughout the war Jean worked alongside her husband in Swiss leftist film productions, using her writing skills as a script girl. At times, she taught English. Although an American citizen, she also took a job as a reporter for the enemy. She worked for Japan's news organization in Switzerland.

That job, and her and Leiberman's continued contacts with the Communist and exile communities in Switzerland, made her a "perfect spy." There is conclusive evidence that she worked for the American OSS and that her work was appreciated by Dulles and by top-level OSS men, such as Paul Blum, James Murphy, and Tracy Barnes, who became leaders in the Cold War's CIA. There is, at yet, no indication that she worked directly with the Communist intelligence apparatus during the war, as did Noel.[4]

Sali Lieberman had a long history of party connections, but he also worked for the Allies before the war and perhaps during the 1940s. He was a well-paid contractor for American-financed refugee rescue committees. Working with the U.S. State Department, he aided the European families of those who had already been brought to America and Mexico by such groups as the Emergency Rescue Committee. That did not mean that he refrained from doing Soviet work before or during the war. For one thing, he acted as an informant for the Tito faction in Yugoslavia while the Allies still regarded it as too radical.[5]

Hermann and the Radical Side of the Thornycrofts: The Czech Connection

Despite her separation and remarriage, Jean and all the Fields kept in friendly contact. Meanwhile, in the late 1930s, Hermann had been on his own in England where he met Kate Thornycroft. They were perfectly matched. Both were highly educated liberals from respectable families with left-leaning political beliefs.

Kate's family was distinguished. Her paternal grandfather was the famous English sculptor Sir William "Hamo" Thornycroft. Her maternal grandfather was the renowned playwright Edward Rose. Her father, an engineer, did well financially and could support five children in a more-than-middle-class style. There were also family connections to greater wealth and power. Kate's great uncle was Sir John Isaac Thornycroft of the huge engineering, shipbuilding, and truck companies. His factories were vital to the British military during World War I.

Sir John and other family members were a solid part of the British establishment. He even funded a chair in economics at Cambridge University. However, there were signs that parts of the family were moving in a new direction. Kate's cousin was the left-leaning Siegfried Sassoon, an heir of the great imperialist family, a war hero, a deserter, a famed antiwar poet, and an open homosexual.

Kate's life reflected the leftist side of the family. After graduating from Cambridge University with a degree in economics and politics, taking classes from Picasso's friend, the renowned advocate of economic planning John Maynard Keynes, Kate received a special honor. She was selected to spend a year at Smith College, one of the elite American Seven Sisters. Perhaps she was one of the exchange students sponsored by Duggan's Institute of International Education (IIE). She focused her studies on the plight of the workers in Lowell, Massachusetts. Lowell was America's first planned industrial community, founded in the early nineteenth century by the Harvard-centered Boston merchant group that invested its and Harvard University's funds to create a textile manufacturing center. Using the new machines of the era (a technology "borrowed" from England) they replaced home-based manufactures with a well-disciplined factory labor force housed in a utopian company town. Unfortunately, that utopia had disintegrated by the late nineteenth century. In the 1920s and 1930s, the New England textile mills became the target of some of the nation's most widespread and violent strikes by immigrant workers.

Kate returned to England just as her mother established a home for children displaced by the Spanish Civil War, and just as her brother Christopher returned from fighting for the leftists. His stories helped lead his brother Bill to becoming a party member, and to his sister Priscilla marrying a devoted German Communist, Hans

Siebert.[6] Kate soon built a reputation as a militant party member, and the Thornycrofts were believed to be part of a Communist group that was gaining control over a British relief effort for refugees.

Following her mother's lead, Kate became involved with a British Quaker group organizing aid for the anticipated flow of refugees from Czechoslovakia if Europe caved in to German demands. Her relief work began just as Robert Dexter and the American Unitarians were seeking advice on how they could help with the consequences of the Munich accord. With a focus on thousands of Czech Unitarians, Dexter negotiated an alliance with the English Quakers. A new temporary relief organization began operations, with Kate Thornycroft on its staff. Soon, Kate, with feminine persuasiveness, convinced her new friend Hermann Field to become a volunteer and to work under her direction.[7]

Hermann made an aid tour through Switzerland, Poland, Hungary, and Czechoslovakia, telling the wary British government that he was just on holiday travel. Then, the British Czech Trust (the successor to the Quaker effort), which the British government believed had been taken over by the Communists, rushed him back to Poland and its camps for refugees. His first tasks were administrative ones, arranging aid and visas in coordination with Polish and British officials who had become involved with the relief work. Hermann, as did Noel, encountered many who would return to their home countries in the after World War II to participate in the recovery and reconstitution of what they hoped would be socialist, if not Communist, nations.[8]

His assignment suddenly became dangerous. In September 1939, the trust asked him to escort some five hundred destitute Czech anti-Fascists to neutral Rumania. There were few resources and the refugees had to beg for shoes and food. They were bombed; only a handful lived. Hermann, who had traveled ahead of the group in a car while surveying routes, survived.

Hermann's experiences made him rather famous among liberals, as well as a source of information for the British and U.S. governments. However, he also gained a reputation among the suspicious as a man who favored people from the extreme Left. British intelligence soon began investigating him and Kate.[9]

Hermann in America: On the "Left" Side of Architecture

Just before England declared war against Germany, Hermann returned to London, spoke on the BBC, and made two important decisions. He would marry Kate and return to America. Perhaps fearful of the outcome of the British investigations, they sailed to New York on the last refugee ship from Ireland. Hermann immediately went public, even telling the State Department he was the official American representative of the Czech Trust. He then offered his expertise as a relief-refugee administrator to the government. The offer was not accepted.

From Celebrity to Frugality: Bending the Story but Not Abandoning the Ideology

Hermann and Kate became minor celebrities. As soon as they landed in America, Eleanor Roosevelt invited them to her Hyde Park estate in New York. They had a poolside lunch and a two-on-one talk. Roosevelt had been important to the work of the International Rescue and the Emergency Rescue Committees, and she knew of the Czech Trust. In addition, in her younger days, she was associated with organizations Hermann's ex-mother-in-law, Susan Ainslie Clark, worked with, such as the Women's Trade Union League.[10]

After the meeting in summer 1940, Hermann turned away from European problems. He and Kate settled into his Shirley, Massachusetts, farm retreat at 110 Center Road and began to establish a new life. He published an emotion-charged book about his escape from Poland. Then, he attended an avant-garde cooperative summer architectural program run by the famed holistic modernist architect and artist, Antonin Raymond. The workshop was at the New Hope Colony in Bucks County, Pennsylvania, near Nina and the Massing's farm. Raymond quickly became an intellectual and career mentor to Hermann.

When the summer course ended, Hermann and Kate settled in New York City. They again decided to let others fight the military battles against fascism. He did not serve in the U.S. armed forces, nor did he or Kate act as volunteers. He somehow obtained a draft deferment based not on being a conscientious objector, but on his being in a "critical occupation," architecture. That allowed him to avoid mandatory noncombatant military service. His and Kate's cultural, ideological, and personal explorations continued. Meanwhile, Noel Field waited for a year after he escaped from France to Switzerland in late 1942 before registering for the draft, although the Selective Service Act had required registration more than three years before. The government had lowered the draft cutoff age by the time he returned the form.[11]

In America, Hermann and Kate were finding the new science of psychology better than any religion. They looked to psychoanalysis to deal with the after-effects of their refugee experiences and their own emotional difficulties.[12] Those problems did not deter them from beginning a family, however. Their first child was born in 1942, the next, two years later. Kate soon obtained a job as a reporter for a British press service, and, later, as a researcher and special feature writer for the influential *Time* magazine that was then torn between a pro- and anti-Soviet policy. One of its editors was Whittaker Chambers, the ex-Communist whose warnings about Reds in the State Department in the 1930s had endangered Noel and his friends.

Hermann had received, like Noel, a generous inheritance and knew that Nina was financially very well off (with an estate worth more than a million dollars), but he became a frugal bohemian. He, Kate, and the children lived in a small New York apartment, then a somewhat larger one in Brooklyn—just two miles from the old family townhome. At times, they sought financial help from Nina, and they spent summers at the secluded Shirley farm and at Nina's Pennsylvania retreat, rather than taking, as the Fields had usually done, regular and rather costly vacations. Hermann, who had met the Massings in Switzerland, became reacquainted with them through Nina.

Hermann had given himself a year or two to recover from the Czech adventure. He did not use the horrors of the Czechoslovakian experience as a pretext for a lifetime of self-pity. However, despite Raymond's help, he did not hold any permanent jobs and was frequently "at liberty" between assignments designing military bases.[13] Hermann had other interests. He wrote more about his 1939 escape from Poland, bending the story to emphasize his role in helping refugees reach Rumania and deemphasizing his connection with the refugee Communists and socialists. He said little about his luck in having the Polish government allow him to ride on the government train that was rushing Poland's gold reserve away from the invading German army. However, Hermann did let the press know he made the last long-distance phone call from Poland to inform Noel of the terrors of the Nazi invasion before the Polish telephone system was cut off from the outside world.[14]

Kate and Hermann did not abandon their left-wing beliefs or involvements during the war. His writings reflected that. While the Soviet's adopted far more regimented workspaces, and imposed rigid architectural and artistic templates on its people, Hermann reserved his criticisms to capitalist architecture and industry. He wrote that a dictatorship ran American architecture, while the Soviets had a democratic system. He also penned criticisms of over-rational architecture such as the Ford Motor Company's huge Willow Run assembly plant.[15]

Hermann remained an idealistic, perhaps unrealistic, leftist, and someone with a heavy touch of moral superiority because of it. He was more than a fellow traveler, but there are no direct indications that he had any day-to-day contact with clandestine activity in the United States. One indicator of his sympathies was his decision to become a member of the Federation of Architects, Engineers, Chemists, and Technicians (FAECT), a union that was part of the Congress of Industrial Organizations(CIO). The CIO was left-leaning but the FAECT was Communist controlled. The FAECT proved useful in obtaining technical information, including chemical, electronic, and atomic secrets for the Soviets during World War II. Of importance, Hermann was central to the founding of the Architects Committee for Soviet-American Friendship, something that came to the attention of the authorities because of its known Communist membership. He and Kate also frequented many left-wing social events in New York City. Yet, perhaps trying to further his career, Hermann joined the conservative Society of Military Engineers.[16]

FBI on the Watch: Hermann, Kate, and Nina Arouse Suspicion

Although Russia was an ally of the U.S. government, the Fields' Communist connections were more than worrisome. Kate and Hermann soon became suspects—as did Nina. They were not aware that the Federal Bureau of Investigation (FBI) had a mail watch on them during 1941. The FBI opted for the politically sensitive monitoring of the addresses on their mail for several reasons. Among them were the Fields' ongoing contacts with Spanish refugees and the concerns about Kate's family in England. It had been suspect for several years. British intelligence asked the FBI to keep a continuing surveillance on all of the Thornycroft's American relatives. The bureau conducted several investigations during the war in addition to the 1941 efforts.[17]

Hermann was of special interest. British intelligence believed he had been an agent since his 1934 visit to Russia; that he had aided only Communists while he worked in Poland in 1939; and that he might have been the courier who transported the Polish Communist Party's funds to safety after the German invasion. He was also suspect because the British knew he tried to gain U.S. visas for Communist reporters and authors.[18]

Hermann's mother's adventures gained FBI attention before her son's did. With help from nephew Henry Field, and an inheritance from her own family, Nina Field had done quite well financially. She had accumulated a small fortune. By 1940, she had over a million dollars in liquid assets, most of which were held in England. She had intelligently planned for her children's welfare. But she also kept much for herself and decided to use that money for adventures. In 1937–1938, she made another grand tour of Europe, including Russia. She also toured the United States by car—by herself. Nina made it safely back to New York, despite her health and intensifying emotional problems. Later, in 1941, she made an excursion the FBI interpreted as more than an old lady's thrilling foray. Her quick trip to Mexico, the spy haven (and refuge for exiled European Communists such as Paul Merker) would have aroused the concern of any counterintelligence agent.[19]

Although she had backed away from the Massings, Nina's behavior had continued to worry the FBI. Neighbors in Shirley commented on her eccentric behavior, her refusal to socialize with them; the many strangers who drove into the isolated farm suggested that she was doing more than providing shelter to men from the Abraham Lincoln Brigade. Nina's attorney became alarmed about her generosity. She was giving money to all the down-and-outs she met and was donating to all sorts of causes, he told an investigator.[20]

Where Is Noel Field? Idealism Trumps Safety: The Unitarian Service Committee and Red Aid

Meanwhile, by late 1942 Noel and Herta (shown in figure 19.1) had become wandering refugees. They had spent almost two years helping internees in unoccupied France with medical aid and resettlement papers, then, as it appeared the Germans would occupy Marseilles, the Unitarians ordered the couple to Portugal where other refugees awaited sanctuary.[21]

Figure 19.1
Noel and Herta Field, circa 1940s, from Library of Congress

Very soon, Noel and Herta decided they were not in a secure base of operations. Although there was still a chance of returning to America from Portugal and employment opportunities in Boston and New York, they opted to rush to Switzerland (on the last train from Vichy France leaving all their possessions behind)—but without telling the Unitarian Service Committee. Its Boston headquarters became worried when it lost contact with Noel. Its staff sent frantic telegrams to the State Department. Finally, the department sent word that Noel and Herta had arrived at the Swiss border, but with only their clothes and a few suitcases![22]

At the end of 1942, Noel and Herta settled in Geneva and then Berne. Their foster daughter, Erica, soon joined them and then moved in with Jean and Sali Liebermann for several months while awaiting her enrollment in college. Noel let the anxious USC operations administrator, Charles Joy, know they were safe and declared they remained committed to the cause of peace. The USC, however, had no (and did not plan to have) extensive programs in Switzerland, and it had little money. When Noel arrived, the USC begged the State Department and the OSS to support him.[23]

Noel and Herta were frightened. They had to begin another life. Noel was not sure how he could make a living. He had to pinch pennies. Noel quickly discovered the Concilium was permanently dead. The Swiss had no interest in what remained of Herbert Field and Johannes Strohl's work, and the disruptions of the war made any ideas about reviving it impractical. As well, after his mid-1930s attempt to help Strohl, Noel had lost most of his interest in his father's old dream, even in science information in general. Noel also found that diplomatic work was not as attractive as it once seemed. He wanted to continue his more direct Red Aid efforts. So, he began sending pleas to Boston to have the USC rethink its policies.

Noel and Herta finally persuaded the USC to support them in establishing a new refugee center, as well as a forward-looking post-war relief program. This Swiss center would carry on the older USC functions, such as medical aid, Noel promised. However, he insisted it had to have an important new capability. He wanted to train social workers who were to be dispatched to manage recovery efforts as the Allies liberated their nations. The USC somewhat apprehensively approved his plan. Noel then began selecting and preparing his "social workers." He failed to point out to Boston that recruits for the program would have to have correct political views.

Allen Dulles and the OSS, and a Dual Role in a Family of Agents

Noel did not tell his managers in America some other things. He was becoming an intelligence asset for Russia and America at the same time. For Noel, such a dual role seemed destined. His rejoining and working with the outlawed Swiss Communist Party (as did his foster-daughter Erica) channeled him into undercover work for left-wing causes. His ties to the USC's Robert Dexter, who had long been a "source" for Allen Dulles, made working with the American intelligence inescapable.[24]

Herbert Field's World War I contact, Allen Dulles, had been chosen to head the Swiss branch of the new American World War II intelligence service, the Office of Strategic Services (OSS). Dulles had been among the first asked to serve, and his original OSS identification card carried serial number "55." Switzerland again was the perfect setting for a spy operation. Although isolated, Dulles had an effective but small staff. Because of closed borders, Dulles had to rely on locals in 1943 and much of 1944. Later, his group included many American agents, including a future key player in the postwar CIA's clandestine work, operative "679," C. Tracy Barnes. The Barnes family would become linked to what became known as the Open Access scientific information movement sixty years later.[25]

Among the Dulles group's first tasks was the creation of a network of informants. Dulles knew that relief agencies could be invaluable sources. So, when preparing for his Swiss assignment, he wasted no time in contacting many of those agencies. He found that Robert Dexter, the head of the Unitarian's relief work, was among the most willing to bend the rules of impartiality. Dexter and his Unitarian colleagues working in Europe had already become operatives for the OSS. In late 1942, Dexter recommended that the OSS recruit Noel. Dulles was a bit hesitant, as he had not regarded Noel as especially adept when he knew him at the State Department. In addition, he had some concerns about Noel's politics. But Dexter persisted and, with a nudge from higher-ups in America, Dulles contacted Noel. Dulles's offer to him was not, however, for the kind of official position Herbert Field had accepted at the end of World War I. Noel's role was less formal, almost unofficial, although he was assigned an OSS registration number of "394."[26]

In early 1943, unaware of future consequences of Dulles's wartime operations, Noel began providing information to the OSS. He also became a channel for OSS funds and messages for leftist resistance groups in occupied nations. Noel had contacts with important exiled Communists from Germany, Poland, France, and many other nations, and he brought them together with OSS agents. Noel was also giving Dulles information and advice on how to restructure postwar Europe. For Noel, as well as for Sali Liebermann, the Communist Tito, not the moderate socialist or royal groups, was the choice for Yugoslavia. Regarding Greece, Noel advised it needed a new socialist government free of royal and imperialist British influence.[27] Noel and Erica also distributed secret OSS subsidies that were laundered through the U.S. government's War Refugee Board which supported the Unitarian efforts and, it seems, Raul Wallenberg.[28]

Noel's Swiss Communist Party controller, Jules Humbert Droz, who was receiving orders from Moscow, directed many of Noel's activities. When Noel and Erica went to the Swiss border to meet with resistance fighters, to shuttle Communist agents from France into Germany, to deliver OSS money, arms, and requests, and to bring refugees across, they were also acting as couriers for the Communists, providing communications between groups throughout occupied Europe, as well as within Germany. Noel's activities seem to have included some trips into Nazi Germany during the war.

In addition, Noel frequently went behind Dulles's back and aided Communist groups the OSS had put on its blacklist. Simultaneously, Noel was using his Unitarian post to help his displaced ideological kin as they waited in Switzerland to return to their homelands. Politically correct leftists were allocated more and better housing, more jobs, and more aid than others. Noel arranged early releases from Swiss internment camps for comrades from many countries, even Rumania. Noel was also putting them at the top of the lists of those who would be the first allowed into liberated territories.

Dulles seemed unaware of, or unconcerned about, the Fields' Red work. While Erica was an active member of the Swiss Communist underground, she was serving as a secretary to one of Dulles's subordinates, a man who was in charge of relations with German Communist labor groups. Such connections may not have bothered Dulles, but they raised the ire of others. The Swiss police had several "talks" with Erica, and they monitored Noel's activities. Noel, and representatives of the USC who seemed overly sympathetic to the Red and Jewish causes, had also came under suspicion by France's Vichy government.

Despite some resurfacing concerns voiced by factions within the State Department, and despite the FBI's newest worries over Noel's possible subversive activities, Dulles did not interfere with Noel or Erica. He should have been more attentive to Noel's Red actions and less willing to follow Noel's advice to bring Erica into the OSS's inner chambers. Erica's secretarial job with the OSS gave her access to much sensitive information that she and Noel passed on to their Communist contacts. Some believe that Noel was the informant that kept Stalin abreast of Dulles's secret negotiations to end the war in Italy—negotiations that did not include Russia.

The Unitarians placed little importance on indications that Noel was not acting as a good neutral. But they did notice that some of the senior men in the Unitarian Service Committee (USC) had fallen into trouble with the French government and the U.S. State Department. The USC's groups in France and Portugal also had been conducting rescue efforts for leftists and they had been using less than strictly legal methods.[29]

The concerns over the USC grew so intense that after an internal Unitarian battle in mid-1944, Robert Dexter had to resign from his USC position. Noel surprisingly escaped much of the 1944 fallout over the USC's actions and, despite all of his leftist work, he was gaining a solid reputation in America as a relief administrator, as well as an American intelligence asset.

In late 1944, Dulles was sent a coded telegram instructing him to smuggle Noel Field into France. Washington wanted Noel to use his contacts to coordinate the old resistance groups and arrange for their participation in the nation's reconstruction. The OSS guided Noel across the border, and then the Maquis (guerrilla bands of French Resistance fighters) put him on its secret underground railroad. After almost being killed during a Nazi bombardment, and after being injured in a car accident, Noel worked in southeastern France. He was the first relief worker into the area. He forged relations with several, usually Communist, groups. After a brief return to Switzerland, Noel was sent to Marseilles. He did much more than establish USC agencies there. He appointed Marxists to head new offices throughout France and other liberated areas. He ordered them to cater to the needs of the ideologically pure.[30]

Then, as the Allied armies were approaching the German border, Dulles dispatched Noel to Paris to have one of his ideas reviewed by a young OSS political officer, Arthur Schlesinger Jr. Noel outlined his plan to form men in refugee camps into cadres ready to take control of governments, perhaps through assassinations, in the soon-to-be-liberated countries in Eastern Europe. The young OSS officer became worried as he began to sense that he had heard of the plan before and that Noel was a not-too-bright ideologue. He tried to squash the idea. Schlesinger, a Harvard student of Hermann's era, concluded that Noel's plan was not new but was just an extension of the Communist activities already in operation in the camps. Schlesinger refused to vet the project and the OSS soon reversed its earlier approval.[31]

Schlesinger's and others cautions about Noel went unheeded during the war, partly because of the inexperience and disorganization in the American intelligence community. The OSS was immature, so much so that Dulles's coded messages from Switzerland were intercepted and read by the Swiss, by the Germans, and probably the Russians.

Librarians at War: The Left at the Library of Congress

In addition to naivety, the OSS had too few experienced people for too many responsibilities. On the top of its demanding task list was information gathering. More than any other American intelligence group, the OSS had responsibility for producing high-level information about foreign nations' economies, leaders, and industrial and scientific capabilities. While it borrowed knowledgeable personnel from the FBI and the military, there were not enough skilled agents to fill its information departments' ranks.

In reaction, the OSS turned to students and faculty at America's elite universities as it sought information specialists and intelligence analysts, as well as those who might do well as covert agents. The agency soon had in its ranks the generation of scholars that succeeded the progressive elite's Kelloggs and Vincents. They quickly made such institutions as Princeton University and the Library of Congress war-related research resources. The OSS used other libraries and bibliographic centers for recruitment because it sought information specialists and Special Librarians to work in its other centers.

The OSS badly needed the librarians and academics. From its inception, the agency had to meet a special version of the Documentalists' challenge. It was responsible for locating and integrating published and secret data on the strategic intentions and capabilities of the enemy. One important step the OSS took was to establish an information hub at the Library of Congress (LC). As in the military research centers at the library (where Adolph Voge worked), the OSS staff at the LC was assigned to gather and analyze open-source public information, as well as that sent from agents or gathered through censorship posts.

The OSS center was located at the LC for good reasons. The library was America's largest repository of international documents. It did not hold all foreign publications, however. Other major American research libraries acted as adjuncts, and the OSS began a major program to acquire additional foreign publications on its own.[32]

The agency also needed to create a unique breed of Special Librarians. As well as bringing practicing librarians and archivists to Washington, it trained its recruits in analytical and data handling methods, and it made translators sensitive to intelligence needs. In addition, many of the European scholars and political figures brought to America by organizations such as the International Rescue Committee worked on OSS projects at the library.

There was much more than simple data dredging at the center. Archibald MacLeish—the poet-politician who was Voge's protagonist, a friend of May Sarton, and the new leader of the Library of Congress—called on the liberal academic establishment in order to staff a high-level think-tank for the OSS. East Coast elite universities supplied the key men. Several of the Marxist scholars rescued by groups such as the Emergency Rescue Committee were recruited. MacLeish's group became known as the Research and Analysis Branch and gained a reputation as being more than liberal in its political orientation. The Communist informant, Philip Keeney, for example, worked as the librarian for the analysis center. But not all at the wartime library were high-level intellectuals "of the Left." The OSS and later CIA dirty-tricks point man of the Cold War, the very politically conservative Edward Lansdale, also worked at center during the war.[33]

Bits of Information, New Methodologies, and New Library Leaders

The OSS also had a data center. Its specialists had a vital and challenging assignment. Central to the intelligence task was, and is, "open source" information. The painstaking verification, correlation, and analysis of thousands of isolated bits of data contained in published as well as captured enemy documents are vital to intelligence work. "Open sources" included radio broadcasts, newspapers, magazines, and journals. They provided an avalanche of data.

Significant data can come from unexpected places. Before the era of spy satellites, old picture postcards of a bridge or a government building could provide important target information. Railroad timetables were invaluable. Travel guidebooks, national histories, old business reports, and even sales catalogs also were helpful. Such seemingly trivial information was so important that the OSS called the travel-guide expert Eugene Fodor into the LC's service. The OSS also had expert scientists in the center to analyze open source information because even closed societies with stringent classification rules leak important information through scientific papers and meetings—even about super-secret issues such as atomic energy.[34]

Because of their "bits of information" challenge, and their large budgets, the OSS and its successors became leaders in information science methods and technology. The OSS soon built files larger than Paul Otlet's or the Concilium's. Retrieval from them called for innovations that went far beyond Herbert Field's cards or the UDC's classifications. For example, a 1940s OSS technician was one of the first to turn the Hollerith-IBM electric tabulator into a word-based information machine. Combining microfilmed documents attached to the Hollerith cards (aperture cards) and with multiple keyword codes punched in each, the tabulator recovered information from unstructured files. Such card systems were not the only advances. Soon after the war, the OSS/Central Intelligence Agency spin-off think-tank at Yale University invested in the refinement of a microfilm automatic retrieval machine, Vannevar Bush's Rapid Selector.[35]

The secret intelligence organizations generated more than just technological innovations. Future leaders of a redoing of American information science, such as Frederick Kilgour, Joseph Becker, Vernon Tate, and Jesse Shera apprenticed at intelligence information centers during and after the war. As with the employees at other agencies dealing with wartime information problems (such as the U.S. Navy project at the LC) the intelligence groups' men were discovering that traditional classification approaches, even Herbert's, were ill suited to the retrieval of the types of information they encountered.

The war's impact on information methods was not confined to agencies such as the OSS, however. Private contractors involved with massive scientific and technological efforts, such as the Manhattan Project, found a need for vastly expanded special libraries. Corporations like DuPont, where Adolph Voge had his short-lived dream job, faced the need for new methods and techniques as wartime projects drove it, and parts of academia, into truly Big Science.[36]

Sidestepping Science Information Barriers in the OSS

The OSS and wartime industry needed more than technology, methods, and the help of America's libraries. The new information specialists had to acquire data that enemies tried to keep to themselves. While OSS Special Librarian–Documentalists searched through the records in America's great libraries; while OSS-aligned radio centers tapped open broadcasts; while codebreakers attempted to read encrypted messages; and, while analysts labored to piece all that information together, operatives in the OSS, including Allen Dulles, were told to fill another important information gap left by the war. A special division of the OSS Research and Analysis branch was ordered to acquire foreign library materials through regular channels. At the same time, the covert action side of the agency developed devious methods to sneak new items out of Axis and occupied countries.

Looking to Libraries for Help

The challenge for the inexperienced Americans was how to obtain both the old and new documents their enemies kept locked away. Scientific information was one of the most protected and coveted types of information as warfare became more high-tech. There were embargoes on all types of science information, including academic journals. Unfortunately, the OSS had few experienced people to overcome the barriers. It had to turn to others.

It first looked to the library community for help. The OSS knew the American Library Association (ALA) had a program during World War I that used booksellers and intermediaries in neutral countries to secure a flow of books and journals for major American libraries. The program was reborn in 1939 as World War II began. Its base of operation was at Nicholas Murray Butler's Columbia University. Unfortunately, the ALA's work was unequal to World War II's challenges, partly because Herbert Field's ally, Clement Walker Andrews, who led the earlier effort, had died in 1930. That left the Americans with few means to acquire embargoed materials. Yet, there were great questions that might be resolved through gaining access to technological and scientific publications. How close were the Axis powers to atomic weaponry? Had they built a new gas and biological warfare capacity? Could they make enough synthetic fuels?[37]

Linking the U.S. IDC and England's ASLIB

More government involvement was required. President Roosevelt first turned to a predecessor of the OSS, the Office of the Coordinator of Information. Although that agency was weak and already overburdened, it got to work. It helped to establish the U.S. government's Interdepartmental Committee for the Acquisition of Foreign Publications (IDC). The committee sent teams abroad to acquire books and journals. Science information immediately became a priority for the IDC—and for the agents of what became the OSS.

The OSS soon created secret acquisition and microfilming centers in many countries, even China, and, importantly, in England. England already had an established acquisitions operation tied to its own Special Librarian–Documentalists group, the Association of Special Libraries and Information Bureaux (ASLIB). ASLIB had been formed after World War I's applied science needs demanded attention. However, the Americans wanted their own operation in England. A young future library innovator, Frederick Kilgour of Harvard University's library (later of the Online Computer Library Center (OCLC) international bibliographic service), became the operating head of the American IDC-OSS program. He personally led a team of ninety at the Library of Congress while fifty worked in foreign nations, including countries occupied by the enemy. His agents obtained and copied books and newspapers, and sent some 270,000 documents back to his abstracting and indexing centers each year.[38]

The task of microfilming and bringing order to the materials that were supplied by, at first, the British, and then the American services, was daunting. The filming responsibility fell on the London staff of an American, Eugene Power, the founder of America's University Microfilms, Inc. He had been working on a project to copy European library materials for the huge Rockefeller-sponsored program for rescuing the classics in the humanities. Taken over by the U.S. government, Power's team shipped millions of pages of filmed intelligence-related documents to America and did so even after the war. His center quickly repaid the great cost of its work. Documents provided by the British spy agencies to the OSS, for example, proved critical to the U.S. atomic bomb program.

Smuggling Secrets: Journal Subscribers with False Identities, and Help from *Chemical Abstracts*

England's stock of materials was not enough to satisfy the intelligence analysts. America's "spies" needed their own supply. One of the OSS's toughest assignments was to devise clever ways to gain access to such restricted items as German scientists' articles. The Axis powers forbid all shipments of publications, even to neutral countries, without prior government approval. There were few established channels for obtaining the publications. For a time, the League of Nation's library was a major source. When America's entry into the war ended its neutral status, access to many other repositories was stopped.

Because the science information embargo was not absolute for neutral countries, one of the duties of OSS operatives was to scour their bookstores for publications and to purchase subscriptions to journals and book series under false identities. The agents were also instructed to pay refugees for items they brought with them. As a result, Portugal, Spain, and Sweden were major centers for the OSS / IDC's operations. In some cases, the IDC operated in the open because it appeared to be just an extension of America's library community. In others, the work had to be disguised. In Sweden, the OSS took great chances when its agents gained covert access to mailbags and copied the scientific documents they contained.

Great care had to be exercised. Most neutral countries feared to be suspected of helping any intelligence operation. Switzerland was especially sensitive about its neutral status because of its location. The need to control espionage within its borders was one of the reasons why the Swiss police frequently took Noel's foster daughter, Erica, to their headquarters for questioning.[39]

Although the Swiss government did not allow the IDC to operate within its borders, the OSS could not afford to ignore the Swiss information potential. Switzerland hosted German businessmen and exiles, and its libraries and bookstores continued to receive publications from Germany, as well as from Axis-occupied nations. The OSS found ways to use all those sources. In addition, the OSS frequently ordered undercover operatives in the occupied areas close to Switzerland to bring scientific as well as politically relevant documents when they rendezvoused with their control agents. Noel Field and Erica were likely to have carried much back from the Swiss border to Dulles's men.

The OSS had other helpers in its quest for science information. *Chemical Abstracts* had established a branch in Switzerland just after America entered the war. It needed to sidestep the transportation problems and the embargoes on chemical publications. The center employed some seventy-five translator-abstractors and shipped their work back to America by sometimes rather devious routes. It did not take long for the OSS to enlist the center's help. The Swiss abstractors' microfilms of OSS-obtained documents were covertly sent on with the *Abstracts*'s results.[40]

The National Defense Research Committee: A New Organization for Big Science

American academia's activities were related to more than information gathering during World War II. They included creating a new organization that embodied, and exceeded, the hopes of the Big Science and cooperative applied science advocates of World War I's National Research Council. The new organization had the potential to profoundly alter America's universities and the nation's science information systems.

During the 1930s, one of the NRC's younger alumni, Vannevar Bush of MIT, had ascended in science's hierarchy. He became the head of the prestigious Carnegie Institution of Washington that Herbert Field had visited twenty years before. (Perhaps ironic, it was Bush who cut off Carnegie grants to Charles Davenport's eugenics research project.)

In 1940, Bush bypassed the old science organizations to create the National Defense Research Committee (the NDRC, later, the Office of Scientific Research and Development, the OSRD). The White House granted it enormous government funding and the power to link academics and their institutions to war time technical needs—while promising to allow significant degrees of academic independence. With previously unimagined amounts of money, groups of scientists in all fields, but especially physics and electrical engineering, applied their knowledge to vital projects ranging from the atomic bomb and radar to computing and health issues.

As a result, academics, and science itself, soon gained levels of recognition that even Watson Davis had not envisioned. Bush and his colleagues hoped that the munificence and prestige would continue during the postwar era and that Big Science would not mean a loss of academic self-determination or the turning of universities from their core intellectual mission.[41]

While America Spent Millions, the Soviets Got Science Information for Free

There was something important that went unnoticed during the war. While America and England were spending billions of dollars on research and the acquisition of scientific information, Russia was using its networks of unpaid ideological converts to steal its allies' most precious and highly guarded technical and scientific secrets. Devoted comrades stole the secrets of the atomic bomb, jet airplanes, radar—even the codebreakers' Ultra Secret. But well before the late 1940s discovery of those techno-espionage triumphs, revelations about other Soviet espionage in the West began to impact the lives the Fields and their allies.

20 Postwar, Looking Forward: More Intrigue, More Big Science and Big Information Stories, and More Ideology

The war's conclusion in August 1945 signaled an opportunity to rebuild, perhaps restructure, the world—including its science information systems. However, the Allies' triumph had been costly. All the major nations except one lay in ruin. It seemed that only the United States had the resources to support luxuries such as the dream of Herbert Field for an international science information system for academics.

Mismatched Desires of the New Scientists

Hopes for some type of worldwide system arose, but not even theoretically oriented European scientists called for a rebirth of the great efforts made in Zurich and Brussels. Europe's destitution was not the central reason for the abandonment of the Concilium. The U.S. libraries could have funded the rebirth of the Zurich center's bi-weekly offering. The cost of coding, sorting, and maintaining the hundreds of thousands of Concilium cards was bearable. By the 1940s, major American research libraries were willing to maintain up to six copies of the cards for each of their regular books and to carry the burden of the sorting and filing by long alphabetical subject terms as well as by author, title, and call number.[1]

The Concilium's fate was the result of its mismatch with the information desires of the new scientists. The card technology seemed unnecessary and, more fundamentally, its numeric UDC complex hierarchical indexing-classification system did not fit the needs of the newer generations of "pep and push" researchers who would find less intellectually complex systems better adapted to the practice of "useful" science. As a result, the Concilium was without any American advocates in the postwar years, even at the Rockefeller Foundation or Columbia University.

In contrast, the war ended as a triumph for the type of large, well-funded, and cooperative science that the National Research Council's Robert Yerkes had advocated. The atomic bomb, radar, computer, and other projects of World War II made scientists, even academic ones, hope they could remain relevant in the peacetime economy. They perhaps naively anticipated that permanent Big (and expensive) Science with private and government funding would allow independence in selecting and conducting projects. The war also made scientific information seem more valuable than ever before. Leaders such as Vannevar Bush believed the time had come to invest in radically new types of science information systems and technologies, ones tailored to the needs of applied work. Another important, but then implicit, corollary of the Big Science vision was making sure the Flexner/Carnegie high-powered research university remained the favored type in American higher education.

Another Chance for the Grand Dreams of the UDC and Abstracts?

At the same time, however, Europe's inheritors of Otlet's grand dream of a universal hierarchical classification system for all types of knowledge believed they had an opportunity to remake and expand the Universal Decimal Classification system. They revitalized the Federation for International Documentation (FID) and planned to launch a new era of truly international intellectual and bibliographic cooperation through the United Nations' body for social and educational cooperation, UNESCO. The FID's receipt of one of the few early postwar era grants for science information made by a U.S. private foundation bolstered its optimism.

However, that Rockefeller grant to the FID was not intended for the exploration of hierarchical classification systems but for the rejuvenation, through UNESCO, of the earlier international abstracts program that Vernon Kellogg had supported. Even the hierarchical system for general libraries, the Dewey Decimal classification, faced difficulties in the postwar era. Although it had a large customer base and the help of the Library of Congress, it would be decades before it was made financially and institutionally secure. Meanwhile, postwar budget reductions slowed the development of classification systems at the Library of Congress.

Far-Reaching Liberal Dreams for Peace, Harmony, Education, and Health

America's scientists and European bibliographers were not alone in having broad visions that were frustrated. World War II ended with what idealistic American liberal internationalists, such as Stephen Duggan and Alger Hiss, believed were grand leaps for peace and world harmony—the reduced influence of American isolationists and the creation of the United Nations. Unlike the League of Nations, the new organization was to have the full support of the rich and newly world-dominant U.S. government.

There were other far-reaching dreams. The 1940s "progressives" thought they could extend the New Deal's economic and social programs and avoid the horrendous strikes and class conflicts that had led to so many Americans turning to the Red cause during the 1930s. Former vice president Henry Wallace and his supporters, for example, laid plans for free college tuition and national health programs. In addition, progressive educators, such as Carmelita Hinton, had their own visions for America's colleges. They looked not to Flexner-Carnegie universities or to community colleges for vocational preparation, but to institutions for moral training for a new era of equality as they labored to redefine humanities education to fit their leftist philosophy.

The Fields Keep the Faith: Red Aid, Socialist Cities, Leftist Sympathies, and Hollywood Strikes

After the war, although many American leftists turned away from communism and ultraliberal causes, the Fields remained faithful. In 1945, Noel and Herta decided to stay in Europe to continue their relief work and to further their Red Aid objectives. They sought an idealistic Marxist future in Central and Eastern Europe. Meanwhile, Hermann Field worked for a holistic peoples' architecture in the United States, watched over Nina, looked forward to helping Europe's leftists rebuild with true socialist cities, and worked as a peace activist. He was also involved with an educational crusade. He became entangled in a battle for another alternative to the Flexner university model: nighttime adult education.

Erica, Noel's foster daughter, married an American army officer but continued with her sympathies for the left. She helped with Communist publications in France where her new husband was studying under the GI Bill. Elsie Field stayed in Illinois with her mathematician husband to be a mother and housewife but continued her efforts to further Margaret Sanger's great causes.

Others connected to the Fields had deeper leftist ambitions. They thought they had the mandate to make the entire world Red. The Hinton children moved to China and became committed to its revolution. Jean Clark (Hermann's ex-wife) and her new husband, Sali Liebermann, returned to America to find that the other members of her family had remained party stalwarts. The Boston Clarks were part of the American Communist network that supported the party's return to obeying Soviet orders. They aided aggressive and disruptive labor and racial-equality campaigns; they encouraged the postwar wave of strikes hoping they would be a first step to at least a socialist America; they and Jean Clark were in touch with the Hollywood leftist strikers of 1946; and they approved the purging of American Communist Party leaders who did not wish to go back to the hardcore practices of the 1930s. The deposed included the party's head, Earl Browder, who had played a central role in the espionage of the 1930s and 1940s, including Hede Massing's network.

An Era of Ambivalence and Frustration, then Cold War

Hopes, whether technical or liberal or radical, did not mean realization. Indecision, frustration, and inaction marked the immediate postwar years. The goal of continued Big Science in academia was difficult to attain. The private sector did not step forward to replace the massive government funding of academics during the war years, and

Washington seemed unwilling to provide support to nonmilitary activities. Vannevar Bush's dream of a great fund for academic scientists made little headway. Faith in the United Nations' type of internationalism lasted but a short time. There was a resurgence of isolationism in America.

Then, there was a turn to policies based on a balance of power to maintain the peace. George F. Kennan's ideas dominated, not those of Henry Wallace. The new international power politics had an important twist. For the first time, the United States played a significant part.

At the same time, the desire for the extension of New Deal domestic policies encountered a revival of conservative politics. Measures to weaken the power of radicals in the labor movement had a major impact. Higher education expanded, but it did not receive the influx of money required to realize the Flexner dream in any but a few already rich universities. Most of the flood of new American students attended low-cost institutions, not research-capable universities.

In Western Europe, the Universal Decimal Classification campaign faltered when it received little support from the United States. In America, traditional academic science information systems showed little progress. No major private foundation seemed willing to provide funding to resuscitate the traditional science journals, indexes, and abstracts injured by the Great Depression and the war—nor did the government.

However, applied science information was beginning to change and to receive a new type of attention, but not because of its purely scientific value. National defense needs and ideology were altering and politicizing it.

The traumatic story of a proposed postwar government information initiative was one of the first indicators that science information was becoming more valuable—and more deeply entwined in political struggles. Henry Wallace, who would soon lead the left-leaning, Communist-supported, and Soviet-friendly national Progressive Party, proposed a government system to coordinate and disseminate applied science literature to the "common man" rather than just to favored academic and corporate research centers. The issue soon politicized the issue of sci-tech information. The proposal confronted more than the traditional American hostility to socialism. It suffered from the beginnings of the separation of the Soviet and American "blocs" and a reinvigorated and emotion-charged anti-Red movement in America.

What contemporaries soon called the Cold War did more than defeat Wallace's information scheme. It almost tore-apart the usually harmonious Unitarian denomination. It nearly destroyed the lives of the Fields and their friends such as Alger Hiss and Robert Oppenheimer. In one case, that of Noel's close friend Laurence Duggan, a life was ended. The Cold War also redirected American science, faculty careers, and education—and it altered the nature of science information systems and the information professions.

Interacting with American beliefs, the Cold War's military expenditures created the foundations for a multibillion dollar, for-profit science information industry in the

West. In the Soviet sphere, it led to the establishment of the closest the world came to a repeat of Herbert Field's centralized information service. Russia's VINITI also became perhaps the UDC's most loyal user.

Big Science, Big Information Science, and a Tiny Soviet Satellite

Propelled by national security concerns and the need to develop a high-tech military, the American government spent billions of dollars on research, with much of it flowing to government contractors and a significant portion to the largest universities. The research demanded new information methods, ones better suited to the needs of time-conscious applied science. Elegant systems such as the UDC gave way to increasingly pragmatic and flexible approaches and to a reliance on expensive computers that few nations could afford. The clients, methods, and machines created a novel type of information specialist with few ties to older librarianship or Documentalism.

As a result of American polices supporting the private sector, some of the new "information scientists" formed companies that would dwarf traditional for-profit suppliers such as Halsey Wilson. Eugene Garfield, for example, sidestepped all the old bibliographic rules to create radically new systems and a huge company that better served applied science and the increasingly bureaucratic higher education markets. Mortimer Taube based his huge business on his simple "Uniterm" method, one that bypassed classification, reduced the need for expensive skilled classifiers and indexers, and that led to the first examples of quantitative information science.[2]

The 1950s seemed to be the beginning of an American "Information Century." However, the early Cold War information initiatives did not translate into the rescue or modernization of the sector Burton Livingston had fought for in the 1920s, academia's traditional information systems. It was not until the fearful surprise of the Soviet's space flight successes in the late 1950s that the American government paid attention to the condition of nonprofit journals, abstracts, and indexes. The resulting policies reflected American politics and values. Instead of creating a recommended centralized federal information capability, the government helped the older nonprofit sector through various types of subsidies and agencies such as the National Science Foundation.

By the 1960s, policies such as the Defense Education Act and the revitalization of the National Science Foundation spread the Flexner model of higher education.[3] Combined with a vast increase in college enrollments, and political demands that the government distribute research contracts throughout academia, higher education was becoming one of the nation's largest industries and markets. Increased academic specialization and widespread demands on faculty to publish (or perish) led to more and larger journals and other publications, and to escalating costs for libraries and other information systems. At the time, however, it did not seem that a crisis in the cost of science publications would arise because the United States had been in an era of unprecedented economic prosperity since the late 1940s. There was also a hope, backed by millions in government subsidies, for the immediate development of a true "science of information," one that could compensate for any deluge of scientific publication.

The Traumas and Continued Morphing of the American Left

The Cold War had a profound impact on the Red cause in America and on its supporters such as Jean and the other Clarks. Even before the postwar East-West split, confessions by Soviet agents such as Nina Field's friend Hede Massing made it clear that Russia had been ordering and paying the Communist Party to spy on the United States since the 1920s, even using many Americans in high-level government positions. With those revelations and the threats to democracy in Europe and Asia, the U.S. government began investigations and prosecutions of party members and fellow travelers. Communist Party leaders were jailed, radical aliens deported.

Leftists sought low visibility. The Clarks in Massachusetts kept their ideological faith and party contacts but became, as much as they could, inconspicuous. Jean Clark and her husband Sali Liebermann moved from Boston to Hollywood and then to the hills above a small fishing village across the bay from San Francisco. Once there, they remained in the background but helped build a welcoming ultraliberal refuge for others of the Left, such as Hollywood Ten movie writers and directors who had served prison time for refusing to discuss their Communist connections during U.S. congressional hearings. Jean's sisters also moved to California to help the cause while keeping a low profile.

The Fields and the Wilderness of Mirrors

The Fields, however, were unable to stay in the shadows as the East–West tensions of the 1940s led to one of the most bizarre episodes in modern history. Beginning in mid-1949, Noel Field, then Herta, then Hermann, and then Erica disappeared after traveling to Central and Eastern Europe. For years, it was unclear what happened to them. Had they defected? Had an American intelligence agency kidnapped them because of their ties to European Communism? Had the Soviets taken them?

It was some five years before the West learned they had all been tortured for information the Stalinists sought for a new and brutal round of purges in the Soviet bloc. Soviet intelligence accused them of being American spies despite their decades of contributions to the Red cause. As bizarrely, some thought the entire affair was part of an American plot to destabilize the Communist bloc.

Shortsighted Predictions about Ideology and an American Information Century

By the mid-1950s, news of the treatment of the Fields, the revelations of Stalin's horrors, and economic prosperity in the West combined to produce a decline of a belief in old-line communism in Europe and America. One scholar, Daniel Bell, even declared that America was experiencing an "end to ideology" and domestic conflict—and a permanent commitment to a moderate version of East Coast "establishment" cultural-political liberalism.[4] Meanwhile, the extraordinary investments in maintaining a high-tech international military balance of power continued to fuel the expansion of research, higher education, and science information systems.

The investments helped avert a third world war, but not America's involvement in smaller but costly conflicts. As a result, by the mid-1960s it was evident that Daniel Bell was wrong. There had not been an end to ideology. As John Patrick Diggins has so well portrayed in works such as *The Rise and Fall of the American Left,* ideology did not die, it ebbed and flowed and continually changed. The Vietnam War of the 1960s led to a rebirth and another reworking of radical ideas.[5] The New Left took its cues from the likes of Paul Massing's friends the Frankfurt scholars, and the Hintons' hero Mao, rather than Lenin. The New Left also bypassed old allies such as labor unions. Its adherents were oriented to counter-culture movements, even challenging the worth of science and the validity of any "objective" education. Its more radical members were extremely violent, even attacking university libraries and destroying catalogs as they attempted to turn American higher education into their version of a Carmelita Hinton moral training school.

After the Vietnam War the New Left retreated, left-wing ideology softened, and "activist" became a favored term. At the same time, the United States faced new challenges, ones stemming from unexpected consequences of its new internationalism—as well as the successes of its programs to spur economic recovery in the world.

Beginning in the 1970s, America's economy faced major threats including declines in private and government income. In response, its states began to retrench support for higher education and thus for science information, although the need for both grew as financially hard-pressed colleges struggled to remain Flexner-compliant. Competing for income from a shrinking pool of traditional federal grants and private industry contracts became a financial necessity for faculty and institutions. At the same time, there was growing acceptance of the commercialization of science as a means of generating income for higher education and, in some instances, faculty members. As well, the old style of Herbert Field's "gentlemen scientists," who unreservedly contributed their discoveries to the world, seemed quaint as an era of "intellectual property" began.

New Threats from Information Conglomerates

By the 1990s, the European Communist bloc had collapsed and China welcomed capitalism. The Cold War seemed to have ended. But new threats to America's science and educational infrastructures appeared. For-profit European information conglomerates began purchasing American academic publishers, including indexing and abstracting services. At the same time, the price of science information was escalating faster than the budgets of academic and research institutions. There were protests and even congressional hearings as unanticipated and very threatening "serials crises" emerged.

A first round of counteractions fit the older progressive-liberal, Vernon Kellogg–like response to the information problem. Foundations stepped forward to encourage new types of Efficient and hopefully self-supporting nonprofit organizations to use the startling progress of computer and Internet technologies to bypass traditional publishers. Some hoped to use the latest technologies to create, finally, a world science bibliography centered on voluntary cooperation.

As well, reacting to protests and pressures, for-profit companies altered their policies to allow some reduced-cost access to publications. At the same time, governments in Europe and America reexamined their policies and began to consider proposals that more types of government-sponsored publications be freely available—but in ways that cost the governments little.

New Information Movements, and a Greater New Link between Information and Intrigue

None of that satisfied a new breed of information radicals. Quite soon, more ideologically informed crusades for Open Societies, Open Access, and Creative Commons arose. They were after much more than a Progressive-like tweaking of the infrastructure of science information. Their rhetoric sounded more like Henry Wallace or Desmond Bernal's than Vernon Kellogg's. They demanded cost-free access to scientific information, limits to intellectual property rights, and an end to commercialized science. There was a call for a boycott of publishers until they reduced their prices.

The most energized voices within the more extreme segments of the "green" Open Access movement had a familiar ring. Their denunciation of "the academic establishment," institutions such as peer review, publishing monopolies, and intellectual property rights paralleled older left-wing arguments. There were even claims that new technologies such as full-text searching and the Internet made journals, and indexing-abstracting services, unnecessary.

The new information movements had parallels, if not ties, to other reform goals, especially to the undoing of European Communist dictatorships, to techno-equality for poor nations, and to minority empowerment in the United States through organizations such as the Association of Community Organizers for Reform Now (ACORN). One of the financial angels for such causes and the Open Access movement was, ironically, one of the world's most aggressive capitalists. George Soros also was a new type of populist internationalist. Among his many interests was finding ways to compensate for the United Nations' failure to secure low-cost science information for less-developed countries.[6]

Meanwhile, the importance of science and research and the computer-communication and Internet revolutions had led to a new level and type of information intrigue—the theft of science information at a distance. The new cyber-spying was not confined to military secrets or wartime needs. International economic competition, combined with ideological conflict, led to ongoing attacks on corporate and university sci-tech information systems.[7]

Notes

Introduction

1. Unless specified otherwise, dollar amounts have been converted to constant U.S. dollar values for 2006 using the Federal Reserve Inflation Calculator.

Chapter 1

1. Thomas D. Hamm, *The Quakers in America* (New York: Columbia University Press, 2003); Tunis Garret Bergen, *Genealogies of the State of New York: Long Island Edition* (New York: Lewis Publishing, 1915).

2. U.S. Manuscript Census, 1860–1920: Cambridge, Mass., newspapers, poll lists, city directories, etc.

3. Frederick Clifton Pierce, *Field Genealogy* (Chicago: W. B. Conkey Co., 1901); Josephine C. Frost, *The Haviland Genealogy: Ancestors and Descendants of William Haviland of Newport Rhode Island, 1653–1688* (New York: The Lyons Genealogical Co., 1914); Bergen, *Genealogies of the State of New York.*

4. On the secularization of America's middle class during the era, see Daniel Walker Howe, *What Has God Wrought, America 1815–1848* (New York: Oxford University Press, 2007); on the Hicksite movement, Hugh Barbour et al., *Quaker Crosscurrents: Three Hundred Years of Friends in the New York Yearly Meetings* (Syracuse, N.Y.: Syracuse University Press, 1995).

5. John W. Oliver, Jr., Charles L. Cherry, and Carline L. Cherry, eds., *Founded by Friends: The Quaker Heritage of Fifteen American Colleges and Universities* (Lanham, Md.: Scarecrow Press, 2007).

6. Nora Travis, *Haviland China: The Age of Elegance* (Altgen, Pa.: Schiffer, 1997).

7. Anonymous, "Lydia Field Obituary," *The Friends Intelligencer* (November 24, 1917): 748–749.

8. Howe, *What Has God Wrought.*

9. Aaron's firm's reputation overcame a multimillion dollar entanglement with an unscrupulous importer and corrupt customs agents in 1875. *New York Times, New York Tribune,* June 11–12,

1875. Aaron went on to become a trusted member of the financial community in New York and Brooklyn.

10. Howard Gotlieb Archival Research Center, Boston University, Hermann Field Collection (BUHF), box 17.

11. On Herbert's health and speech problem: American Philosophical Society, Charles Davenport Papers (APSDP) B-D27, Field to Davenport.

12. Henry Fairfield Osborn, *The American Museum of Natural History, Its Origin, Its History, the Growth of Its Departments to December 31, 1909* (New York: The Irving Press, 1911).

13. Brooklyn Polytechnic Institute, "First Quarter Century of Brooklyn Collegiate and Polytechnic Institute, Presented by the President to the Board of Trustees, June 16, 1880."

14. On the progressive business man, Patrick J. McGrath, *Scientists, Business, and the State, 1890–1960* (Chapel Hill: University of North Carolina Press, 2002).

15. On the history of Unitarians and radicalism, Dan McKanan, *Prophetic Encounters: Religion and the American Radical Tradition* (Boston: Beacon Press, 2011).

16. Daniel Walker Howe, *The Unitarian Conscience: Harvard Moral Philosophy* (Cambridge, Mass.: Harvard University Press, 1988); on the new universities, Laurence R. Vesey, *The Emergence of the American University* (Chicago: University of Chicago Press, 1965); John W. Oliver Jr., Charles L. Cherry, and Carline L. Cherry (eds.), *Founded by Friends: The Quaker Heritage of Fifteen American Colleges and Universities* (Lanham, Md.: Scarecrow Press, 2007), 57. On Eliot "modernizing" Harvard: Bruce Kimball, *The Inception of Modern Professional Education: C.C. Langdell, 1826–1906* (Chapel Hill: University of North Carolina Press, 2009).

17. For example, see George Howard Parker, *The World Expands: Recollections of a Zoologist* (Cambridge, Mass.: Harvard University Press, 1946).

Chapter 2

1. Charles E. Hambrick-Stowe, *Charles G. Finney and the Spirit of American Evangelicalism* (Grand Rapids, Mich.: W. B. Eerdmans, 1996).

2. James McLachlan, "The American College in the Nineteenth Century: Towards a Reappraisal," *Teachers College Record* 86 (December 1978): 28–36, and Colin B. Burke, *American Collegiate Populations: A Test of the Traditional View* (New York: New York University Press, 1982).

3. Thomas Harold André Le Duc, *Piety and Intellect at Amherst College, 1865–1912* (New York: Arno Press, 1969).

4. Whitney R. Cross, *The Burned-Over District: The Social and Intellectual History of Enthusiastic Religion in Western New York, 1800–1850* (New York: Harper & Row 1965).

5. Rebecca Green, "Melvil Dewey's Ingenious Notational System." Presented at North American Symposium on Knowledge Organization, Syracuse University, June 18–19, 2009.

6. Melvil Dewey, "Decimal Classification beginning," *Library Journal* 45, no. 15 (February 1920): 151–154.

7. W. A. Wiegand, "The Amherst Method: The Origins of the Dewey Decimal Classification Scheme," *Libraries and Culture* 33 (Spring 1998): 175–194.

8. Wayne Wiegand, *Irrepressible Reformer: A Biography of Melvil Dewey* (Chicago: ALA, 1996).

9. Useful on classification history: Bruce M. Manzer, *The Abstract Journal, 1790–1920* (Metuchen, N.J.: Scarecrow Press, 1977); W. Boyd Rayward, "The Search for Subject Access to the Catalogue of Scientific Papers 1800–1900," in W. Boyd Rayward, ed., *The Variety of Librarianship: Essays in Honor of John Wallace Metcalfe* (Sandy Bay, Tasmania: Library Association of Australia, 1976); Elaine Svenonius, "Unanswered Questions in the Design of Controlled Vocabularies," *JASIS* 37, no. 5 (1986): 331–342.

10. Jim Ranz, *The Printed Book Catalog in American Libraries 1723–1901* (Chicago: ALA, 1964).

11. Wm. Parker Cutter, *Charles Ammi Cutter* (Boston: Gregg Press, 1972). On objections to card files, M. Freeman, "Pen, Ink, Keys, and Cards: Some Reflections on Library Technology," *College and Research Libraries* 52 (July 1991): 328–335.

12. M. P. Satija, *The Theory and Practice of the Dewey Decimal Classification System* (Oxford: Chandos Publishing, 2007).

13. IIB, "IIV Conference Bibliographique Internationale Londres, 1929, 09, 16/17," in *Bulletin of the IIB*, 163 (1929); Raynard Swank, "Subject Catalogs, Classifications, or Bibliographies: A Review of Critical Discussions 1876–1942," *Library Quarterly* 14 (October 1944): 316–332.

14. Mildred Harlow Downing, *Introduction to Cataloging and Classification* (Jefferson, N.C.: McFarland Press, 1981), 62.

15. Francis Louis Miksa, "Charles Ammi Cutter: Nineteenth Century Systematizer of Libraries," Ph.D. diss., University of Chicago, 1974. On the limits of Dewey's systems, A. C. Foskett, *The Universal Decimal Classification: The History, Present Status and Future Prospects for a Large General Classification Scheme* (London: Linnet Books, 1973, chapters 1 and 2).

16. Satija, *Theory and Practice.*

Chapter 3

1. Howard Gotlieb Archival Research Center, Boston University, Hermann Field Collection (BUHF), box 4, file 2.

2. Biographical materials in Mark Papers, Harvard University Library.

3. Keith R. Benson, "Epilogue: The Development and Expansion of the American Society of Zoologists," in Keith R. Benson et al., eds., *The Expansion of American Biology* (New Brunswick, N.J.: Rutgers University Press, 1991).

4. Mary P. Winsor, *Reading the Shape of Nature: Comparative Zoology at the Agassiz Museum* (Chicago: University of Chicago Press, 1991).

5. Gladys Doolittle, "A History of the Concilium Bibliographicum, Zurich," MS thesis, School of Library Science, New York, June 1929, p. 6.

6. Franz Stafleu, "Fifty Years of International Biology," *Taxon* 21, no. 1 (February 1971): 141–51.

7. George Howard Parker Papers, Harvard University.

8. G. H. Parker and E. L. Mark, *Mark Anniversary Volume* (New York: H. Holt, 1903).

9. His American work began with Herbert Haviland Field, "The Development of the Pronephros and Segmental Duct in Amphibia," *Bulletin of the Museum of Comparative Zoology* 21, no. 5 (1891). For his later zoological publications: Henry B. Ward, "Herbert Haviland Field," *Science* 54, no. 1401 (November 4, 1921): 424–428.

10. For insight into his early activities, Herbert H. Field, "An International Zoological Record," *Nature* 47 (April 27, 1893): 606–607; Field, "Bibliographical Reform and the Zoological Record," *Natural Science: A Monthly Review of Scientific Progress* 6 (1895): 333–335.

11. Field, "An International Zoological Record."

12. Robert B. Downs, "Problems of Bibliographic Control," https://www.ideals.illinois.edu/bitstream/handle/2142/5548/librarytrendsv2i4D_opt.pdf?sequence=1.

13. Useful surveys of the problems are Stafleu, "Fifty Years of International Biology," and Eckhardt Fuchs, "The International Catalogue of Scientific Literature," in Christopher Clark, Jurgen Schtiewer, and Peter Wagner, eds., *Transnational Intellectual Networks*.

14. Herbert Field, "Bibliographical Reform."

15. Fuchs, "International Catalogue."

16. For a list of Herbert's writings on the subject, see my web-based bibliography (http://userpages.umbc.edu/~burke). A general survey of the problem is Alex Wright, *Glut: Mastering Information through the Ages* (Ithaca, N.Y.: Cornell University Press, 2008). A more focused work is Katherine O. Murra, "History of Some Attempts to Organize Bibliography Internationally," in Jesse Shera and Margaret Egan, eds., *Bibliographic Organization* (Chicago: University of Chicago Press, 1950), 24–53. A magnificent overview is Brian Vickery, "A Century of Scientific and Technical Information," *Journal of Documentation* 33, no. 3 (December 1999): 476–527. For a longer perspective, D. Rosenberg, ed., "Early Modern Information Overload, Special Issue," *Journal of the History of Ideas* 64, no. 1 (2003).

17. Jack Meadows, "The Growth of Journal Literature: A Historical Perspective," in *The Web of Knowledge: A Festschrift in Honor of Eugene Garfield,* by Blaise Cronin and Helen Barsky Atki, eds. (Medford, N.J.: ASIS-Information Today, 2000).

18. C. A. Minchin, "A Plea for an International Zoological Record," *Nature* 46 (August 18, 1892): 367–368; Alex Csiszar, "Seriality and the Search for Order: Scientific Print and Its Problems during the Late Nineteenth Century" *History of Science* 48 (2010): 399–434.

19. Stafleu, "Fifty Years of International Biology." On the general problem of national bibliography: Katherine Oliver Murra, "Notes on the Development of the Concept of Current Complete National Bibliography," Appendix to The UNESCO Library of Congress Bibliographical Survey, *Bibliographical Services: Their Present State and Possibilities of Improvement* (Washington, D.C.: n.p., 1950).

20. Robert L. Collision, *Abstracts and Abstracting Services* (Santa Barbara, Calif.: ABC-CLIO, 1971); Verner Clapp, "Research in Problems of Scientific Information—Retrospect and Prospect," *American Documentation* 14, no. 1 (1963): 1–19.

21. Bruce M. Manzer, *The Abstract Journal, 1790-1920: Origin, Development and Diffusion* (Metuchen, N.J.: Scarecrow Press, 1977).

22. W. Boyd Rayward, "The Search for Subject Access"; William Hoyle, "Concilium Bibliographicum at Zurich and Its Work," *Library Association Record,* 1899: 709–718.

23. Manzer, *The Abstract Journal.*

Chapter 4

1. Herbert H. Field, "An International Zoological Record," *Nature* 47 (April 27, 1893): 606–607.

2. T. D. A. Cockerel, "A Suggestion for the Indexing of Zoological Literature," *Nature* 46 (September 8, 1892): 442–443.

3. Manzer, *The Abstract Journal;* Laurent Rollett and Phillipe Nabonnand, "An Answer to the Growth of Mathematical Knowledge? The Repertoire Bibliographique des Science Mathematiques," *European Mathematical Newsletter* 4 (March 2003): 2–17.

4. Gladys Doolittle, "A History of the Concilium Bibliographicum, Zurich," MS thesis, School of Library Science, New York, June 1929, p. 6.

5. Field, "An International Zoological Record."

6. Henry F. Nachtrieb papers, University of Minnesota Archives.

7. W. Boyd Rayward, ed., *European Modernism and the Information Society* (Burlington, Vt.: Ashgate, 2008), 91; Field, "An International Zoological Record."

8. Charles Atwood Kofoid, "Cooperation in Scientific Bibliography," *Science* 27, no. 692 (April 3, 1908): 543–545.

9. Howard Gotlieb Archival Research Center, Boston University, Hermann Field Collection, various Herbert Field items for 1895–1914. Hereafter, BUHF.

10. W. Boyd Rayward, "The Search for Subject Access to the Catalogue of Scientific Papers," in W. Boyd Rayward (ed.), *The Variety of Librarianship: Essays in Honour of John Wallace Metcalfe* (Sandy Bay, Tas.: Library Association of Australia, 1976). William Hoyle, "Concilium Bibliographicum at Zurich and Its Work," *Library Association Record,* 1899: 709–718; J. Victor Carus, "On the International Catalogue of Scientific Literature," *Science* 9, no. 233 (June 16, 1899): 825–835.

11. Herbert H. Field, "A Brief Account of the Foundation of the Concilium, Bibliographicum," *Annotationes Concilium Bibliographici* (also *Concilium Bibliographicum*), December 1907; BUHF, Field to Davenport, September 29, 1895.

12. The details of Herbert's 1890s actions were gathered from facts scattered in his many writings as well as items in the BUHF collection. See the bibliography cited on my web page (http://userpages.umbc.edu/~burke).

13. Nachtrieb papers, correspondence and items from Concilium.

14. Charles Atwood Kofoid, "Cooperation in Scientific Bibliography," *Science* 27, 692 (April 3, 1908): 543–545.

15. John Lawler, *The H. W. Wilson Company* (Boston: Gregg Press, 1972).

16. Field, "A Brief Account."

17. Doolittle, "A History of the Concilium Bibliographicum."

18. BUHF, box 4, file 1.

19. Hoyle, "Concilium Bibliographicum at Zurich and Its Work"; Herbert Haviland Field, "The Concilium Bibliographicum in Zurich," *Proceedings of the 28th Meeting of the American Library Association* (St. Louis, 1905).

20. Alva T. Stone, *One Hundred Years with the Library of Congress Subject Headings* (New York: Haworth Press, 2000).

21. Kenneth E. Carpenter, *The First 350 Years of the Harvard University Library* (Cambridge, Mass.: Harvard University Library, 1986); Francis Louis Miksa, "Charles Ammi Cutter: Nineteenth Century Systematizer of Libraries," Ph.D. diss., University of Chicago, 1974.

22. Francis L. Miksa, *The Development of Classification at the Library of Congress* (Champaign Urbana, Ill.: University of Illinois, Graduate School of Library and Information Science, circa 1984); Andrew D. Osborn, "From Cutter to Dewey to Mortimer Taube and Beyond: A Complete Century of Change in Cataloging and Classification," *Cataloging and Classification Quarterly* 12, 3/4 (1991): 35–50.

Chapter 5

1. Gladys Doolittle, "A History of the Concilium Bibliographicum, Zurich," MS thesis, School of Library Science, New York, June 1929; Herbert H. Field, "A Brief Account of the Foundation of the

Concilium, Bibliographicum," *Annotationes Concilium Bibliographici* (also *Concilium Bibliographicum*), December 1907.

2. "Bibliographica Zoologica and Anatomica," *Science* 2, no. 46 (November 15, 1895): 656–657.

3. A useful summary of the many features and codes of the UDC is in Shirley F. Harper, "The Universal Decimal Classification," *American Documentation* 5, no. 4 (October 1954): 195–213.

4. *Annotationes.*

5. Helpful on Otlet's systems, I. C. McIlwaine, "The Universal Decimal Classification: Some Factors Concerning its Origins, Development, and Influence," in Trudi Bellardo Hahn and Michael Buckland, eds., *Historical Studies in Information Science* (Medford, N.J.: Information Today, 1998), 94–106. On the need for more complexity given new customers and subjects, Adolf Law Voge, *Systematic Code and Index for the Relative Decimal Classification of a Card Bibliography of Electrochemistry and Allied Subjects* (Zurich: Concilium Bibliographicum, 1905–1906).

6. For example, Concilium Bibliographicum, *Conspectus methodicus et alphabeticus numerorum classificationis bibliographici* (International Institute of Bibliography, 1902).

7. On the post WWII file organization revolutions: Mortimer Taube, *Studies in Coordinate Indexing*, vol. 1 (Washington, D.C.: Documentation Inc., 1953).

8. American Philosophical Society, Charles Davenport Papers (APSDP) B-D27, Field to Davenport, circa 1895.

9. Concilium Bibliographicum, *Conspectus methodicus et alphabeticus numerorum classificationis bibliographici* (International Institute of Bibliography, 1898).

10. Useful on the catalog, J. Victor Carus, 'On the International Catalogue of Scientific Literature," *Science* 9, 233 (June 16, 1899): 825–835; Wm. E. Hoyle, *The Dewey Decimal Classification and the International Catalogue of Science* (London: Page and Pratt, c. 1900).

11. Eckhardt Fuchs, "The International Catalogue of Scientific Literature," in Christopher Clark, Jurgen Schtiewer, and Peter Wagner, eds., *Transnational Intellectual Networks*; W. Boyd Rayward, "The Search for Subject Access to the Catalogue of Scientific Papers 1800–1900," in W. Boyd Rayward, ed., *The Variety of Librarianship: Essays in Honor of John Wallace Metcalfe* (Sandy Bay, Tasmania: Library Association of Australia, 1976); W. Boyd Rayward, *The Universe of Information: The Work of Paul Otlet for Documentation and International Organization,* FID 520 (Moscow: FID, 1975); H. P. Bowditch et al., "Report of the Committee of the University Council to Consider the Communication of the Royal Society Relating to a Catalogue of Scientific Papers to be Made by International Cooperation, *Science* 1, no. 7 (February 15, 1895): 184–186; Hoover Institution, Vernon Kellogg Papers, 56004.

12. Doolittle, "A History of the Concilium Bibliographicum"; Katherine O. Murra, "History of Some Attempts to Organize Bibliography Internationally," in Jesse Shera and Margaret Egan, eds., *Bibliographic Organization* (Chicago: University of Chicago Press, 1950), 24–53.

13. *Annotationes*, February 1907; Henry B. Ward, "Herbert Haviland Field," *Science* 54, no. 1401 (November 4, 1921): 424–428.

14. H. H. Field, "The New Bibliography Bureau for Zoology," *Science* 2, no. 34 (August 1895): 234–237.

15. Herbert Haviland Field, "A Brief Account of the Foundation of the Concilium Bibliographicum," in *Annotationes*, December 1907; *Conspectus*, 1902.

16. Doolittle, "A History of the Concilium Bibliographicum," 10; William Hoyle, "Concilium Bibliographicum at Zurich and Its Work," *Library Association Record*, 1899: 709–718."

17. Concilium Bibliographicum, *Conspectus*, 1898 seems to be the first available edition.

18. H. H. Field, "The New Bibliography Bureau."

19. Fuchs, "The International Catalogue of Scientific Literature."

20. Doolittle, "A History of the Concilium Bibliographicum," 40.

21. Herbert Field, "The International Catalog of Scientific Literature," *Science* 10, no. 240 (August 4, 1899): 133–143; Elaine Svenonius, "Unanswered Questions in the Design of Controlled Vocabularies," *JASIS* 37, no. 5 (September 1986): 331–340.

22. Wm. E. Hoyle, "The Dewey Decimal Classification."

23. Henry F. Nachtrieb Papers, University of Minnesota.

24. Among W. Boyd Rayward's many works on Otlet, *The Universe of Information: The Work of Paul Otlet for Documentation and International Organization,* FID 520 (Moscow: FID, 1975).

25. E. Scott, "IFLA and FID History and Program," *Library Quarterly* 32 (1962): 1–18; S. Fayete-Scribe, "The Cross Fertilization of the U.S. Public Library Model and the French Documentation Model (IIB, French Correspondent of FID) through the French Professional Associations between World War I and World War II," *JASIS* 48, no. 9 (1997): 782–793; R. R. Bowker, "The Institut international de bibliographie, Brussels," *Library Journal* 25 (1900): 273–274.

Chapter 6

1. Concilium Bibliographicum, *Conspectus*.

2. *Annotationes*.

3. G. Brown Goode, "The Ideal Index to Scientific Literature," *Science* 1, no. 16 (April 19, 1895): 433–437.

4. H. P. Bowditch et al., "Report of the Committee of the University Council to Consider the Communication of the Royal Society Relating to a Catalogue of Scientific Papers to be Made by International Cooperation, *Science* 1, no. 7 (February 15, 1895): 184–186; John Shaw Billings, "A Card Catalogue of Scientific Literature," *Science* 1 (April 12, 1895): 406–408.

5. Henry F. Nachtrieb Papers, University of Minnesota; Concilium Bibliographicum, Physiological Division of the Zoological Bibliography (Zurich: Concilium Bibliographicum, n.d.).

6. H. H. Field, "The New Bibliography Bureau for Zoology," *Science,* 2, no. 34 (August 1895): 234–237.

7. Ch. Richert, *Bibliographia Physiologia* 1893–1894 (Paris: Felix Alcan Editor, 108 blvd. St-Germain, c. 1898).

8. Gladys Doolittle, "A History of the Concilium Bibliographicum, Zurich," MS thesis, School of Library Science, New York, June 1929; *Annotationes.*

9. *Annotationes,* 1907; American Philosophical Society, Charles Davenport Papers B-D27 (APSDP), Field to Davenport, June 1896.

10. "Doolittle, "A History of the Concilium Bibliographicum."

11. *Conspectus,* 1898–1902.

12. *Annotationes,* 1909; Concilium *Annual Report,* 1903.

13. *Annotationes,* 1907.

14. *Conspectus.*

15. Concilium Bibliographicum, "To the Committee Appointed by the 7th International Zoological Congress in Boston," 1907.

16. C. C. McClung et al., "The Biological Record and Biological Abstracts," *Science,* 61, no. 1574 (February 27, 1925): 232–234. A. J. Meadows, ed., *The Origins of Information Science* (London: Taylor-Graham, 1987), p. 144.

17. George M. Eberhart, *The Whole Library Handbook 4* (Chicago: American Library Association, 2006).

18. *Annotationes,* 1905.

19. APSDP B-D27, Field to Davenport, November 1914; Harold Fowler McCormick, "Via Pacis: Further Considerations Concerning How Terms of Peace Can Be Automatically Prepared while the War is Still Going In" (A suggestion Offered by an American), Harold F. McCormick (Zurich: Concilium Bibliographicum, 1917).

20. *Annotationes,* 1907.

21. T. D. Cockerel, "A Suggestion for the Indexing of Zoological Literature," *Nature,* 46 (September 8, 1892): 442–443; Herbert Field, "New Departures in the Concilium Bibliographicum II—The Supplementary Bibliography." *Science* 17, no. 443 (June 26, 1903): 1014–1016; Field, "Report to Seventh International Congress."

22. *Annotationes.*

Chapter 7

1. *Annotationes*; John Lawler, *The H. W. Wilson Company* (Boston: Gregg Press, 1972), 41.

2. *Conspectus*.

3. Herbert H. Field, "Statement to the Committee Appointed by the 7th International Zoological Congress in Boston," August 23, 1907.

4. Gladys Doolittle, "A History of the Concilium Bibliographicum, Zurich," MS thesis, School of Library Science, New York, June 1929; A. L. Voge, "Indexing of Periodical Literature and the Work of the Concilium Bibliographicum, Zurich," *Papers of the Bibliographical Society of America* 2 (1907–1908).

5. *Annotationes*, 1908.

6. *Annotationes*; Harvard University class books.

7. Based on tuition at the University of Pennsylvania.

8. Ulrich Periodical Register.

9. *Annotationes*.

10. Useful on evolutions history in America, Edward J. Larson, *Trial and Error: The American Controversy Over Creation and Evolution,* 3rd ed. (New York: Oxford University Press, 2003).

11. Gregory A. Kimble et al., *Portraits in Psychology,* vol. II, *Yerkes* (Hillsdale, N.J.: L. Erlbaum Associates: American Psychological Association, 1991); Daniel J. Kevles, "Testing the Army's Intelligence: Psychologists and the Military in World War I," *The Journal of American History* 55, no. 3 (1968): 565–581.

12. Doolittle, "A History of the Concilium Bibliographicum," 86. The Smithsonian continued its subscription in the 1930s.

13. E. Carlton MacDowell, "Charles Benedict Davenport, 1866–1944: A Study in Conflicting Influences," 1946, reprinted from *Bios* 17, no. 1 (March 1946); "Bibliography of Charles B. Davenport," 37–50.

14. Oscar Riddle, *Biographical Memoir of Charles Benedict Davenport, 1866–1944* (Washington, D.C.: National Academy of Sciences, 1947).

15. Mark B. Adams et al., "Human Heredity and Politics: A Comparative Study of the Eugenics Record Office at Cold Spring Harbor (United States), the Kaiser Wilhelm Institute for Anthropology, Human Heredity, and Eugenics (Germany), and the Maxim Gorky Medical Genetics Institute (USSR)," *Osiris* 20 (2005): 232–262; Jane Maienschein et al., eds., *Centennial History of The Carnegie Institution of Washington:* Vol. V, *The Department of Embryology* (New York: Cambridge University Press, 2004).

16. *Annotationes*.

17. Henry Fairfield Osborn, *The American Museum of Natural History, Its Origin, Its History, the Growth of Its Departments to December 31, 1909* (New York: The Irving Press, 1911).

18. Ronald L. Numbers, *Darwinism Comes to America* (Cambridge: Harvard University Press, 1998), 152.

19. Ronald Rainger, *An Agenda for Antiquity: Henry Fairfield Osborn & Vertebrate Paleontology at the American Museum of Natural History, 1890–1935* (Tuscaloosa: University of Alabama Press, 1991). On the emergence of modern science and liberalism at Columbia, Robert A. McCaughy, *Stand: Columbia: A History of Columbia University* (New York: Columbia University Press, 2003).

20. Roger Lewin, *Bones of Contention: Controversies in the Search for Human Origins* 2d ed. (Chicago: University of Chicago Press, 1997); Howard Gotlieb Archival Research Center, Boston University, Hermann Field Collection (BUHF), Field to Davenport, August 1895; Henry Fairfield Osborn, "Progress of the Concilium Bibliographicum," *Science* 17, no. 41 (June 12, 1903): 951–952; National Academy of Sciences, *Biographical Memoirs*, vol. 19 "Osborn" (Washington, D.C.: NAS, 1918).

21. William Hoyle, "Concilium Bibliographicum at Zurich and Its Work," *Library Association Record,* 1899: 709–718; Wm. E. Hoyle, *The Dewey Decimal Classification and the International Catalogue of Science* (London: Page and Pratt, c. 1900).

22. Adolf Voge, "The Concilium," *Public Libraries* (1908), as in Voge Papers at Library of Congress; Doolittle, "A History of the Concilium Bibliographicum."

23. Doolittle, "A History of the Concilium Bibliographicum," 36.

24. Voge, "Indexing of Periodical Literature"; American Philosophical Society, Charles Davenport Papers (APSDP) B-D27, Field to Davenport, 1907; *American Ornithologist*, 1907.

25. W. Boyd Rayward, *The Universe of Information: The Work of Paul Otlet for Documentation and International Organization,* FID 520 (Moscow: FID, 1975); S. Fayete-Scribe, "The Cross Fertilization of the U.S. Public Library Model and the French Documentation Model (IIB, French Correspondent of FID) through the French Professional Associations between World War I and World War II," *JASIS* 48, no. 9 (1997): 782–793.

26. Herbert Field, "The Concilium Bibliographicum in Zurich," *Proceedings of the 28 Meeting American Library Association* (St. Louis, 1905).

27. Voge Papers, Library of Congress.

28. Robert Kanigel, *The One Best Way: Frederick Winslow Taylor and the Enigma of Efficiency* (New York: Viking, 1997).

29. U.S. Manuscript Census, 1860–1920: Cambridge, Mass., newspapers, poll lists, city directories, etc.; Voge Papers, Library of Congress.

30. General Electric Co., *Relative Index: An Extension of the Dewey Decimal System, for the Electric Industry* (Schenectady, N.Y.: General Electric Company, 1903).

31. Dave Muddiman, "A New History of ASLIB 1924–1950," *Journal of Documentation* 61, no. 3 (2005): 402–428; May Wilson et al., *Special Libraries Directory* (New York: SLA, 1925); Anthony Thomas Kruzas, "The Development of Special Libraries for American Business and Industry," PhD. diss., University of Michigan, 1960.

32. Kruzas, "The Development of Special Libraries," 278; Elaine Svenonius, *The Intellectual Foundations of Information Organization* (Cambridge, Mass.: MIT Press, 2000), 173.

33. Bruce M. Manzer, *The Abstract Journal, 1790–1920* (Metuchen, N.J.: Scarecrow Press, 1977).

34. Voge, "Indexing of Periodical Literature"; Rayward, *The Universe of Information*, 101.

35. Voge, *Systematic Code and Index for the Relative Decimal Classification of a Card Bibliography of Electrochemistry and Allied Subjects* (Zurich: Concilium Bibliographicum, 1905–1906) and "Indexing of Periodical Literature."

36. *Annotationes*.

37. Voge, "Indexing of Periodical Literature."

38. Charles Atwood Kofoid, "Cooperation in Scientific Bibliography," *Science* 27, no. 692 (April 3, 1908): 543–545.

39. Herbert Haviland Field, "The Relations of the Concilium Bibliographicum to Publishers and Editors," in *Annotationes*, December 1901.

40. Useful on Putnam, Jane Aiken Rosenberg, *The Nation's Great Library, Herbert Putnam and the Library of Congress 1899–1939* (Urbana: University of Illinois Press, 1993).

41. Voge Papers, Library of Congress.

42. Rayward, *The Universe of Information*, 101.

43. On Voge's personal life, Voge Papers, Library of Congress; U.S. Census; State Department Documents; *New York Times*; *Washington Post*, passim.

44. A. L. Voge, *Bibliography of Periodicals & Serials and Society Publications in Electrochemistry* (Zurich, 1909).

45. Adolf Law Voge, *Systematic Code*.

46. E. J. Crane, ed., *The History of Chemical Abstracts* (n.p.: ACS, 1952).

47. Adele Hoskins, "A History of Chemical Bibliography in the United States Prior to Chemical Abstracts," Ph.D. diss., University of Chicago, 1971.

48. BUHF, box 3, 1.

49. Thomas Hapke, "Wilhelm Ostwald, the 'Brucke' (Bridge) and Connections to Other Bibliographic Activity at the Beginning of the Early Twentieth Century," in Mary Ellen Bowden, Trudi Bellardo Hahn, and Robert V. Williams, eds., *Proceeding of the 1998 Conference on the History and*

Heritage of Science Information Systems, ASIS-CHF (Medford, N.J.: Information Today, 1999), 139–147.

50. Voge Papers; U.S. Census (passim); ALA Proceedings, 1915–17; San Francisco City Directories.

Chapter 8

1. For Hamilton's general biography, Howard Gotlieb Archival Research Center, Boston University, Hermann Field Collection (BUHF), passim and especially box 17; Doreen Bolger, "Hamilton Easter Field and His Contribution to American Modernism," *American Art Journal* 20, no. 2 (1998): 78–107; Wendy Jeffers, "Hamilton Easter: The Benefactor from Boston," *Archives of American Art Journal* 50, no. 1–2 (2011): 26–37.

2. "Paul Haviland," http://en.wikipedia.org/wiki/Paul_Haviland.

3. Hamilton's great hopes for his boarding school were not fulfilled, see U.S. Manuscript Census, 1860–1920: Cambridge, Mass., newspapers, poll lists, city directories, etc.

4. Thomas D. Hamm, *The Quakers in America* (New York: Columbia University Press, 2003); Hans A. Schmidt, *Quakers and Nazis: Inner Light and Outer Darkness* (Columbia: University of Missouri Press, 1997).

5. *New York Times*, June 14, 1916; Hamilton Easter Field, "The Answers to a Letter in Which Are Expressed Doubts About My Loyalty . . . Made in the Prospectus of the Thurnscoe Forum of the Season of 1918" (Sowbery Farm, Cape Neddick, Maine, August 31, 1918: Hamilton Easter Field).

6. Useful on Dos Passos, Virginia Spencer Carr, *Dos Passos: Life* (Garden City, N.Y.: Doubleday, 1984).

7. BUHF, Passports; useful is "Noel Field," at www.Documentstalk.com/field-noel haviland-1904-1970.

8. Flora Lewis, *Red Pawn: The Story of Noel Field* (Garden City, N.Y.: Doubleday, 1965).

9. BUHF, box 3, file 1, February 20, 1904; American Philosophical Society, Charles Davenport Papers (APSDP) B-D27, Strohl to Davenport, February 14, 1929.

10. *Annotationes*; Adolf Voge, "The Concilium," *Public Libraries* (1908), as in Voge Papers at Library of Congress; Katherine O. Murra, "History of Some Attempts to Organize Bibliography Internationally," in Jesse Shera and Margaret Egan, eds., *Bibliographic Organization* (Chicago: University of Chicago Press, 1950).

11. BUHF, box 3, file 1; box 4, file 1; APSDP, Field to Davenport, September 1909.

12. "W. Bush," *New York Times,* February 11, 1941; *Who's Who* 1943.

13. Murra, "History of Some Attempts to Organize Bibliography Internationally"; *Annotationes*.

14. APSDP, Field to Davenport, 1902–1910.

15. Larry L. Fabian, *Carnegie's Peace Endowment* (Washington, D.C.: Carnegie Endowment for International Peace, 1984).

16. BUHF, box 24, file 17.

17. Concilium, Annual Reports, 1904–.

18. APSDP, Field to Davenport, December 16, 1902, October 30, 1905.

19. Eckhardt Fuchs, "The International Catalogue of Scientific Literature," in Christopher Clark, Jurgen Schtiewer, and Peter Wagner, eds., *Transnational Intellectual Networks* (Chicago: University of Chicago Press, 2004).

20. Fuchs, "The International Catalog," 187; American Museum of Natural History archives collection (AMNHC), Davenport to Osborn, various.

21. APSDP, Field to Davenport, May 1909.

22. Herbert Field, "Progress of the Concilium Bibliographicum," *Science* 19, no. 490 (May 20, 1904): 802–805; APSDP, Field to Davenport, November 1909.

23. Herbert Field, "The Concilium Bibliographicum in Zurich," *Proceedings of the 28 Meeting American Library Association* (St. Louis, 1905); Henry Fairfield Osborn, "Progress of the Concilium Bibliographicum," *Science* 17, no. 41 (June 12, 1903): 951-952.

24. APSDP Field to Davenport, October 1905.

25. Concilium, *Annual Report*, 1907; APSDP, Field to Davenport, November 1909.

26. *Annotationes*, 1906–1907.

27. APSDP, Field to Davenport, 1907–1908.

28. Concilium Bibliographicum, "To the Committee Appointed by the 7th International Zoological Congress in Boston, 1907"; Concilium, *Annual Report*, 1907.

29. *Annotationes*, 1907; "Preliminary List of Periodicals Excepted by the Concilium Bibliographicum."

30. Henry Fairfield Osborn, "Concilium Bibliographicum," *Science* 26, no. 660 (August 23, 1907): 260–261; American Museum of Natural History Collection (hereafter, AMNHC), "Concilium Notes," 1907–1908.

31. Robert F. Bruner, *The Panic of 1907: Lessons Learned From the Market's Perfect Storm* (Hoboken, N.J.: John Wiley & Sons, c. 2007); Concilium Bibliographicum, "To the Committee Appointed by the 7th International Zoological Congress in Boston," 1907; Herbert H. Field, "An International Zoological Record," *Nature* 47 (April 7, 1893): 606-607; Herbert H. Field, "Statement to the Committee Appointed by the 7th International Zoological Congress in Boston," August 23, 1907.

32. APSDP, Concilium matters, June 1908.

33. AMNHC, Bumpas to Davenport, January 28, 1908; AMNHC, February 1, 1908.

34. APSDP, "AAAS Committee," circa 1909–1910.

35. Gladys Doolittle, "A History of the Concilium Bibliographicum, Zurich," MS thesis, School of Library Science, New York, June 1929, 61; AMNHC, "Concilium Constitution," June 1909; *Annotationes,* 1909.

36. APSDP, Field to Davenport, January–May 1910.

37. Concilium, *Annual Report,* 1907.

38. John Lawler, *The H. W. Wilson Company* (Boston: Gregg Press, 1972).

39. AMNHC, October 1907–May 1908.

40. AMNHC, May–August 1908; APSDP, June 3, 1908.

41. APSDP, Field-Davenport letters, June 1908.

42. AMNHC, Bumpas to Davenport, January 1908.

43. AMNHC, Field-Osborn, April 1911.

44. BUHF, 17.

45. *Annotationes.*

Chapter 9

1. Howard Gotlieb Archival Research Center, Boston University, Hermann Field Collection (BUHF), special folder.

2. Federal Bureau of Investigation (FBI) Freedom of Information Act (FOIA), http://vault.fbi.gov/American Friends Service Committee.

3. American Philosophical Society, Charles Davenport Papers (APSDP), Field to Davenport, 1914.

4. Useful on WWI U.S. intelligence, Rhodi Jeffrey-Jones, *Cloak and Dollar: A History of American Secret Intelligence* (New Haven: Yale University Press, 2002).

5. R. Harris Smith, *OSS the Secret History of America's First Central Intelligence Agency* (n.p., 1973), 205; Robert Murphy, *Diplomat Among Warriors* (New York: Doubleday, 1964), 7–8.

6. Murphy, *Diplomat Among Warriors,* passim; and, useful for background, Klaus Schwabe, "United States Secret War Diplomacy, Intelligence, and the Coming of the German Revolution in 1918," *Diplomatic History* 16, no. 2 (Spring 1992): 175–200.

7. Hoover Institution, Vernon Kellogg Papers 56004, Kellogg memos circa 1919; also Herron Papers; Ralph H. Lutz, ed., *Fall of the German Empire, 1914–1918,* vol. 1 (New York: Octagon Books, 1969; originally published 1932). Dr. de Fiori was not cited with a first name in the sources. It was probably W. Robert. See, for example, George G. Bruntz, Allied Propaganda and the Collapse of the German Empire in 1918 (Stanford, Calif.: Hoover Institution, 1972; originally published 1938).

8. Leonard Mosley, *Dulles: A Biography of Eleanor, Allen, and John Foster Dulles and Their Family Network* (New York: Dial Press, 1978), 46.

9. BUHF, box 3, file 2; box 24, file 17; Mosley, Leonard, *Dulles*.

10. BUHF, August 8, 1918.

11. BUHF, October 10, 1918; Hoover Archives, Herron Papers, Finding Aid; Mitchell Pirie Briggs, *George D. Herron and the European Settlement* (Stanford, Calif.: Stanford University Press, 1932).

12. BUHF, box 3, file 4; *New York Times*, November 11, 1917.

13. BUHF, 17; Hoover Institution, Herbert Haviland Field folders.

14. BUHF, 1; Hoover Institution, Herbert Haviland Field folder.

15. BUHF, 4; David Clay Large, *Where Ghosts Walked: Munich's Road to the Third Reich* (New York: W.W Norton, 1997).

16. BUHF, special folders; Oswald Garrison Villard, *Fighting Years: The Memoirs of a Liberal Editor* (New York: Harcourt Brace and Co., 1939); Hoover Institution, Herbert Haviland Field folder.

17. Dominic Streatfield, *Brainwash: The Secret History of Mind Control* (New York: St. Martin's Press, 2007).

18. BUHF, special folders. The "Herplant" Herbert identified as a Bolshevik agent was most probably Alexander Israel Helpland (Parvus). See Z. Zeman, Z. and Winfried Scharlau, *The Merchant of Revolution: The life of Alexander Israel Helpland (Parvus) 1867–1924* (London: Oxford University Press, 1965).

19. BUHF, box 3, file 5.

20. Lawrence Emerson Gelfand, *The Inquiry: American Preparations for Peace, 1917–1919* (New Haven: Yale University Press, 1963).

21. BUHF, box 3, file 1.

22. Katherine Oliver Murra, "Notes on the Development of the Concept of Current Complete National Bibliography," Appendix to The UNESCO Library of Congress Bibliographical Survey, *Bibliographical Services: Their Present State and Possibilities of Improvement* (Washington, D.C.: n.p., 1950), 18–19.

Chapter 10

1. Howard Gotlieb Archival Research Center, Boston University, Hermann Field Collection (BUHF), box 3, file 1; Melvin I. Urofsky, *Louis D. Brandeis: A Life* (New York: Pantheon, 2009).

2. U.S. manuscript census.

3. BUHF, box 3, file 1.

4. Mark B. Adams et al., "Human Heredity and Politics: A Comparative Study of the Eugenics Record Office at Cold Spring Harbor (United States), the Kaiser Wilhelm Institute for Anthropology, Human Heredity, and Eugenics (Germany), and the Maxim Gorky Medical Genetics Institute (USSR), *Osiris* 20 (2005): 232–262.

5. American Philosophical Society, Charles Davenport Papers (APSDP), Field-Davenport 1919; BUHF, box 3, file 1.

6. National Research Council, *A History of the National Research Council 1919–1933* (Washington, D.C.: The Council, 1933); George Ellery Hale, *A National Focus of Science and Research*, Reprint and Circular series of the National Research Council, no. 39, circa 1922.

7. APSDP, Davenport Papers, August 16, 1919, re from NRC.

8. BUHF, box 3, file 1.

9. U.S. Census 1920; BUHF, box 3, file 1. Only one resident at number 104 was an artist. At the Roebling's old townhouse (number 110), which Hamilton now called his primary gallery and school, there were ten boarders. A mulatto widow from South Carolina acted as the "keeper," helped by her four adult children. Of the ten residents, four called themselves artists while a maiden "aunt" Haviland made her living as a dressmaker.

10. G. H. Parker Papers, Harvard University; Federal Bureau of Investigation Freedom of Information Act request (FOIA) 1098 300–900–000 and 1081306–000 on Noel Field, Director Mass. Industrial Defense Association, to FBI, November 1926. Hereafter, Noel's FBI file = FOIANF.

11. BUHF, box 3, file 1.

12. National Academy of Sciences / National Research Council archive collection (NRCC) Kellogg to Morgan, June 5, 1926.

13. BUHF, box 3, file 1; NRCC, Mark to Yerkes, May 1920.

14. BUHF, box 3, file 1.

15. Peter Mosk and Henry Hope, *The Robert Laurent Memorial Exhibition 1972–1973* (Durham: University of New Hampshire, n.d.), 161.

16. BUHF, box 3, file 1.

17. The Negro boy was probably Charles "Keene," the woodcutter who was boarding with his family at number 110; U.S. Census 1920, 1930.

18. *New York Times*, May 9, 1907.

19. Adolph Voge Papers, Library of Congress.

20. BUHF, box 3, file 1.

Chapter 11

1. Howard Gotlieb Archival Research Center, Boston University, Hermann Field Collection (BUHF), box 3, file 11, September 1919; National Academy of Sciences / National Research Council archive collection (NRCC), Field to McClung, November 12, 1919.

2. American Philosophical Society, Charles Davenport Papers (APSDP), Field to Davenport, November 1914, re McCormick.

3. APSDP, Vincent to Davenport, December 26, 1919.

4. APSDP, Davenport to Vincent, January 1920.

5. NRCC, Field to NRC, January 1920.

6. Information on the membership is scattered in the APSDP and NRCC documents for the early 1920s.

7. An informative sketch of Yerkes's career is in the introduction to his papers at Yale University. On the initial contacts with Field in 1919–1920: APSDP, Yerkes Papers, Davenport to Herbert, October 1919; BUHF, box 3, file 1.

8. Gregory A. Kimble et al. *Portraits in Psychology,* vol. 2, "Yerkes" (Hillsdale, N.J.: L. Erlbaum Associates: American Psychological Association, 1991); Sophie Bledsoe de Aberle and George W. Corner, *Twenty-Five Years of Sex Research; History of the National Research Council, Committee for Research in Problems of Sex, 1922–1947* (Philadelphia: Saunders, 1953).

9. For insight into the Council's ambitions: Robert M. Yerkes, ed., *The New World of Science, Its Development During the War* (Freeport, N.Y.: Books for Libraries Press, 1969, circa 1920); Sally Gregory Kohlstat et al., *The Establishment of Science in America: 150 years of the American Association for the Advancement of Science* (Rutgers University Press, 1999).

10. BUHF, box 3, file 1.

11. Useful on McClung: Anonymous, "Charles Erwin McClung," http://www.encyclopedia.com/doc/1G2-2830902724.html; on his plans for biological information systems, in addition to the separate items in BUHF, APSDP, and NRCC. On the McClung et al. abstracts plans: National Research Council, *A History of the National Research Council 1919–1933* (Washington, D.C.: The Council, 1933); Anonymous, "Biological Abstracts," *Science* 60, no. 1561 (November 28, 1924); J. R. Schramm, "The Abstracting and Indexing of Biological Literature," *Science,* 56, no. 1453 (November 3, 1922): 495–501; Alvin Leroy Bennett, "The Development of Intellectual Cooperation Under the League of Nations and United Nations," PhD diss., University of Illinois, Urbana, 1950; and W. C. Steere, *Biological Abstracts/BIOSIS: The First Fifty Years* (New York: Plenum Press, 1976).

12. Kohlstat, *The Establishment of Science in America.*

13. Yerkes, *New World of Science*; National Research Council, *A History of the National Research Council.*

14. Vernon Kellogg, "Isolation or Cooperation in Research," *Science* 63, no. 1626 (February 26, 1926).

15. Daniel J. Kevles, "George Ellery Hale, the First World War, and the Advancement of Science in America," *Isis* 59, no. 4 (Winter 1968): 427–437.

16. Yerkes, *New World of Science*.

17. Smithsonian Institution, "Science Service," http://siarchives.si.edu/findingaids/FARU7091.htm#FARU7091h.

18. Helpful for overviews of the foundations: Raymond B. Fosdick, *The Story of the Rockefeller Foundation* (New York: Harper & Brothers, 1952); Howard J. Savage, *Fruit of an Impulse: Forty-Five Years of the Carnegie Foundation, 1905–1950* (New York: Harcourt-Brace, 1953); Gerald Jonas, *The Circuit Riders: Rockefeller Money and the Rise of Modern Science* (New York: W.W. Norton, 1989).

19. Roger. L. Geiger, *To Advance Knowledge: The Growth of American Research Universities, 1900–1940* (New York: Oxford University Press, 1986); Sophie Bledsoe D. Aberle and George Washington Corner, *1922–1947* (Philadelphia: Saunders, 1953).

20. National Academy of the Sciences, *The National Academy of Sciences, The First Hundred Years, 1863–1963* (Washington, D.C.: NAS, 1978).

21. Thomas M. Spaulding, *The Cosmos Club on Lafayette Square* (Washington, D.C.: Cosmos Club, 1949).

22. Burke, Colin, *Information and Secrecy, Vannevar Bush, Ultra, and the Other Memex* (Metuchen, N.J.: Scarecrow Press, 1994).

23. A. L. Kroeber et al., *Franz Boas 1858–1942* (New York: Kraus, 1969).

24. APSDP, Davenport to Field, October 18, 1919; C. J. West, "A History of the National Research Council's Research Information Service, 1919–1933," *Science,* 78, no. 2019: 203–206; Charles L. Reese, *Information Needs in Science and Technology*, Reprint and Circular series, of the National Research Council #33, circa 1922.

25. APSDP, Memoranda of March 23, 1920.

26. Yerkes, *New World of Science*, 142.

27. BUHF, box 3, file 1, November 1919 letters; NRCC November 1919 documents.

28. Insights into the NRC international ambitions, including science information can be found in Yerkes, *New World of Science*, especially the essays by Hale and Angell.

29. NRCC, Field to Vincent, March 6, 1920.

30. Nathan Reingold, *Science, American Style* (New Brunswick: Rutgers University Press, 1991); NAS, *National Academy of Science*; NRCC, Cattell letters, June 1920.

31. Franz Stafleu, "Fifty Years of International Biology," *Taxon* 21, no. 1 (February 1971): 141–51; Daniel J. Kevles, "Into Hostile Political Camps: The Reorganization of International Science in

WWI," *ISIS* 62, no. 1 (Spring 1971): 47–60; Anonymous, "The International Research Council," *Science* 56, no. 1444 (January 1933): 243–244.

32. NRCC, November 1919, Field-NRC items; BUHF, Herbert to Nina letters, November 1919.

33. Phillip C. Jessup, *Elihu Root* (Hamden, Conn.: Archon Books, 1964 (c. 1938).

34. Larry L. Fabian, *Carnegie's Peace Endowment* (Washington, D.C.: Carnegie Endowment for International Peace, 1984).

35. BUHF, box 3, file 1.

36. BUHF, box 17, September 1919, Herbert to State Department; November 1919 letters re State Department visit.

37. BUHF, box 3, file 1, on letters re world peace and the League.

38. George B. Noble, *Christian Herter* (New York: Cooper Square Publishers, 1970).

39. W. John Hutchins, *Early Years of Machine Translation: Memoirs and Biographies of Pioneers* (Philadelphia: John Benjamin, 2000); Patrick H. McNamara, *A Catholic Cold War: Edmund A. Walsh, S.J., and the Politics of American Anticommunism* (New York: Fordham University Press, 2005); Christopher Simpson, *Blowback: America's Recruitment of Nazis and Its Effects on the Cold War* (New York: Collier, 1989).

40. BUHF, box 3, file 1, November–December 1919, Herbert to Nina.

41. *New York Times*, various articles on Hamilton, 1919–1922, especially December 8, 1922.

42. BUHF, box 3, file 1, December 1919.

43. American Museum of Natural History Collection (AMNHC), on committee meeting, October 16, 1907.

Chapter 12

1. Thomas Neville Bonner, *Iconoclast: Abraham Flexner and a Life in Learning* (Baltimore: Johns Hopkins University Press, 2002): Helen Thomas Flexner, *A Quaker Childhood* (New Haven, Conn.: Yale University Press, 1940).

2. Joan Marie Johnson, *Southern Women at the Seven Sister Colleges: Feminist Values and Social Activism* (Athens: University of Georgia Press, 2008).

3. "Eleanor Flexner," http://en.wikipedia.org/wiki/Eleanor_Flexner.

4. Howard Gotlieb Archival Research Center, Boston University, Hermann Field Collection (BUHF), box 3, file 1; American Philosophical Society, Charles Davenport Papers (APSDP), Cattell to Flexner, January 1920.

5. National Academy of Sciences / National Research Council archives collection (NRCC). Field to McClung, December 27, 1919.

6. NRCC, various January 1920.

7. APSDP, Davenport to Vincent, January 2, 1920.

8. APSDP, February 19, 1920; NRCC, Angell to Vincent, February 19, 1920.

9. Michael Rosenthal, *Nicholas Miraculous: The Amazing Career of the Redoubtable Dr. Nicholas Murray Butler* (New York: Farrar Straus Giroux, 2006; Lawrence A. Cremin, *American Education, the Metropolitan Experience, 1876–1980* (New York: Harper & Row, circa 1988); Lawrence A. Cremin, *The Transformation of the School: Progressivism in American Education, 1876–1957* (New York: Knopf, 1964, circa 1961).

10. New School History, http://www.newschool.edu/nssr/subpage.aspx?id=9064.

11. BUHF, box 3, file 1, Herbert Field to Nina Field, December 1919.

12. Stephen Duggan, *A Professor at Large* (Freeport, N.Y.: Books for Libraries Press, 1943); Richard Arndt, *The First Resort of Kings: American Cultural Diplomacy in the Twentieth Century* (Washington, D.C.: Potomac Books, 2005).

13. Stephen Duggan, *The Eastern Question: A Study in Diplomacy,* Columbia University Studies in History, Economics, and Public Law 14, no. 3 (New York: AMS Press, n.d.); Stephen Duggan, "The Inspiration of UNESCO," *Journal of Higher Education* 18, no. 3: 124–127.

14. Duggan, *Professor at Large*; U.S. House of Representatives, "Guide to Subversive Organizations and Publications (and appendix) March 3, 1951," Committee on Un-American Activities (Washington, D.C., 1951).

15. Joseph E. Persico, *Edward R. Murrow: An American Original* (New York: McGraw-Hill, circa 1988); Arndt, *First Resort of Kings.*

16. S. J. Taylor, *Stalin's Apologist: Walter Duranty, The New York Times' Man in Moscow* (New York: Oxford University Press, 1990).

17. Stephen Duggan and Betty Drury, *The Rescue of Science and Learning; the Story of the Emergency Committee of Displaced Foreign Scholars* (New York: Macmillan, 1948).

18. APSDP, Duggan File.

19. Henry Evelyn Bliss, "The System of the Sciences and the Organization of Knowledge," *Philosophy of Science* 2, no. 1 (January 1935): 86–103.

20. APSDP, Davenport to Field, January 1–14, 1920.

21. APSDP, Cattell to Flexner, January 24, 1920; BUHF, box 3, file 1, January 10, 1920.

22. Franz Stafleu, "Fifty Years of International Biology," *Taxon* 21, no. 1 (February 1971): 141–151.

23. J. R. Schramm, "The Abstracting and Indexing of Biological Literature," *Science,* no. 56, no. 1453 (November 3, 1922): 495–501; Vernon Kellogg, "Isolation or Cooperation in Research," *Science* 63, no. 1626 (February 26, 1926).

24. Edward P. Cheyney, *History of the University of Pennsylvania, 1740–1940* (Philadelphia: University of Pennsylvania Press, 1940); Wistar Institute of Anatomy and Biology, *Organization and Work of the Wistar Institute* (Philadelphia: The Institute, 1916); Wistar Institute, *Twentieth Anniversary of the Organization of the Advisory Board of the Wistar Institute of Anatomy and Biology, April 13, 1925* (Philadelphia, 1925).

25. Gordon S. Fulcher, *Indexing of Scientific Articles*, Reprint and Circular series of the National Research Council, no. 34, circa 1921.

26. NRCC, various January 1922, re Wistar, abstracts, and value of indexes.

27. BUHF, box 3, file 1, January 1920.

28. BUHF, box 3, file 1, February 1920.

29. BUHF, box 3, file 1.

30. Richard J. Storr, *Harper's University: The Beginnings: A History of the University of Chicago* (Chicago: University of Chicago Press, 1966).

31. APSDP, Vincent Folder; *New York Times*, May 10, 1920.

32. E. W. Burgess, "Obituary: George Edgar Vincent: 1864–1941," *American Journal of Sociology* 46, no. 6 (May 1941): 887.

33. NRCC, Angell to Vincent, February 13, 1920; Wissler to Angell, February 19, 1920.

Chapter 13

1. National Academy of Sciences / National Research Council archives collection (NRCC), January 13, 1920, McClung to Field.

2. American Philosophical Society, Charles Davenport Papers (APSDP), Davenport to Field, October 18, 1919; February 13, 1920, re Vincent to Angell; NRCC, Angell to Vincent February 13, 1920; Wissler to Angell, February 19, 1920.

3. Walter Miles, "James Rowland Angell, 1869-1949, Psychologist-Educator," *Science* 110, no. 2844 (July 1, 1949), 1–4; J. R. Angell, "Autobiography of James Roland Angell," in Carl Murchison, ed., *A History of Psychology in Autobiography*, vol. III (Worcester, Mass.: Clark University Press, 1932), 1–38; Dzuback, Mary Ann, *Robert M. Hutchins, Portrait of an Educator* (Chicago: University of Chicago Press, 1991), 38–39.

4. On the initial stages of the controversy": NRCC, Angell to Vincent, February 13, 1920, Vincent to Davenport, February 18, 1920; APSDP Davenport to NRC February 13, 1920, Vincent to Davenport, February 18, 1920.

5. NRCC, Wissler to Angell, February 19, 1920.

6. NRCC, Davenport to Vincent, February 24, 1920; Angell to Vincent, February 19, 1920.

7. NRCC, Angell to Wissler, February 20, 1920.

8. NRCC, Angell letters of February 21, 23, 25, 1920.

9. Howard Gotlieb Archival Research Center, Boston University, Hermann Field Collection (BUHF), box 3, file 1, February 21, 22, 1920; NRCC Angell to Vincent, February 25, 1920; Vincent to Field re Vincent accusations, February 28, 1920.

10. NRCC, Angell to Vincent, February 25, 1920.

11. BUHF, box 3, file 1, February 21. 1920; NRCC February 22–25 1921.

12. APSDP, Angell to Vincent, February 29, 1920; NRCC, Vincent to Davenport, February 29, 1920.

13. BUHF, box 3, file 1, March 1920.

14. APSDP, Vincent to Davenport, February 29, 1920; BUHF, box 3, file 1, March 5, 1920; NRCC, Field to Vincent, March 6, 1920.

15. NRCC, March 1, 1920.

16. NRCC, Vincent to Angell, March 5; 1920 and Angell to Davenport; NRCC, March 12, 1920.

17. NRCC, Livingston letters, March 1920; Joint Committee reports, March 1920; On Kellogg, C. C. McClung, *Biographical Memoir of Vernon Lyman Kellogg* (Washington, D.C.: NAS, 1938).

18. NRCC, Angell to RIS group, March 19, 1920.

19. BUHF, box 3, file 1, February 25, 1920; NRCC March 1, 1920.

20. BUHF, box 3, file 1, Herbert Field to Nina Field, March 1, 1920; Gladys Doolittle, "A History of the Concilium Bibliographicum, Zurich," MS thesis, School of Library Science, New York, June 1929, 64.

21. NRCC, Field to Angell, March 8, 1920.

22. NRCC, Field to Vincent, March 6, 1920.

23. APSDP, Livingston to Davenport, February 9, 1920; NRCC Livingston to Kellogg, March 1920.

24. BUHF, box 3, file 1, March 9, 1920: On Imperator, *New York Times*, March 1920.

25. NRCC, March 11, 1920, Angell to Field in Zurich.

26. NRCC, September 1921, Kellogg reports.

27. BUHF, box 1, file7.

28. APSDP, Vincent folder, June 16, 1920; NRCC, Yerkes to H. Wilson, March 16, 1922; Wilson to Kellogg, May 1922; Lawler, *The H. H. Wilson Company,* 40.

29. Verner Clapp, "Indexing and Abstracting Services for Serial Literature," *Library Trends* 2, no. 4 (1954): 509–521.

30. BUHF, box 3, file 1, Special folders; Richardson, Ernest Cushing, *Some Aspects of International Library Cooperation* (Radley, Pa.: F. S. Cook and Son, 1928).

31. BUHF, box 3, file 1, September.1920; APSDP, Yerkes folder, August 8, 1920.

32. APSDP, Yerkes file, Davenport letters, March 1920.

33. APSDP, Davenport to committee members, April 1920.

34. NRCC, various March–April 1920, Livingston to Kellogg, April 1920.

35. APSDP, Davenport to Pearl, March 20, 1920; Committee Report May 1920; NRCC Committee Report, May 1920.

36. W. Boyd Rayward, *The Universe of Information: The Work of Paul Otlet for Documentation and International Organization,* FID 520 (Moscow: FID, 1975).

37. NRCC, Angell memoranda, May 1920.

38. NRCC, Mark to Yerkes, May 19, 1920.

39. For example, The Institute, League of Nations International Institute of Intellectual Cooperation (Paris, 1933).

40. NRCC Franz report(s) 1920; E. Scott, "IFLA and FID History and Program," *Library Quarterly* 32 (1962): 1–18; F. Donker Duyvis, *F. Donker Duyvis: His Life and Work* (The Hague, Netherlands: NIDER, 1964); A. C. Foskett, *The Universal Decimal Classification: The History, Present Status and Future Prospects for a Large General Classification Scheme* (London: Linnet Books, 1973), 32.

41. Temperance Elizabeth Smith, "A Study of the International Organization for Intellectual Cooperation within the League of Nations," MS thesis, Claremont College, 1935; Alvin Leroy Bennett, "The Development of Intellectual Cooperation Under the League of Nations and United Nations," PhD diss., University of Illinois, Urbana, 1950.

42. APSDP, Davenport June 16, 1920.

43. Henry B. Ward, "Herbert Haviland Field," *Science* 54, no. 1401 (November 4, 1921): 424–428.

Chapter 14

1. National Archives and Records Administration (NARA) Record Group (RG) 59, Central Decimal File (CDF), Herbert-Nina Field entries 354.113, 411 .

2. Howard Gotlieb Archival Research Center, Boston University, Hermann Field Collection (BUHF), box 3, file 1, the foundation to Nina Field, May 1921; NRCC, Nina Field to Kellogg, May 1921, Nina Field to Embree May 1921.

3. The best sources on the state of the Concilium's finances are Kellogg's letters and reports, see National Academy of Sciences / National Research Council archives collection (NRCC) from April 1921. Also helpful is Gladys Doolittle, "A History of the Concilium Bibliographicum, Zurich," MS thesis, School of Library Science, New York, June 1929.

4. For example, NRCC, Livingston letters of 1920, letters to Kellogg January 1922, Yerkes, March 16, 1922, Greenman to Kellogg, May 17, 1922.

5. American Philosophical Society, Charles Davenport Papers (APSDP), Livingston to Davenport, February 9, 1920 and Ward to Davenport, March 15, 1920.

6. "Dr. Clarence Jay West," in *American Men and Women of Science: Physical and Biological Sciences*, ed. the Jacques Cattell Press (New York: Bowker, 1971–circa 1986).

7. Clarence J. West, "A Survey of Biological Chemical Literature" *Science* 47, no. 1226 (June 28, 1918): 623–626.

8. Amos Alfred Fries and Clarence J. West, *Chemical Warfare* (New York: McGraw-Hill, 1921).

9. E. J. Kahn, Jr., *The Problem Solvers: A History of Arthur D. Little Inc.* (Boston: Little-Brown, 1986).

10. NRCC, Kellogg reports, July 1921–September. 1921.

11. For example, Vernon Kellogg, *International Biology* (reprint and circular series of the NRC, 1925); Vernon Kellogg, *Vernon Kellogg 1867–1937* (Washington, D.C.: Anderson House, 1939); and C. C. McClung, *Biographical Memoir of Vernon Lyman Kellogg* (Washington, D.C.: NAS, 1938).

12. Vernon Kellogg, *Mind and Heredity* (Princeton: Princeton University Press, 1923).

13. Vernon Kellogg, *Headquarters Nights* (Boston: Atlantic Monthly Press, 1917); Mark Aaron Larson, "These Are Times of Scientific Ideals": Vernon Lyman Kellogg and Scientific Activism, 1890–1930," PhD diss., University of Minnesota, 2000.

14. Dorothy B. Lilley and Ronald W. Trice, *A History of Information Science 1945–1985* (San Diego, Calif.: Academic Press Inc.; 1989), and Dave Muddiman, "Red Information Scientist: The Information Career of J. D. Bernal," *Journal of Documentation* 59, no. 4 (2003): 387–409.

15. Smithsonian Institution files on the Science Service; Peter J. Kuznick, *Beyond the Laboratory: Scientists as Political Activists in 1930s America* (Chicago: University of Chicago Press, 1987).

16. American Museum of Natural History Collection (AMNHC), Tower to Osborn, April 17, 1920 NRCC, Tower to Kellogg, May 11, 1920.

17. NRCC, Kellogg re Zoological Record, February 1922.

18. NRCC, Yerkes to H. H. Wilson, March 16, 1922; Greenman to Kellogg, May 17, 1922; NRCC, Kellogg to Barrows, September 8, 1921; APSDP, Field letters September 1909.

20. NRCC, Kellogg letters to NRC re Russia, September 1921; Kellogg Congressional testimony, December 1921; on Herbert Hoover and relief, George H. Nash, *The Life of Herbert Hoover, Master of Emergencies, 1917–1918* (New York: W.W. Norton circa 1996); Benjamin M. Weissman, *Herbert*

Hoover and Famine Relief to Soviet Russia, 1921–1923 (Stanford, Calif.: Hoover Institution Press, 1974).

21. National Academy of the Sciences, *The National Academy of Sciences, The First Hundred Years, 1863–1963* (Washington, D.C.: NAS, 1978); NRCC, Kellogg Russia letters, September 1921.

22. G. J. A. O'Toole, *Honorable Treachery: A History of U.S. Intelligence, Espionage, and Covert Action from the American Revolution to the CIA* (New York: Atlantic Monthly Press, 1991), 299; David J. Foglesong, "Xenophon Kalamantiano: An American Spy in Russia," *Intelligence and National Security* 6, no. 1 (1999): 154–195; Marguerite E. Harrison, *Marooned in Moscow: The Story of an American Woman Imprisoned in Russia* (New York: George H. Doran Co., 1921).

23. Foreign Policy Association Pamphlets, *Russia the Obscure: A Report from the Viewpoint of Two American Observers*, pamphlet 22, series of 1923–1924.

24. Vernon Kellogg, *Mind and Heredity*, 18.

25. Steve Weinberg, *Armand Hammer: The Untold Story* (Boston: Little Brown, 1989), 58.

26. NRCC, Kellogg letters and reports, December 1921.

27. For example, see Robert M. Yerkes, ed., *The New World of Science, Its Development During the War* (Freeport, N.Y.: Books for Libraries Press, 1969, circa 1920).

28. National Academy of the Sciences, *The National Academy of Sciences, The First Hundred Years, 1863–1963* (Washington, D.C.: NAS, 1978); C. C. McClung et al., "The Biological Record and Biological Abstracts," *Science* 61, no. 1574 (February 27, 1925): 232–234.

29. For example, APSDP, Livingston to Davenport, February 9, 1920; J. R. Schramm, "The Abstracting and Indexing of Biological Literature," *Science* 56, no. 1453 (November 3, 1922): 495–501.

30. NRCC, NRC report on abstracting, March 6, 1921.

31. Franz Stafleu, "Fifty Years of International Biology," *Taxon* 21, no. 1 (February 1971): 147.

32. Alvin Leroy Bennett, "The Development of Intellectual Cooperation Under the League of Nations and United Nations," Ph.D. diss., University of Illinois, Urbana, 1950; J. R. Schramm, "The Abstracting and Indexing of Biological Literature," *Science,* 56, no. 1453 (November 3, 1922): 495–501; NRCC, Schramm, January 1, 1922.

33. W. C. Steere, *Biological Abstracts/BIOSIS: The First Fifty Years* (New York: Plenum Press, 1976); Cynthia Steinke, ed., *Information Seeking and Communications Behavior of Scientists and Engineers* (New York: Haworth Press, 1999).

34. "A Century of CAS," *Chemical and Engineering News* (June 11, 2007): 41–53.

35. Bennett, "The Development of Intellectual Cooperation "; NRCC, Schramm to Kellogg, October 31, 1922; Kellogg-Strohl correspondence 1924.

36. APSDP, Livingston to Davenport, February 9, 1920; NRCC,, Livingston plan, March 1920.

37. James McKeen Cattell and Albert Theodor Poffenberger, *James Mc Keen Cattell, Man of Science 1860–1944* (Lancaster, Pa. The Science Press, 1948).

38. NRCC, Cattell letters of May–June 1920 to Angell, June 1922 re NRC plans. Ellen Condliffe Lagemann, *The Politics of Knowledge: The Carnegie Corporation, Philanthropy, and Public Policy* (Middletown, Conn.: Wesleyan University Press, 1989), 40-.

39. NRCC, Kellogg reports on Concilium agreements, February 1922.

40. NRCC, Strohl to Kellogg, February–April 1922.

41. Documents from the Mondaneum Archive, Mons, Belgium, as supplied by W. Boyd Rayward; Strohl letters to/from Otlet from Otlet files.

42. NRCC, Strohl letters of 1924.

43. NRCC, "Statement of the Activities of the NRC for the Years July 1, 1922 to June 30, 1923"; Kellogg letter to Nina Field re compensation 1922. All the Rockefeller grants, and payments to Nina, 1920–1926, totaled to close to $1,400,000.

44. BUHF, box 22, file 12.

45. Anonymous, "The Concilium Bibliographicum," *Science Monthly* 14, no. 3 (March 1922): 201–202; NRCC, Kellogg to Strohl, February 1922, April 14, 1924; Strohl's protest, May 18, 1922.

46. *Conspectus,* 1925.

47. Concilium, "Conditions of Subscription," May 1924.

48. NRCC, Strohl's protest May 18, 1922.

49. NRCC, Yerkes letter of March 16, 1922; Fulcher, Mrs. Jane R. "Card Form Publications in Medicine and Related Fields," (National Library of Medicine, circa 1949): 48–53.

50. For example, NRCC, Greenman letters of 1922.

51. NRCC, Greenman to Kellogg, May 17, 1922, Schramm to Kellogg, October 31, 1922.

52. NRCC, Yerkes to Wilson, March 16, 1922; Jimena Canales, "Einstein, Bergson, and the Experiment that Failed: Intellectual Cooperation at the League of Nations," *MLN* 120, no. 5 (December 2005): 1168–1191.

53. J. E. de Vos van Steenwijk, "Intellectual Cooperation in Biology," *Science* 67, no. 1747 (June 22, 1928): 633–634.

54. NRCC Strohl to Kellogg, August 22, 1922; Meeting report, March 31, 1923.

Chapter 15

1. Howard Gotlieb Archival Research Center, Boston University, Hermann Field Collection (BUHF), box 3, file 1, box17, passim, and special folders; Flora Lewis, *Red Pawn: The Story of Noel Field* (Garden City, NY: Doubleday, 1965); Shiplists and passport data.

2. Cambridge city directories, voting lists, etc.; U.S. Census; BUHF, box 31, file 11.

3. BUHF, box 17.

4. Doreen Bolger, "Hamilton Easter Field and His Contribution to American Modernism," *American Art Journal* 20, no. 2 (1998): 78–107; *New York Times*, December 8, 1922; Peter V. Moak and Henry Hope, *The Robert Laurent Memorial Exhibition 1972–1973* (Durham: University of New Hampshire, nd.), especially p. 17; Smithsonian Institution, Archives of American Art, Papers of Robert Laurent, TN278561, reels 2065–2067, especially 2066.

5. U.S. manuscript census, 1930.

6. Lewis, *Red Pawn*; Craig Thompson, "What Has Stalin Done with Noel Field," *Saturday Evening Post* 224, no. 24 (December 15, 1951): 17; Michael Kazin, "The Agony and Romance of the American Left," *American Historical Review* 100 (December 1995): 1488–1512.

7. Robert A. Rosenstone, *Romantic Revolutionary: A Biography of John Reed* (New York: Vintage Books, 1975).

8. Don S. Kirschner, *Cold War Exile: The Unclosed Case of Maurice Halperin* (Columbia: University of Missouri Press, 1995). When and if Noel joined the American Party is contested but there are some credible indications that he and Herta gained formal but irregular status in the United States in 1933 and in Switzerland in 1936. See http://www.documentstalk.com/wp/field-noel-haviland-1904-1970; Thomas Sakmyster, *Red Conspirator: J. Peters and the American Communist Underground* (Urbana: University of Illinois Press, 2011), 93.

9. BUHF, box 17.

10. For later investigations of the need for deep indexing: Eugene Garfield, *Citation Indexing: Its Theory and Application in Science, Technology, and Humanities* (Philadelphia: ISI Press, 1983), especially pp. 6–18.

11. NRCC Strohl-Kellogg, August 8, 1922; *New York Times,* Kellogg, October 4, 1925.

12. Concilium Financial Report, 1923.

13. Serpil De Costa, "Foundation and Development of IFLA, 1926–1946, *Library Quarterly* 52, no. 1 (1982): 41–45; Alvin Leroy Bennett, "The Development of Intellectual Cooperation Under the League of Nations and United Nations," PhDdiss., University of Illinois, Urbana, 1950.

14. NRCC, Kellogg memoranda, July–September 1923.

15. Bennett, "The Development of Intellectual Cooperation"; Temperance Elizabeth Smith, "A Study of the International Organization for Intellectual Cooperation within the League of Nations," MS thesis, Claremont College, 1935; on the unfilled ambitions of the IIIC, Edwin Williams and Ruth V. Noble, *Conference of International Cultural, Educational, and Scientific Exchanges* (Chicago: American Library Association, 1947), especially pp. 16–33.

16. New York Times on IIIC, October 4, 1925; The Institute, *League of Nations International Institute of Intellectual Cooperation* (Paris: 1933).

17. NRCC, Meeting report(s), 1924.

18. NRCC, Strohl to Kellogg, 1926.

19. NRCC, Nauer to NRC, February 13, 1926.

20. NRCC, Kellogg correspondence 1924; NRCC reports, December 1924.

21. For example, NRCC, Strohl to Kellogg, May 19, 1924, November 1924.

22. NRCC, Davenport-Kellogg, Kellogg memoranda, March–June 1924.

23. NRCC, November–December 28, 1924, re abstracts.

24. NRCC, Strohl; to Vincent, December 15, 1925.

25. NRCC, Strohl to NRC, December 1926; Davenport et al., to Kellogg, June–October 1926.

26. NRCC, Strohl letters, December 1925; Kellogg, February 1925.

27. NRCC, Strohl to Kellogg, June 1925; *New York Times*, "On International Cooperation," October 4, 1925.

28. American Philosophical Society, Charles Davenport Papers (APSDP), McClung article and Strohl letters circa February 1925; Kellogg-Davenport letters June 1925; re Davenport trip, May 1925.

29. NRCC, Davenport to Strohl, August 1925.

30. J. Strohl, "The Scope of Bibliographies," *Science* 63, no. 1626 (March 26, 1926): 218–221.

31. Vernon Kellogg, "Isolation or Cooperation in Research," *Science* 63, no. 1626 (February 26, 1926); NRCC Kellogg to Strohl June 1925; Kellogg memorandum November 18, 1926.

32. Raymond B. Fosdick, *The Story of the Rockefeller Foundation* (New York: Harper & Brothers, 1952).

33. APSDP, Davenport to Strohl, January 17, 1927.

34. NRCC, letters to Kellogg, June–October 1926.

35. NRCC, Parker to Strohl, March 22, 1927.

36. NRCC, Kellogg to Morgan, June 5, 1926; American Museum of Natural History Collection (AMNHC), passim, on space problems; Shirley F. Harper, "The Universal Decimal Classification," *American Documentation* 5, no. 4 (October 1954): 195–213.

37. W. Boyd Rayward, "The International Exposition and the World Documentation Congress, Paris 1937," *Library Quarterly* 53, no. 3 (July 1983): 254–268.

38. *Annotationes,* 1930s.

39. Concilium, Conditions of Subscription, May 1924.

40. De Costa, "Foundation and Development of IFLA."

41. NRCC, Strohl, November 15, 1929; Noel Field to Kellogg, August 15, 1931.

42. BUHF, 17; ASPDP, McClung to Davenport, August 20, 1932; NRCC, Noel Field to Kellogg, August 15, 1931.

43. For example, NRCC, Noel Field to Lewis of NRC December 1933.

44. APSDP, Noel to Davenport, April 12, 1933.

Chapter 16

1. *Historical Statistics of the United States, Millennial Edition* (New York: Cambridge University Press, 2006).

2. David Henry Wenrich, "Biological Abstracts, a Condensed History," 1956, as from *Biological Abstracts*, vol. 30 (circa 1957); Donald Reddick, "International Union of Biological Sciences," *Science* 81, no. 2105 (May 3, 1935): 434–435.

3. Richard T. Kaser and Victoria Coz Kaser, *BIOSIS Championing the Cause: The FIRST 75 YEARS* (Philadelphia: NFAIS, 2001); American Philosophical Society Charles Davenport Papers (APSDP), NRC to Davenport, October 4, 1934; Herbert B. Landau, "Engineering Index 1994–1984," *Special Libraries* 75, no. 4 (1984): 12–18; ALA Periodicals Section, "Preliminary Survey of Indexes and Abstracting Services," *Library Journal* 63 (April 1, 1938): 265–271.

4. Wenrich, "Biological Abstracts"

5. W. C. Steere, *Biological Abstracts/BIOSIS: The First Fifty Years* (New York: Plenum Press, 1976).

6. Franz Stafleu, "Fifty Years of International Biology," *Taxon* 21, no. 1 (February 1971): 141–151.

7. NRCC, Concilium advertisements, June 8, 1940, and 1942.

8. W. Boyd Rayward, *The Universe of Information: The Work of Paul Otlet for Documentation and International Organization,* FID 520 (Moscow: FID, 1975). Katherine O. Murra, "History of Some Attempts to Organize Bibliography Internationally," in Jesse Shera and Margaret Egan, eds., *Bibliographic Organization* (Chicago: University of Chicago Press, 1950); 205. (The nations in the Soviet and French spheres were the most frequent users of the UDC at the time.)

9. W. Boyd Rayward, "The International Federation for Information and Documentation (FID)," in *Encyclopedia of Library History* (New York: Garland, 1999), 290–294; E. Scott, "IFLA and FID History and Program," *Library Quarterly* 32 (1962): 1–18.

10. League of Nations, "The Aims and Organization of the League of Nations" (Geneva: The League, 1929); *New York Times*, Kellogg on IIIC, October 4, 1925.

11. David Muddiman, "Public Science in Britain, The Origins of Documentation and Information Science," in W. Boyd Rayward, ed., *European Modernism and the Information Society: Informing the Present, Understanding the Past (*Burlington, Vt.: Ashgate, 2008), 201–222; I. C. McIlwaine, "The Universal Decimal Classification: Some Factors Concerning its Origins, Development, and Influ-

ence," in Trudi Bellardo Hahn and Michael Buckland, eds., *Historical Studies in Information Science* (Medford, N.J.: Information Today, 1998), 94–106; A. J. Meadows, ed., *The Origins of Information Science* (London: Taylor-Graham, 1987), 144; S. C. Bradford, "A National Bureau of Information," *Nature* 120, no. 3015 (August 12, 1927): 231–232.

12. A. C. Foskett, *The Subject Approach to Information* (London: Clive Bingley, 1969), 307; S. C. Bradford, *Documentation* (London: C. Lockwood, 1948). His files included a full set of Concilium cards and a half-million items obtained from a continental "Bibliographical Institute" that ceased operation in the 1920s. S. C. Bradford, "Classification for Bibliography of Science," *Nature* 128, no. 3224 (1931): 268–269; Association of Special Libraries and Information Bureaux, *Report of the Proceedings of the Fourth Conference* (London: ASLIB, 1927), 49.

13. S. C. Bradford, "Classification for Bibliography of Science"; Dr. S.C. Bradford, "Why the Science Library Adopted the Universal Decimal Classification," *Library Journal* 55 (1930): 1000–1005.

14. W. Boyd Rayward, "The UDC and FID—a Historical Perspective," *Library Quarterly* 37, no.3 (July 1967): 259–278; Shirley F. Harper, "The Universal Decimal Classification," *American Documentation* 5, no. 4 (October 1954): 195–213.

15. Library of Congress, Voge Papers, passim.

16. Some insight into the navy service is in Library of Congress. Science and Technology Division, *The Navy Research Section, Science Division, Library of Congress* (Washington: 1951). On users of the UDC, Harper, "The Universal Decimal Classification." Voge's work would have a rather ironic relationship to post-WWII information systems. His rather traditional DDC/UDC-like classification system for science and technology was the basis for Vannevar Bush's secret Joint Research and Development Board for the military and Voge's classification was used by the Central Intelligence Agency for its main information system.

17. Mark B., Adams et al., "Human Heredity and Politics: a Comparative Study of the Eugenics Record Office at Cold Spring Harbor . . .," *OSIRIS* 2nd Series 20 1 (n.d.): 232–262.

18. A. L. Kroeber et al., "Franz Boas 1848–1942," *American Anthropologist* 3, no. 2 (July–September 1943).

19. U.S. Congress, "Un-American Activities Committee Hearing," 1953, vol. 2, 1764.

20. Margaret Sanger, *Margaret Sanger: An Autobiography* (New York Cooper Square Press 1999); Adams, "Human Heredity."

21. Edward J. Larson, *Sex, Race, and Science: Eugenics in the Deep South* (Baltimore: Johns Hopkins University Press, 1995).

22. Edward J. Larson, *Summer for the Gods: The Scopes Trial and America's Continuing Debate Over Science and Religion* (Cambridge: Harvard University Press, 1997); Joseph W. Bendersky, *The Jewish Threat: Anti-Semitic Politics of the U.S. Army* (New York: Basic Books, 2000), 197.

23. Peter J. Kuznick, *Beyond the Laboratory: Scientists as Political Activists in 1930s America* (Chicago: University of Chicago Press, 1987), 142.

24. Henry Fairfield Osborn, *The Origin and Evolution of Life: On the Theory of Action and Reaction and Interaction of Energy* (New York: Charles Scribner's, 1918); George Howard Parker, *Biology and Social Problems* (Boston: Houghton-Mifflin, 1914).

25. Larson, *Summer for the Gods*; Adam R. Shapiro, "Civic Biology and the Origin of the School Antievolution Movement," *Journal of the History of Biology* 41 (2008): 409–433.

Chapter 17

1. Federal Bureau of Investigation Freedom of Information Act file on Hermann Field #1062382–000. Hereafter, FOIAHF and FBI FOIA 1122171, 1122172; Howard Gotlieb Archival Research Center, Boston University, Hermann Field Collection (BUHF), boxes 28–24, 32–38; www.documentstalk.com/wp/field-noel-haviland-1904-1970, Biography and Bibliography on Noel Field and Family, accessed April, 2011.

2. FBI FOIA file on Noel Field, 1081306-000, hereafter FOIANF, November 1926, "Industrial Defense Association"; FOIANF, especially November 1926; National Archives and Records Administration (NARA), Record Group (RG) 59 Noel Field, 123F 451 series and 123, Noel Field (box 403).

3. Noel Field, *Banishing War Through Arbitration: A Brief Sketch of Post-war Arbitration Treaties*, 2nd ed. Washington, D.C.: The National Council for Prevention of War, 1927); U.S. House Committee on Un-American Activities. *The Erica Wallach Story* (Washington: U.S. Government Printing Office, 1958); David Caute, T*he Fellow Travelers: Intellectual Friends of Communism,* rev. ed. (New Haven: Yale University Press, 1988).

4. Hede Massing, *This Deception* (New York: Duell, Sloan, and Pearce, 1951); Maria Schmidt, "Noel Field—The American Communist at the Center of Stalin's East European Purges: From the Hungarian Archives," *American Communist History* 3, no. 2 (2004): 215–245; Igor Lukes, "The Rudolf Slansky Affair: New Evidence." *Slavic Review* 58, no. 1 (Spring 1999): 167, states Noel aligned with Soviet intelligence as early as 1927 and that he had some role in the assassination of a defector.

5. Joseph C. Harsch, *At the Hinge of History, A Reporter's Story* (Athens: University of Georgia Press, 1993).

6. David McKnight, *Espionage and the Roots of the Cold War: The Conspiratorial Heritage* (London: Frank Cass, 2002).

7. S. J. Taylor, *Stalin's Apologist: Walter Duranty, The* New York Times' *Man in Moscow* (New York: Oxford University Press, 1990.

8. Arthur Schlesinger, Jr., *A Life in the Twentieth Century* (Boston: Houghton-Mifflin, 2000).

9. BUHF, box 17.

10. BUHF, box 17; The Mundaneum Archive, Mons, Belgium, Box HLF 064 Correspondence: Folder: Lettres de Henry (sic) Haviland Field; lettres de Noel Field (fils) 1922–1930.

11. "Elsie Field: Portrait of a Physician," *Women and Health* 9, 1 (Spring 1984); BUHF, 17.

12. BUHF, box 32, file 18.

13. FOIANF.

14. Massing, *This Deception*, 167; Ship passenger lists, July 29, 1933, July 24, 1941; Susan Elisabeth Subak, *Rescue and Flight: American Relief Workers Who Defied the Nazis* (Lincoln: University of Nebraska Press, 2010), xxii.

15. BUHF, box 28, file 4; Igor Lukes, "The Rudolf Slansky Affair: New Evidence," 167, claims that Noel was working for Soviet intelligence by 1927.

16. BUHF, box 17; Margot Peters, *May Sarton, A Biography* (New York: Knopf, 1997), 83.

17. NARA, RG59 F123, F4521/21-October 6, 1934; *New York Times,* October 25, 1934.

18. FOIAHF, circa 1949 rpt.

19. BUHF, box 28, file 4.

20. C. Edward White, *Alger Hiss's Looking-Glass Wars* (Oxford: Oxford University Press, 2004), 230; Massing, *This Deception.*

21. Later in life, Noel stated that he became a Communist Party member in late 1933. It seems that it was a secret or irregular American membership. Thomas Sakmyster, *Red Conspirator: J. Peters and the American Communist Underground* (Urbana: University of Illinois Press, 2011), 93, 215–245; Igor Lukes, "The Rudolf Slansky Affair: New Evidence." *Slavic Review* 58, no. 1 (Spring 1999), cites statements that Noel was recruited to help Soviet intelligence as early as 1927.

22. Harvey Klehr, John Earl Haynes, and F. I. Firsov, *The Secret World of American Communism* (New Haven: Yale University Press, 1995); CIA CREST files re Noel Field.

23. BUHF, box 32, file 7. Schmidt Manuscript; Schmidt "Noel Field," 233; U.S. State Dept. FOIA 200802520 file on Hermann Field; hereafter, SDFOIAHF, passim.; FOIANF, Fletcher Warren to FBI, December 31, 1940.

24. NARA, RG59 123F 45/21–28 and 121.67/1450.

25. BUHF, box 17; May Sarton, *I Knew a Phoenix* (New York: Rhinehart & Co. 1954), 142.

26. U.S. Census; Harvard Class Books; *New York Times.*

27. Sue Ainslie Clark and Edith Franklin Wyatt, *Making Both Ends Meet the Income and Outlay of New York Working Girls* (New York: Macmillan Co., 1911).

28. Nancy Schrom, *As Equals and as Sisters: Feminism, the Labor Movement, and the Women's Trade Union League of New York* (Columbia: University of Missouri Press, 1980); FOIAHF and FBI FOIA 1122171, 1122172, Sali and Jean Lieberman, passim. Hereafter, FOIASL.

29. FOIASL, June 3, 1951.

30. Gary Dorrien, *The Making of American Liberal Theology…* (Louisville, Ky.: Westminster John Knox Press, 2001), 374.

31. Herbert A. Philbrick, *I Led Three Lives: Citizen, Communist, Counterspy* (Washington, D.C.: Capitol Hill Press, 1972), 261; Alan R. Sadovnik and Susan F. Stemel, *Founding Mothers and Others, Women Educational Leaders During the Progressive Era* (New York: Palgrave, 2002); Dao-yuan Chou, *Silage Choppers & Snake Spirits: The Lives and Struggles of Two Americans In China* (Queson City, Philippines: IBON, 2009).

32. John A. Beinkere, *There Were Giants in the Land: The Life of William Heard Kilpatrick* (New York: Peter Land, 1998).

33. May Sarton Papers, Berg Collection, New York Public Library, letters from Clark sisters to May, circa 1928–1930.

34. FOIAHF; FOIANF.

35. BUHF, 31, 11; FOIASL June 23, 1948; *New York Times*, 1934–1935.

36. NARA, RG59 811.4276/61/43-Duggan file, December 28, 1934; *New York Times,* July 30, 1934.

37. FBI FOIA, Joseph Figueiredo, NARA case # nw 39746.

38. BUHF, box 26, file 6; Whaley, Barton, *Soviet Clandestine Communication Nets* (Cambridge, Mass.: Center for International Studies, MIT, 1969). BUHF, box 26, file 6; Michael Gelb, ed., *An American Engineer in Stalin's Russia, the Memoirs of Zara Witkin 1932–4* (Berkeley: University of California Press, 1991).

Chapter 18

1. Federal Bureau of Investigation (FBI) Freedom of Information Act (FOIA) file on Hermann Field #1062382–000, hereafter FOIAHF. FBI FOIA 1122171, 1122172, Sali and Jean Lieberman, January 31, 1948, hereafter, FOIASL.

2. FOIASL, June 23, 1948, LA 100 23969 (FBI document)

3. Howard Gotlieb Archival Research Center, Boston University, Hermann Field Collection (BUHF), Hermann to Nina, June 2, 1935.

4. BUHF, box 28, file 5.

5. Dave Muddiman, "Red Information Scientist: The Information Career of J.D. Bernal," *Journal of Documentation* 59, no. 4 (2003): 387–409; Mary Catherine Bateson, *With a Daughter's Eye: A Memoir of Margaret Mead and Gregory Bateson* (New York: Murrow, 1984).

6. FBI FOIA Act, Noel Field 1081306-000, hereafter FOIANF.

7. U.S. Congress, "Un-American Activities Committee Hearing," 1953, vol. 2, 1764; Biography and bibliography on Noel Field and family at Documents Talk.com (hereafter, Documents Talk, http://www.documentstalk.com/wp/field-noel-haviland-1904-1970. Hede Massing, *This Decep-*

tion (New York: Duell, Sloan, and Pearce, 1951), 177; BUHF, file 28, box 4; Andrew Webster, "The Transnational Dream: Politicians, Diplomats and Soldiers in the League of Nations' Pursuit of International Disarmament, 1920–1938," *Contemporary European History* 14, no. 4 (November 2005): 493–518; Elisabeth K. Poretsky, *Our Own People: A Memoir of "Ignace Reiss" and His Friends* (Ann Arbor: University of Michigan Press, 1969), 144.

9. Maria Schmidt, "The Hiss Dossier: A Historian's Report," *New Republic*, November 8, 1993, 17–18, 20; *New York Times*, "CBS MAN," June 20, 1955, p. 1.

10. Rosalee McReynolds and Louise S. Robbins, *The Librarian Spies: Philip and Mary Jane Kenney and Cold War Espionage* (Westport, Conn.: Praeger, 2009).

11. NARA, RG59, Central Decimal File index, April 1936.

12. FOIASL, 100-29196; http://en.wikipedia.org/wiki/Hede-Massing.

13. Harvey Klehr, John Earl Haynes, and F. I. Firsov, *The Secret World of American Communism* (New Haven: Yale University Press, 1995); FOIANF.

14. NARA, RG59, 123F 451 Noel Field, March 10, 1936.

15. Barry M. Katz, *Foreign Intelligence: Research and Analysis in the Office of Strategic Services* (London: Oxford University Press, 1989); Lewis A. Coser, *Refugee Scholars in America: Their Impact and Their Experiences* (New Haven: Yale University Press, 1984), especially pp. 85–120.

16. FOIANF; *New York Times*, December 1, 1938 re Noel; Kurt Singer, *The Men on the Trojan Horse* (Boston: Beacon Press, 1953).

17. BUHF, box 33, file 11; May Sarton, *At Seventy: A Journal* (New York: Norton, 1987); Mundaneum Archive, Mons, Belgium, Box HLF 064.

18. BUHF, box 28, file 4; FOIASL passim circa 1935–50.

19. BUHF, box 32, file 8; Unitarian-Universalist Archives bms 16031/1 (5).

20. Patricia Riddle, "Elsie Field: Portrait of a Physician," *Women & Health* 9, no. 1 (Spring 1984): 1–4.

21. Paul R. Halmos, *I Have a Photographic Memory* (Providence, R.I.: AMS, 1987); e-mail interview with James Reeds; FBI FOIA release, Joseph Doob NARA case nw 39746 ; Thomas R. Johnson, *American Cryptology During the Cold War* (Ft. George G. Meade, Md.: Center for Cryptologic History, National Security Agency, 1995, vol. 2.).

22. FOIAHF re 1939; FOIASL, re 1935.

23. Ship passenger lists, July 24, 1941, July 29, 1933.

24. Karel Kaplan, *Report on the Murder of the General Secretary* (Columbus: Ohio State University Press, 1990), 20; Herbert Romerstein and Eric Breindel, *The Venona Secrets* (New York: Regency Press, 2000).

25. Romerstein and Breindel, *The Venona Secrets*, 134; Massing, *This Deception*; FOIAHF, 100-32889-21.

26. FOIASL, reports re Jean Clark, 1938; Veronica A. Wilson, "Now You Are Alone": Anticommunism, Gender, and the Cold War Myths of Hede Massing and Whittaker Chambers," *Diplomatic History* 36, no. 4 (2012): 699–722.

27. U.S. Congress, "Un-American Activities Committee Hearing," 1953, re Noel letter, June 1939.

28. NARA, RG59 F123F451/21–28, February 18 and 28, 1939; *New York Times*, December 30, 1937.

29. Erica Wallach, *Light at Midnight* (New York: Doubleday, 1967); Robert A. Rosenstone, *Crusade of the Left: The Lincoln Battalion in the Spanish Civil War* (Cambridge: Harvard University Press, 1990, circa 1975). On the frightful death toll due to the war and, especially, the brutal political cleansings by both left and right, Julius Ruiz, "Seventy Years on: Historians and Repression during and after the Spanish Civil War," *Journal of Contemporary History* 44, no. 3 (July 2009): 449–472. The infamous Bela Kun had been sent to Spain to ensure "purity." There are indications he was helped by Archie Brown, an American who became linked to the Fields after World War II.

30. BUHF, box 32, file 8.

31. *The Erica Wallach Story* (Washington: U.S. Government Printing Office, 1958).

32. Lawrence Emerson Gelfand, *The Inquiry: American Preparations for Peace, 1917–1919* (New Haven: Yale University Press, 1963).

33. "Documents Talk."

34. Ships' passenger list, March 15, 1939.

35. Noel Field items sent from Unitarian archives at Harvard University; see also the recent release of thousands of pages of documents concerning the USC, Noel, Fry, and others at the Harvard University websites for the Unitarian/Universalist organizations. A pivotal starting address is http://oasis.lib.harvard.edu/oasis/deliver/~div16146.

36. NARA, RG59 123F451/2–28 and 123 Noel Field (Box 403).

37. Adolf Augustus Berle, *Navigating the Rapids 1918–1971: From the Papers of Adolf Berle* (New York: Harcourt Brace Jovanovich, 1973), 583–586; FOIANF; Jordan A. Schwarz, *Liberal: Adolph A. Berle and the Vision of an American Era* (New York; Free Press, 1987), 24.

38. NARA, RG59 123F451/21–28, 123 Noel Field (Box 403) and 811.10 N/441.

39. Spruille Braden, *Diplomats and Demagogues: The Memoirs of Spruille Braden* (New Rochelle: Arlington House, 1971), 346; A. Berle, Leo Cherne, and Clare Boothe Luce, *Hungary Under Soviet Rule, a Survey of Developments from September 1957 to August 1958* (New York: prepared by friends of the captive nations, circa 1958).

40. BUHF, box 33, file 17; Ghanda Di Figlia, *Roots and Visions: The First Fifty Years of the Unitarian Universalist Service Committee* (Boston: Unitarian-Universalist Service Committee, 1990); Susan

Elisabeth Subak, *Rescue and Flight: American Relief Workers Who Defied the Nazis* (Lincoln: University of Nebraska Press, 2010).

41. James Ford Lewis, "The Unitarian Service Committee," PhD diss., University of California Berkeley, 1952.

42. Clyde Sanger, *Lotta and the Unitarian Service Committee Story* (Toronto: Stoddart Publishing Company, 1986).

43. FOIASL, September 19, 1946 and various.

44. Stephen H. Fritchman, *Heretic, a Partisan Autobiography* (Boston: Skinner House, 1983).

45. FOIAHF, August 8, 1941; Unitarian Universalist Archives, Harvard University, bms 16031/1 (7).

46. BUHF, box 33, file 17.

47. Barsky biography, Robert Wagner Archives, New York University; "Survey of the Work of the Joint Anti-fascist Refugee Committee, to date June 1944"; various items in *New York Times*, 1941–1947, especially September 1, 1944; Phillip Deery, "A Blot upon Liberty: McCarthyism, Dr. Barsky and the Joint Anti-Fascist Refugee Committee," *American Communist History* 8, no. 2 (2009): 167–196.

48. On the Barsky organizations, *New York Times,* January 1, 1944, 11 and passim, and Unitarian/Universalist Archives, BMS16007, BMS16013, and especially 16031/ (6) anonymous to USC 1946. A more detailed story of the "Joint" committee and its complex relations with the USC, socialist critics, and Noel will be found in my upcoming work on the Fields and science information after World War II.

49. Varian Fry, *Surrender on Demand* (Boulder, Colo.: Johnson Books, 1997); Andy Marino, *A Quiet American: The Secret War of Varian Fry* (New York: St. Martin's Press, 1999).

50. Sebastian Farber, "Scenes of Bravery and Determination," *The Volunteer* 26, no. 1 (January 3, 2009): 3–8. The intricate story of Varian Fry's achievement and difficulties, and of the USC's ties to the OSS, will be detailed in my upcoming work on the Fields and science information after World War II.

51. Tommie Sjoberg, *The Powers and the Persecuted: The Refugee Problem and the Intergovernmental Committee on Refugees (IGCR), 1939–1947* (Lund, Sweden: Lund University Press, 1991), 91, 201.

52. Noel soon found the cost of living in his base of operations, Switzerland, so high that he sent rather emotional pleas for raises to Boston. Unitarian-Universalist Archives, Harvard University, 1941–1942. Within a year after World War II, he convinced the USC to pay him a salary close to what he had received at the League.

53. Barth Bernd-Rainer and Werner Schweizer. *Der Fall Noel Field: Schlüsselfigur der Schauprozesse in Osteuropa* [The Case of Noel Field: Key Figure in the Show Trials in Eastern Europe], assisted by Thomas Grimm (Berlin, Germany: Basis Druck, 2005).

54. USC Papers, Unitarian Universalist Archives, Harvard University. Throughout the war and the postwar years, Noel's reports, especially concerning the use of the Barsky funds, emphasized his organizations' strict neutrality in giving aid. However, in a 1945 report to Boston on efforts in Switzerland, the writer stated they would never aid "fascists nor national socialists nor their sympathizers" (USC bms16035).

Chapter 19

1. Federal Bureau of Investigation (FBI), Freedom of Information Act (FOIA) requests, 1122171, 1122172, Sali and Jean Lieberman, January 31, 1948, hereafter FOIASL.

2. FOIASL, LA 100–23969; SF 100–29196, January 1. 1948.

3. Un-American Activities Committee Hearings, New England, 1958; Northern California, 1960; http://www.historia-actual.org/Publicaciones/index.php/haol/article/viewFile/73/71; NewYork University Wagner Archive interview with Joe Figueiredo.

4. FOIASL, July 1, 1959, January 26, 1945, SF 100–29196, Joyce and Blum letters

5. Interview with Andrew Kluger 2010; Barth Bernd-Rainer and Werner Schweizer. *Der Fall Noel Field: Schlüsselfigur der Schauprozesse in Osteuropa* [The Case of Noel Field: Key Figure in the Show Trials in Eastern Europe], assisted by Thomas Grimm (Berlin, Germany: Basis Druck, 2005).

6. U.S. State Department (SD) Freedom of Information Act Request (FOIA), Hermann Field, case # 200802520, hereafter, SDFOIAHF, from "Consulate, Glasgow," December 14, 1954.

7. SDFOIAHF, May 16, 1950.

8. SDFOIAHF, From Consulate Glasgow, December 14, 1954; FOIAHF, June 29, 1940.

9. Hermann Field, "Few of 600 Fleeing Escape Poland: Many Slain, Most Still Wander in Country," *New York Times,* September 20, 1939, 4; Hermann Field, *From Krawkow* [sic] *to Rumania* (London: Drummand, 1940).

10. Howard Gotlieb Archival Research Center, Boston University, Hermann Field Collection (BUHF), box 28, file 5.

11. SDFOIAHF, December 16, 1943.

12. BUHF, box 28, file 5; Antonin Raymond, *Antonin Raymond, an Autobiography* (Rutland, Vt.: Charles Tuttle, 1973).

13. Federal Bureau of Investigation, Freedom of Information Act request 1082382-000, Hermann Field, hereafter, FOIAHF, circa 1940–45; BUHF, box 33, file 12.

14. FOIAHF, September 1, 1949.

15. BUHF, box 28, file 4.

16. BUHF, box 4, file 5; Katherine A. S. Sibley, *Red Spies in America* (Lawrence: University of Kansas Press, 2004), 137; Steven T. Usdin, *Engineering Communism* (New Haven: Yale University

Press, 2005); Ellen Schrecker, *Many Are the Crimes: McCarthyism in America* (Boston: Little, Brown & Co., 1998), 107.

17. BUHF, box 24 and 28, file 5; SDFOIAHF, November 4 and 25, 1954, and May 8, 1950.

18. FOIAHF, April 25, 1949.

19. BUHF, box 33, file 12; box 28, files 4; 24.

20. FOIAHF, September 21–24, 1941.

21. James Ford Lewis, "The Unitarian Service Committee," PhD diss., University of California Berkeley, 1952; Unitarian Universalist Archive, Harvard University, "Noel Herta, and Hermann Field, 1941-50" file & bms 16031/1(7).

22. National Archives and Records Administration (NARA) Record Group (RG) 59, 123F451/24–6, Berle to Joy etc., November 1942.

23. FOIAHF; Elizabeth McIntosh, *Sisterhood of Spies: The Women of the OSS* Annapolis: Naval Institute Press, 1998), 183. FOIASL, August 14, 1958.

24. Susan Elisabeth Subak, *Rescue and Flight: American Relief Workers Who Defied the Nazis* (Lincoln: University of Nebraska Press, 2010); NARA RG226 Dulles file; FOIAHF, April 11, 1950.

25. Patrick O'Donnell, *Operatives, Spies, and Saboteurs* (New York: Free Press, 2004; John Prados, *Safe for Democracy: The Secret Wars of the CIA* (Chicago: Ivan R. Dee, 2007), 82.

26. BUHF, box 24, file 13; Pierre-Th. Braunschweig, *Secret Channel to Berlin: The Masson-Schellenberg Connection and Swiss Intelligence in World War II* (Philadelphia: Casemate, 2004). 394; Joseph C. Harsh, *At the Hinge of History* (Athens: University of Georgia Press, 1993), 60; Neal H. Petersen, *From Hitler's Doorstep: The Wartime Intelligence Reports of Allen Dulles, 1942–1945* (University Park: Pennsylvania State University Press, 1996); Subak, *Rescue & Flight*, 137 & passim.

27. Erica Wallach, *Light at Midnight* (New York: Doubleday, 1967); Ghanda Di Figlia, *Roots and Visions: The First Fifty Years of the Unitarian Universalist Service Committee* (Boston: Unitarian-Universalist Service Committee, 1990), 29.

28. R. Harris Smith, *OSS the Secret History of America's First Central Intelligence Agency* (n.p., 1973), 222.

29. NARA RG59, 123F451/21–28, 1943 FBI rpt. to State; Subak, *Rescue and Flight*.

30. BUHF, box 24, file 13; FOIANF, September 1, 1944. The details of Noel's trips in 1942–1944 will be covered in a future volume. Some new evidence indicates that he may not have suffered as much on this 1944 trip into France as previously thought.

31. Arthur Schlesinger Jr., *A Life in the Twentieth Century* (Boston: Houghton-Mifflin, 2000), 333–334; Nelson MacPherson, *American Intelligence in War Time London* (London: Frank Cass Publishers, 2003). Schlesinger later became one of the most important U.S. intellectuals fighting Communism. The OSS did recruit some of those Communist "Calpo" refugees to infiltrate Germany during the war but soon stopped using them. A future volume will cover the details of the Calpo involvements and Noel's other OSS related activities.

32. Ralph D. Warner, *A History of the Farmington Plan* (Lanham, Md.: The Scarecrow Press, 2002); Barry M. Katz, *Foreign Intelligence: Research and Analysis in the Office of Strategic Services* (London: Oxford University Press, 1989).

33. Scott Donaldson, *Archibald MacLeish, an American Life* (Boston: Houghton-Mifflin, 1992).

34. Stephen C. Mercado, "Sailing the Sea of OSINT in the Information Age," https://www.cia.gov/library/center-for-the-study-of-intelligence/csi-publications/csi-studies/studies/vol48no3/article05.html; Rosalee McReynolds and Louise S. Robbins, *The Librarian Spies: Philip and Mary Jane Kenney and Cold War Espionage* (Westport, Conn.: Praeger, 2009).

35. Robin Winks, *Cloak and Gown: Scholars in the Secret War, 1939–1961* (New Haven: Yale University Press, 1996).

36. NARA RG226, Personnel files; Frederick G. Kilgour, "The Origins of Coordinate Searching," in Hahn and Buckland, *Historical Studies in Information Science*, 107–115.

37. Pamela Spence Richards, *Scientific Information in Wartime: The Allied German Rivalry, 1939–1945* (Westport, Conn.: Greenwood, 1994); Kathy Lee Peiss, "Cultural Policy in a Time of War: The American Response to Endangered Books in World War I," *Library Trends* 55, no. 3 (Winter 2007): 370–386.

38. Barry M. Katz, *Foreign Intelligence: Research and Analysis in the Office of Strategic Services* (London: Oxford University Press, 1989); NARA RG226, Kilgour personnel file, box 404; E. M. R. Ditmas, "Coordination of Information: A Survey of Schemes Put Forward in the Last Fifty Years," *Journal of Documentation* 3, no. 4 (March 1948): 209–221; Robin Winks, *Cloak and Gown,* 103–110.

39. *The Erica Wallach Story* (Washington: U.S. Government Printing Office, 1958); Irene S. Farkas-Conn, *From Documentation to Information Science* (New York: Greenwood Press, 1990). The Swiss were not always neutral. One of their diplomats informed a German agent that the Allies were reading the most secret German codes. James DeBrosse and Colin Burke, *The Secret in Building 26: The Untold Story of Ameirca's Ultra War Against the U-boat Enigma Codes* (NY: Random House, 2004), 124.

40. E. J. Crane, ed., *The History of Chemical Abstracts;* Robert K. Stewart, "The Office of Technical Services: A New Deal in the Cold War," *Knowledge Creation, Utilization, and Diffusion* 15, no. 1 (September 1993): 46–77.

41. James Phinney Baxter, *Scientists Against Time* (Wilmington, Del.: Scholarly Resources, 1983); Vannevar Bush, *Science, the Endless Frontier* (Washington, D.C.: National Science Foundation, 1980, circa 1945).

Chapter 20

1. The libraries had administrative catalogs as well those used by patrons.

2. Colin Burke, "The 'Term' in the Classifier's Garden," in Toni Carbo and Trudi Bellardo Hahn, eds., *International Perspectives on the History of Information Science and Technology* (Medford, N.J.:

ASIST, 2012), 163–173. All the topics in this concluding chapter, including information methods and technology, are covered in detail in my upcoming work.

3. Trudi Bellardo Hahn and Diane L. Barlow, "The Fortuitous Confluence of the National Science Foundation, the American Society for Information Science & Technology, and Information Science," in Carbo and Bellardo Hahn, *International Perspectives*, 5–17.

4. Daniel Bell, *The End of Ideology; on the Exhaustion of Political Ideas in the Fifties*, new rev. ed. (New York: Collier Books, 1962).

5. John Patrick Diggins, *The Rise and Fall of the American Left* (New York: W.W. Norton, 1992).

6. Michael T. Kaufman, *Soros, the Life and Times of a Messianic Billionaire* (New York: Knopf, 2002); John Atlas, *Seeds of Change: The Story of ACORN, America's Most Controversial Antipoverty Community Organizing Group* (Nashville, Tenn.: Vanderbilt University Press, 2010); Peter Suber, *Open Access* (Cambridge, Mass.: MIT Press, 2012), for legal/technical aspects. For a sampler on the flavor and history of the Open Access movement: Stevan Harnad, "For Whom the Gate Tolls? How and Why to Free the Refereed Research Literature Online Through Author/Institution Self-Archiving, Now," *Historical Social Research / Historische* 29, no. 1, 107 (2004):76–113; http://www.jstor.org/stable/20761931; http://www.opensocietyfoundations.org/voices/ten-years-on-researchers-embrace-open-access; http://www.opensocietyfoundations.org/openaccess/read.

Index

Note: page numbers followed by "f" indicate illustrations. Throughout the index, "HF" refers to Herbert Field.